Springer Theses

Recognizing Outstanding Ph.D. Research

Aims and Scope

The series "Springer Theses" brings together a selection of the very best Ph.D. theses from around the world and across the physical sciences. Nominated and endorsed by two recognized specialists, each published volume has been selected for its scientific excellence and the high impact of its contents for the pertinent field of research. For greater accessibility to non-specialists, the published versions include an extended introduction, as well as a foreword by the student's supervisor explaining the special relevance of the work for the field. As a whole, the series will provide a valuable resource both for newcomers to the research fields described, and for other scientists seeking detailed background information on special questions. Finally, it provides an accredited documentation of the valuable contributions made by today's younger generation of scientists.

Theses are accepted into the series by invited nomination only and must fulfill all of the following criteria

- They must be written in good English.
- The topic should fall within the confines of Chemistry, Physics, Earth Sciences, Engineering and related interdisciplinary fields such as Materials, Nanoscience, Chemical Engineering, Complex Systems and Biophysics.
- The work reported in the thesis must represent a significant scientific advance.
- If the thesis includes previously published material, permission to reproduce this must be gained from the respective copyright holder.
- They must have been examined and passed during the 12 months prior to nomination.
- Each thesis should include a foreword by the supervisor outlining the significance of its content.
- The theses should have a clearly defined structure including an introduction accessible to scientists not expert in that particular field.

More information about this series at http://www.springer.com/series/8790

Abhijeet Alase

Boundary Physics and Bulk-Boundary Correspondence in Topological Phases of Matter

Doctoral Thesis accepted by Dartmouth University, USA

Springer

Abhijeet Alase
Institute for Quantum Science
and Technology
University of Calgary
Calgary, Alberta, Canada

ISSN 2190-5053 ISSN 2190-5061 (electronic)
Springer Theses
ISBN 978-3-030-31959-5 ISBN 978-3-030-31960-1 (eBook)
https://doi.org/10.1007/978-3-030-31960-1

This Springer imprint is published by the registered company Springer Nature Switzerland AG.
The registered company address is: Gewerbestrasse 11, 6330 Cham, Switzerland

Supervisor's Foreword

I am very pleased to introduce this work by my former graduate student, Dr. Abhijeet Alase. In a broad sense, Abhijeet's thesis fits into the rich tradition in theoretical physics of uncovering relevant symmetry properties and leveraging them to obtain mathematically exact results, alongside a deeper physical understanding. Within condensed-matter physics, lattice translation symmetry arguably provides the underpinning for much of modern electronic band and transport theory in crystalline solids: as formally captured by the Bloch's theorem, translation symmetry allows labeling of the one-electron wavefunctions in bulk matter in terms of their crystal momenta and, with that, mandates their organization within the Brillouin zone and ultimately their conduction properties. Yet, even in the absence of any impurity or bulk disorder, the translational symmetry of the infinite lattice, as reflected in the torus topological constraint that periodic (Born–von Karman) boundary conditions impose, is clearly broken in a real, finite crystal. On the one hand, properly accounting for boundary conditions other than periodic—as necessary, for instance, to describe hard-wall (open) boundaries or more complex scenarios (such as boundary impurities or disorder)—is central in the study of surface phenomena and surface states, as systematically initiated in the early investigations by Tamm and Shockley and culminating in the discovery of the quantum Hall effect. On the other hand, new impetus for carefully re-examining the extent to which *bulk* and *boundary* properties are interconnected and influencing each other stems from the challenge of understanding and classifying topological phases of matter: more precisely, to elucidate how, exactly, the presence of a topologically non-trivial bulk in both insulators and superconductors leads to the emergence of electronic states that are localized on the boundary, and can withstand disorder as long as certain discrete symmetries are obeyed. It is precisely this "bulk-boundary correspondence" in *symmetry-protected topological phases* that Abhijeet's thesis aims to explore, focusing on the simplest yet paradigmatic limit of independent electrons.

The central tool for this investigation, and what I regard as a major contribution of this thesis, is a *generalized Bloch's theorem*, which provides exact, analytic expressions for all energy eigenvalues and eigenstates of a lattice system of independent fermions, under the sole assumptions that any coupling entering the relevant (quadratic) Hamiltonian has finite range and lattice translation is broken *only* by boundary conditions. In essence, this theorem makes mathematically precise the intuition that, for such clean (disorder-free) systems, translational symmetry is still obeyed "sufficiently far away" from the boundary: technically, this is achieved by reformulating the single-particle diagonalization problem in terms of corner-modified banded block-Toeplitz matrices, and by recognizing that the resulting eigensystem inherits most of its solutions from an auxiliary, infinite translation-invariant Hamiltonian that allows for *non-unitary representations* of translation symmetry—hence effectively an analytic continuation to the complex plane of the standard Bloch's Hamiltonian off the Brillouin zone.

Abhijeet's results are important from a twofold perspective. One is exact solvability: like the usual Bloch's theorem, its generalization is, first and foremost, a practical tool for calculations, granting direct access to exact (often closed-form) solutions for all energy eigenvalues and eigenstates of the model system of interest. In many ways, such a structural characterization is close in spirit to what Bethe ansatz techniques accomplish for a different class of (interacting) quantum integrable systems: the linear-algebraic task of diagonalizing the single-particle Hamiltonian is mapped to one of solving a "small" system of equations, with the simpler (non-interacting) nature of the problem reflecting in those equations being *polynomial*. From this mathematical-physics standpoint, it is worth noting that the diagonalization techniques for corner-modified banded block-Toeplitz matrices that the generalized Bloch's theorem hinges on are, in fact, more generally applicable; since, in particular, Hermiticity is not a requirement, these same methods are likely to find use in other (fermionic as well as bosonic many-body) settings where mildly broken translational symmetry is at center stage—notably, non-Hermitian effective Hamiltonians or open-system Liouvillian super-operators.

From a condensed-matter physics perspective, the other major implication is, as anticipated, on symmetry-protected topological phases of matter. Here, the generalized Bloch's theorem is applied to a number of paradigmatic models, in both one and higher spatial dimension—describing topological insulators and superconductors, as well as interfaces and junctions. While for many of these models solutions for specific parameter regimes or boundary conditions had been independently derived in the literature, access to an exact, general framework affords two main advantages: not only is a unified understanding between seemingly disparate physical systems but, perhaps most importantly, genuinely *new* phenomena or interpretations may be uncovered, sometimes even while revisiting well-known territory. The discovery that both exponentially decaying edge modes and more exotic modes with *power-law prefactors* can emerge at zero energy in Kitaev's Majorana chain, under open boundary conditions in some *non-generic* parameter regimes, can serve as a good case-in-point in this respect. Precisely in these same non-generic parameter regimes, the use of the generalized Bloch's theorem permits to give physical meaning to generalized eigenvectors in the context of transfer matrix techniques, which have long been successfully employed in statistical mechanics. In short, I believe the results in this thesis will be useful to students and scientists with an interest in, and a taste for, exact solvability and topological quantum matter alike.

In closing, it is also a pleasure for me to acknowledge that, as with the best graduate students, Abhijeet has changed the way I, as his supervisor, have initially thought of and approached these subjects and has, in turn, contributed to shape this very exploration by bringing in altogether new tools and insights. Especially, noteworthy in this sense are the initial investigations of matrix Wiener–Hopf factorization techniques which Abhijeet also reports on here and which, I believe, are poised to deliver a fresh perspective on the bulk-boundary correspondence and the stability of zero-energy edge states in one-dimensional topological phases. To sum up, Abhijeet's thesis provides an excellent reference and entry point in a fascinating area at the crossroad between contemporary condensed-matter and quantum-statistical physics.

Professor of Physics, Dartmouth College
Hanover, NH, USA
July 2019

Lorenza Viola

Preface

The emphasis in this thesis is on the investigation of boundary physics of topological phases of matter. Writing this preface is a good opportunity to document how I ended up working on this topic, and how the work shaped up over the past 4 years. I joined Dartmouth college in the Fall of 2013 as a very young researcher, aspiring to do experiments on Nuclear Magnetic Resonance (NMR) in Prof. Chandrasekhar Ramanathan's laboratory. I enjoyed working with Sekhar for a couple of terms, however also realized that theoretical problems captured my attention a lot more than the experimental challenges. Consequently, I decided to join Prof. Lorenza Viola's research group with the intention of working primarily in the field of quantum information. Accordingly, I started studying entanglement in fermions in the Summer of 2014. In March 2015, one of the postdocs working in Lorenza's group, Amrit Poudel, decided to move for a new position. Amrit was working on topological superconductivity as well as many-body quantum chaos. I was handed the responsibility to retain some of the knowledge that Amrit had generated during his stay at Dartmouth. I got so involved in this work that it evolved into my PhD thesis.

Lorenza and Gerardo Ortiz (Professor of Physics at Indiana University, Bloomington) had been working together on topological superconductivity for quite some time before I joined the collaboration. They both give utmost importance to "practicality" of their research, to which their earlier works stand testimony. I was, in the beginning, much less attracted to the physics part, but took interest in the project due to beautiful mathematics. I was naturally drawn to the question of diagonalizing free-fermionic systems with terminations, the kind we often encounter when investigating topological phases. Lorenza, Gerardo, Emilio Cobanera (a postdoc in Netherlands at that time, now Assistant Professor at SUNY Polytechnic, Utica), and I spent months developing and discussing the solution of Kitaev's p-wave superconductor model, which eventually led to the results that we published in *Physical Review Letters*. Emilio joined Lorenza's group as a postdoc in September 2015. Over the next 2 years, Emilio and I had countless sessions of intense discussions on the specific topic and condensed-matter physics in general, apart from the weekly meetings where all four of us would skype to discuss progress.

In developing the solution of Kitaev's model, we primarily thought of our results as providing an ansatz for eigenvectors of systems under arbitrary boundary conditions. We were aware that this ansatz could capture all the eigenvectors in the cases that we investigated. However, knowing that the underlying mathematics was going to be quite involved, I was not inclined to spend a lot of effort in proving the completeness of the ansatz. It so happened that Emilio was interested in understanding the mathematical underpinnings of our ansatz. He landed on the "Smith normal form" of matrix polynomials and started applying it in the context of our work. It was this approach that allowed us to confirm that localized states that decay exponentially with a power-law prefactor indeed could appear in systems that we considered. Since we had definite progress in this mathematical direction, I changed my mind and decided to prove the completeness of the ansatz. We could integrate these two things nicely in our paper in the *Journal of Physics A: Mathematical and Theoretical*. Thanks to these mathematical foundations we could formulate a generalization of Bloch's theorem in our next paper that appeared in *Physical Review B*. It is funny that Emilio and I had something else in our mind that we referred to as generalized Bloch's theorem. It was Lorenza once again who pushed us in the right direction. Many of the applications presented in the two papers in *Physical Review B* were worked out even before we started working on the mathematical foundations, although many details became clear only after we formulated the generalization of Bloch's theorem.

The fact that matrix factorization can play an important role in deciding localized eigenstates of a Hamiltonian under open boundary conditions was known to us since the time we were working on the Kitaev's model. However, I could focus more on this aspect only after our publication of the first paper in *Physical Review B*. Most of the work in this context was led by me since Emilio was already away for his faculty position in Utica. This experience made me realize how much easier the earlier work was made due to more frequent discussions with collaborators.

I am fortunate to have been guided by two very experienced researchers (Gerardo and Emilio) during my PhD, apart from my advisor. Especially in the beginning, Lorenza, Emilio, and Gerardo were very tolerant of how little I knew in this field. They have also been extremely supportive of my strange theoretical attempts at understanding the topic. This is the main reason that we could land on some unorthodox methods (especially the bulk-boundary separation technique in the generalized Bloch's theorem, and also the use of the Wiener–Hopf factorization) in our research and capitalize on them. When it came to hands-on training, Emilio has always been there for me. My views about physics are therefore heavily influenced by Emilio's. I have rarely met a more passionate and positive personality. I feel grateful for continued support (financial, academic, and mentorship) of Lorenza which has allowed me to contribute to this work. I am amazed by the freedom she has provided me in exploring the problems at hand. She has also steered the work somewhat in the direction of quantum information and quantum control, which may prove very important in future. Gerardo's emphasis on consistency and correctness throughout has been an inspiration for me. His experience has also been extremely valuable for all the work we have done together. I have met some wonderful people in Hanover, at and outside of Dartmouth College. I am very fortunate to have met my wife, Salini Karuvade, at Dartmouth. Salini has been by my side throughout, and has supported me in every possible way in this journey, which would have been simply impossible without her. My parents and sister deserve a great deal of appreciation for the way they have encouraged me to pursue my dreams. It is not possible to name everyone at Dartmouth who has played an important role in my stay here. Sagar Kale, Dhana Nair-Schäf, Aarathi Prasad, Peter Johnson, Ariana Sopher, Christopher Caroll, Javiera Born, Maulik Patel, Parth Sabharwal, Kanav Setia, Hui Wang, William Braasch, Laura Carpenter, Italo Lemos, Nishant Aggarwal, Ashna Sharma, and Dhananjay Beri are only a few names that I would like to mention here. Special thanks to Billy and Laura for letting me stay with them for the few days when I was combatting an allergy during thesis writing! I could not have maintained consistency over the 5 years if not for some quality time spent playing badminton with an amazing bunch of people at Dartmouth Badminton Club. During the summers, I also enjoyed playing ultimate frisbee, to which I was introduced by former Lorenza's graduate student Peter Johnson, and soccer, for which Roberto Onofrio was twice as enthusiastic as me. Finally, I want to say thank you to my committee members, Prof. Chandrasekhar Ramanathan, Prof. James Whitfield, and Prof. Adrian Del Maestro. Sekhar is an excellent guide and a very inspirational teacher and was the first to give me an opportunity to conduct a guest lecture in his class. I will remember my experience as a TA with James, and also many times we played chess in the chess club at Hanover. Sincere thanks to Adrian for accepting my request to be on the thesis committee!

Calgary, Alberta, Canada
January 2019

Abhijeet Alase

Parts of This Thesis Have Been (Will Be) Published in the Following Journal Articles

1. A. Alase, E. Cobanera, G. Ortiz, and L. Viola, "Exact solution of quadratic fermionic hamiltonians for arbitrary boundary conditions", Phys. Rev. Lett. **117**, 076804 (2016).
2. E. Cobanera, A. Alase, G. Ortiz, and L. Viola, "Exact solution of corner-modified banded block- toeplitz eigensystems", J. Phys. A: Math. Theor. **50**, 195204 (2017).
3. A. Alase, E. Cobanera, G. Ortiz, and L. Viola, "Generalization of Bloch's theorem for arbitrary boundary conditions: theory", Phys. Rev. B **96**, 195133 (2017).
4. E. Cobanera, A. Alase, G. Ortiz, and L. Viola, "Generalization of Bloch's theorem for arbitrary boundary conditions: interfaces and topological surface band structure", Phys. Rev. B **98**, 245423 (2018).
5. A. Alase, E. Cobanera, G. Ortiz, and L. Viola, "Matrix factorization approach to bulk-boundary correspondence and stability of zero modes", in preparation.

Contents

Chapter 1
Introduction

The transformation of condensed-matter physics brought about by the discovery and exploration of topological phases of matter has been revolutionary. Before these discoveries, Landau's theory of continuous phase transitions [1], based on spontaneous breaking of symmetries, underpinned many of the successes in the twentieth century, including a microscopic theory of superconductivity. This theory had to be set aside to describe the topological states of matter, making way for an entirely new theoretical framework. Over the last two decades, concepts from topology have populated the condensed-matter literature at an ever increasing rate. Today, not only have we experimentally confirmed the existence of many topological solid-state materials [2], but we have also been able to synthesize systems with topological properties on a wide variety of physical platforms [3, 4]. Beyond their fundamental importance, these developments have also garnered a great deal of attention in view of potentially transformative device applications, notably in the context of spintronics and quantum computation [5, 6]. The Nobel Prize in Physics awarded in 2016 to Michael Kosterlitz, Duncan Haldane, and David Thouless recognized the significance of all the achievements in this field.

1.1 Context and Motivation

We begin this section with a brief review of some of the major developments that have shaped the field of topological phases of matter. We next discuss the theoretical challenges that have motivated the research work reported in this thesis.

A. Alase, *Boundary Physics and Bulk-Boundary Correspondence in Topological Phases of Matter*, Springer Theses, https://doi.org/10.1007/978-3-030-31960-1_1

1.1.1 A Brief Historic Overview

The role of topology in phase transitions in condensed-matter physics was first recognized by Berezinskii, Kosterlitz, and Thouless [7, 8] in the early 1970s. They showed that it is theoretically possible for certain two-dimensional (2D) systems to undergo a rather new kind of phase transition. These systems are capable of hosting stable topological defects, known as "vortices." Below a critical temperature, these defects appear in pairs closely bound together, known as vortex–antivortex pairs. Above the critical temperature, however, it is energetically favorable for the vortices and antivortices to disassociate, which reflects in an exponential decay of correlations. The distinguishing feature of this phase transition is that below and above the critical temperature, the *same set of symmetries* are satisfied by the system. Therefore, it cannot accommodate a description in terms of spontaneously broken symmetries associated with a local order parameter, as in Landau's theory. In fact, it was already known at the time, thanks to a theorem by Mermin and Wagner [9], that 2D systems *cannot* spontaneously break a continuous symmetry. This new mechanism of phase transition paved the way to exotic phases in low-dimensional systems.

In 1980, the discovery of the integer quantum Hall effect provided a major breakthrough in the field of topological phases of matter. In the original experiment [10] by von Klitzing, a 2D gas of electrons was trapped between layers of silicon and silicon dioxide in a silicon field effect transistor. When a strong magnetic field was applied perpendicular to the plane of the trapped electrons, it was observed that the transverse conductance of the sample assumed values that were integral multiples of the universal constant e^2/h, with e and h being the absolute charge of the electron and the Planck's constant, respectively. The measured conductance values were so precise that they were proposed to provide a more accurate figure for the fine structure constant than was available at the time. It is now known that these values of conductance do not depend on the material being used in the experiment, and do not change even if the experimental sample contains impurities. In a landmark paper, Thouless and coworkers explained the quantized conductance steps in terms of a topological invariant, now known as the Thouless–Kohmoto–Nightingale–den Nijs (TKNN) invariant [11], which can be inferred from the quantum mechanical wavefunctions of the electrons in the bulk bands parametrized by the crystal momentum.

The analysis by Thouless and coworkers assumed a 2D system *without* terminations (edges). The physical mechanism behind the integer quantum Hall effect for a more realistic system with edges was made clear by the investigations in subsequent years [12]. In a semi-classical picture, a strong magnetic field forces electrons away from the edges of the sample to revolve in cyclotronic orbits, therefore essentially blocking any flow of electric current. Near the edges, however, the electrons show a peculiar kind of motion, called "chiral" transport: each edge of the sample allows flow of electrons in only one direction, with opposite edges carrying counterpropagating currents. Because the left and the right moving electrons are separated by several hundred atomic layers, they cannot backscatter despite the presence of lattice vibrations or impurities. This ensures a dissipationless transport of electrons along the edges, leading to the quantized conductance values. Yet, it was not clear why the TKNN invariant accurately provides the conductance values of a system with edges. In 1993, Yasuhiro Hatsugai showed that the number of edge states is indeed related to the TKNN invariant, therefore providing the missing link [13].

Around the same time when the quantum Hall effect was being scrutinized, Duncan Haldane showed that topological quantities also have important consequences for spin chains. In particular, he put forth a conjecture [14] that states that the spectrum of spin chains composed of integer spins is gapped, as opposed to those composed of half-integer spins, which are gapless. His ideas were later applied to a model of spin-1 chain now referred to as the Affleck, Lieb, Kennedy, and Tasaki (AKLT) model [15]. It was found that the Hamiltonian of this model remains gapped provided certain symmetries are obeyed by the perturbations. This provided one of the early examples of what are now known as "symmetry-protected topological phases" (or SPT phases for short).

A question that emerged as an aftermath of the quantum Hall effect was whether similar topological states could arise in the absence of an external magnetic field. In 1988, Duncan Haldane indeed proposed a model of the quantum Hall effect in zero field [16]. Nevertheless, the Hamiltonian of this model broke time-reversal symmetry, which was crucial for the presence of chiral edge states. This phase of matter is now called a "Chern insulator," due to the topological invariant at play, which is the so-called "Chern number." However, the model proposed by Haldane did not have a physical realization at that time.

The next important development in this context was the prediction and discovery of topological insulators. In 2005, Kane and Mele proposed the existence of a "quantum spin Hall state" [17]. In this case, thanks to the fact that time-reversal symmetry is preserved, each edge carries electrons in both directions, but their direction of motion is correlated to their spin. This kind of motion of electrons is called "helical" transport. It was soon demonstrated that the quantum spin Hall effect can be realized in the presence of a strong spin–orbit interaction in a thin layer of mercury telluride (HgTe) sandwiched between cadmium telluride (CdTe) [18, 19]. The 2D materials showing the quantum spin Hall effect were the first experimentally discovered members of the family of topological insulators. Spin-correlated dissipationless transport makes them ideal candidates for low power spintronic devices [20, 21].

The quantum spin Hall effect differs from the quantum Hall effect in a crucial way: the dissipationless character of the currents on each edge in the former is destroyed by an applied external magnetic field, or by the presence of magnetic impurities near the surface. In other words, the exotic properties in this case hold *only* as long as time-reversal symmetry is obeyed by the system as a whole. Also, as opposed to the quantum Hall states labeled by integers taking values in all of $\mathbb{Z}$, there are only two quantum spin Hall states, labeled by a $\mathbb{Z}_2$-valued topological invariant. Only a year after the discovery of 2D topological insulator, a 3D topological insulator was theoretically predicted and its topological surface bands were observed experimentally [22, 23]. These discoveries confirmed beyond any doubt the existence of SPT phases [24, 25]. Following these developments, a plethora of topological insulating materials were predicted and discovered, satisfying various symmetry conditions in one, two, and three space dimensions.

Many of the concepts applicable to topological insulators can be carried over to superconducting materials [26]. In fact, some initial theoretical developments in topological superconductors happened before the discovery of topological insulators. The driving force behind these developments was the pursuit of Majorana quasiparticle excitations, which were predicted to be hosted by certain superconducting systems. While Majorana fermions were originally hypothesized by Ettore Majorana in 1937 in the context of particle physics [27], in solid-state systems Majorana excitations are quasiparticles with the properties that they are their own anti-excitations. In 2001, Alexei Kitaev showed theoretically that a p-wave superconducting nanowire can host Majorana quasiparticles of zero energy [28]. Unlike fermions and bosons, the exchange ("braiding") of two such "Majoranas" can lead to more than a simple change of phase in the overall wavefunction, depending on the underlying fermion parity of the system. They are therefore said to exhibit non-Abelian exchange statistics. Among other exciting aspects, Majoranas can be used, in principle, to encode quantum information and be manipulated in a way such that the principles of topology largely protect the information from environment-induced errors [29, 30]. The theoretical search for Majorana fermions continued in 2D [31] and 3D [32] superconductors after Kitaev's proposal, including extension to s-wave topological superconductors that respect time-reversal symmetry [33, 34]. However, their connection to topology did not come to the fore until the discovery of topological insulators [35].

With several topological phases being discovered, there was a need to classify them based on their properties. However, it was not obvious what the organizational principle should be. Altland and Zirnbauer had used particle-hole symmetry, time-reversal symmetry, and their composition to classify random matrix ensembles into ten symmetry classes [36]. Shinsei Ryu and coworkers adapted this symmetry classification to topological phases, leading to what is now known as the "tenfold way" [37]. According to this classification, and assuming that the above are the only discrete symmetries the system possesses, the abovementioned symmetries allow us to place each topological phase of non-interacting fermionic matter into one of the ten symmetry classes. Within the symmetry class, it is labeled by an appropriate topological invariant. As a result, Hamiltonians labeled by different values of the topological invariant cannot be transformed into each other adiabatically, and while respecting the symmetries of the class. The tenfold way can be extended to include the effect of space-group symmetries of crystal lattices, which are responsible in some cases for providing robustness to the edge or surface states [38]. As a result of these developments, it became clear what the defining property of a SPT phase is: namely, the presence of energy states localized on and extending along the boundary, which can sustain disorder as long as certain discrete symmetries are obeyed. The presence and robustness of such localized energy states in each case can be associated with a topological invariant of the wavefunctions of independent electrons in the bulk. The latter property is broadly referred to as the "bulk-boundary correspondence" principle. Although much of the investigation of SPT phases has focused on *gapped* phases of matter (such as insulators and superconductors), gapless phases of matter (such as Dirac and Weyl semimetals as well as nodal and other gapless superconductors) have also been explored more recently [39, 40].

The developments in this field have seen a healthy exchange between theory and experiments. In the characterization of 3D topological insulators, angle-resolved photoemission spectroscopy for detection of surface Dirac cones has emerged as an important experimental technique [23, 41]. The experiments involving detection of Majoranas in artificial superconductors have been in limelight over the last decade [42, 43]. Several experiments have reported strong evidence of the existence of Majoranas, by relying on a variety of signatures such as RF irradiation in a Josephson junction [42], zero-bias conductance peak [44], and scanning tunneling microscopy [45]. In parallel, much progress has been made in realizing topological systems on simulation platforms, in particular on ultracold atomic gases [3]. These platforms allow verification of unexplored topological phenomena by realizing models that are not known to describe solid-state materials. Equally importantly, they can probe some dynamics that are difficult to simulate numerically on existing computers.

We conclude this brief survey with a mention of some works that have aimed to formulate the bulk-boundary correspondence in a mathematically rigorous manner. The aforementioned investigations of quantum Hall effect by Hatsugai were arguably the first attempts in such a direction. Several investigations have now appeared, which unveil various aspects of the exact relations between the bulk and the boundary physics of topological materials. They use a multitude of approaches, including operator K-theory [46–50], Green's function techniques [51, 52], and the transfer matrix formalism [53], to name a few. We point out in particular Ref. [46], which contains a comprehensive treatment of "complex classes" of topological insulators using the tools of operator K-theory.

1.1.2 What Motivates the Work in This Thesis

The overarching goal of the work in this thesis is to contribute to the understanding of the physics of quantum (zero-temperature) phase transitions beyond Landau's paradigm. It is therefore valuable to gain a deeper understanding of the same in SPT phases. One of the immediate objectives is to obtain a characterization of these phases in terms of boundary physics, and to develop theoretical tools required to carry out this task. The characterization of topological phases is intricately woven with the problem of their classification. Establishing exact bulk-boundary relations is necessary for achieving these objectives. A characterization based on boundary physics is also of great interest for experiments, since one needs to identify unambiguous observable signatures of topology [54, 55]. Further, it can provide guiding principles for engineering exotic topological states in quantum simulation platforms and synthetic quantum matter. Finally, this kind of analysis is ultimately of relevance to the quest of developing quantum technologies, such as spintronics and topological quantum computing.

From an operational perspective, one can distinguish topological phases from ordinary ones only on the basis of their distinctive behavior at the boundary. Therefore, systems with a boundary are required to establish this difference qualitatively and quantitatively. Accordingly, it is desirable to have characterizations of topological phases based on the boundary physics, and not (or not only) on the topological invariants of the bulk as is predominantly established currently. It is important to note in this context that, even though topological invariants are computed using the properties of the bulk, they are *global* quantities rigorously defined for infinite systems. Unlike local observables which are well-defined for finite systems (up to finite-size effects), no experiment has *direct* access to a bulk topological invariant. Consider, for example, the TKNN invariant that is proportional to the transverse conductance values in the quantum Hall effect. The derivation of this invariant computes the bulk transverse conductivity while completely disregarding the edge states; whereas in a finite sample, bulk conductivity vanishes and the transverse conductivity depends entirely on the edge states. It can be argued that the measured conductance is equal in value to the quantity predicted by TKNN invariant, but not physically the same. A similar argument can be made for the quantized electromagnetic response established in Ref. [56], the derivation of which is again based on the bulk properties. It is worth mentioning in this context that topological phase transitions do not necessarily leave any thermodynamic signatures unless the non-extensive effects due to the boundary are considered [57].

The main hurdle in theoretical investigation of the boundary physics of topological systems is that a system with boundary is no longer described by the standard band theory. The band theory of solids, which has been phenomenally successful over the last century in explaining and predicting many important properties, rests on Bloch's theorem [58]. The latter describes the structure of energy eigenstates in the independent electron approximation, by leveraging the translational symmetry of systems that are effectively assumed to be of infinite extent (or, mathematically, are subject to Born–von Karman periodic boundary conditions (BCs)). The latter assumption means that band theory is well suited to describe the properties that are least affected by the presence of the boundary. Yet, the characteristic boundary properties of topological materials are determined by the energy states localized near the boundary, which are outside the purview of Bloch's theorem. Much of the work in this thesis is directed towards filling this gap.

In light of the above discussion, it is clear that a correspondence between the bulk properties of the topological materials and their boundary properties is crucial for the very definition of topological phases. The bulk-boundary correspondence principle is generally assumed to hold as long as the classifying symmetries are obeyed by the system as a whole. If these symmetries are broken explicitly, say by disorder on the boundary, then the boundary physics deviates from the expected behavior, even though the bulk invariant is still well-defined. This suggests that a very careful evaluation of not only the very formulation of the bulk-boundary correspondence, but also of the physical conditions under which it holds, is in order. As mentioned, a number of works have attempted a characterization of boundary physics and bulk-boundary correspondence in SPT phases. These include both the identification and verification of correspondences on a case-by-case basis, as well as some attempts at formulating and rigorously proving the correspondence for relevant classes of models. Despite these efforts, a complete picture is far from being reached, even for the well-established "strong" topological phases. For the "weak" topological phases in which the boundary properties are not as robust, and also for gapless topological phases, no consensus about the form of the bulk-boundary correspondence has been reached as yet. In so-called "higher-order topological insulators," which host localized energy states on boundaries of co-dimension two or higher (for example, 1D edge of a 3D sample), whether any form of bulk-boundary correspondence holds at all remains to be seen.

In the context of establishing rigorous bulk-boundary identities, the series of works based on operator K-theory suffer a common drawback. While they have been very successful in establishing some rigorous connections, valid also for disordered systems, they hide or ignore much of the information about the energy states localized on the boundary, other than their topological invariant (boundary invariant). Instead, a framework for explaining the bulk-boundary correspondence, that is based on an approach to explicit calculation of localized energy states, can be expected to provide much more physical insight and theoretical control over the investigation of topological phases of matter.

We emphasize that, in this thesis, we are not concerned with states of matter that are said to have "intrinsic topological order," such as fractional quantum Hall states [59] or the ground states of Kitaev's toric code [60]. Although points of contact with SPT phases exist (notably, in terms of robustness against perturbations), this kind of topological order is known to be associated with long-range entanglement, and the resulting quasiparticles are predicted to exhibit fractional charge, among other exotic features. The topological states we are concerned with here are short-range entangled instead, described by the topology of the wavefunctions of the filled band in the independent fermions approximation.

1.2 Outline

We now describe the main contributions of each of the chapters in this thesis. A more detailed outline of the contents of each chapter is provided in each case at its beginning.

In Chap. 2, we formulate a generalization of Bloch's theorem for systems of independent electrons terminated by two parallel hyperplanes, and subject to arbitrary BCs. This generalization provides an exact description of the energy states of such systems in the tight-binding approximation, by allowing the crystal momenta to take complex values. This result is crucial for getting an analytic grasp on the localized energy states in clean systems, and characterizing their response to disorder on the boundary. We successfully capture the exact interplay between (explicitly) broken symmetry of lattice translation and arbitrary BCs in the formation of energy eigenstates, via a "boundary matrix." The key ingredient in our analysis is the separation of the time-independent Schrödinger equation into simultaneous bulk and boundary equations. The former is independent of the BCs and can be solved algebraically. We show that our analysis can be extended to the limiting case of systems of very large size, which is often used to make exact connections between bulk topological invariants and localized energy states. We also describe the extension of the theorem to systems with multiple bulks, notably, junctions.

In Chap. 3, we apply the generalized Bloch theorem to several prototypical models of topological materials in 1D and 2D. The main motivation is to study the localized energy states and bulk-boundary correspondence in these models. The generalized Bloch theorem allows calculation of the energy and wavefunction of the localized energy states in closed-form for several of these models, which is the most complete characterization one can hope for. Not only does it provides a verification of proposed correspondences in these models, but it also affords some insight into the underlying mechanism. We show that the generalized Bloch theorem is a very useful tool also for studying the Josephson response of topological superconductors, as well as Andreev bound states on an SNS junction. Further, we demonstrate its utility for Hamiltonian engineering with a simple 1D model.

In Chap. 4, we outline an approach based on matrix factorization to prove the bulk-boundary correspondence conjectures in the literature as applicable to SPT phases in 1D, with an emphasis on physical interpretation of the intermediate and final results. Such rigorously established relations are necessary, although not sufficient, to characterize SPT phases on the basis of boundary physics. Our approach is applicable to all five non-trivial classes in 1D. The main tool that we employ in our proofs is the "matrix Wiener–Hopf factorization," which provides an analytic method of solving initial value problems for certain type of difference operators. We also define the concept of stability of zero modes, and explore its connections with topology and symmetries. This is an important step towards bringing clarity to various claims in the literature about the topological "protection" of energy states localized on the boundary. We also discuss a model of s-wave topological superconductor, in which the Majorana modes are protected by multiple sets of symmetries, a scenario not discussed in the ten-fold classification scheme.

In Chap. 5, we provide the mathematical proof of all the lemmas and theorems appearing in Chap. 2. The proof of the generalized Bloch theorem relies entirely on the analysis in this chapter. We consider a more general setting, that of an eigenvalue problem of a "corner-modified banded block-Toeplitz (BBT) matrix," without the assumption of Hermiticity. The key ingredient in the analysis is the use of the so-called "Smith normal form" of matrix polynomials for solving the (possibly generalized) eigenvalue problem of corner-modified BBT matrices.

The main results of all chapters and their implications, along with future directions, are discussed in Chap. 6.

The discussion in Chaps. 2 and 3 is based on publications [61, 62] and [63], except for some calculations in the section on Kitaev's Majorana chain adopted from Ref. [64]. Chapter 4 is based on a manuscript currently under preparation [65]. Chapter 5 presents the relevant results from Ref. [64].

References

1. L.D. Landau, E.M. Lifshitz, Chapter I- the fundamental principles of statistical physics, in *Course of Theoretical Physics*, 3rd edn., ed. by L.D. Landau, E.M. Lifshitz (Pergamon, Oxford, 1980), pp. 1–33
2. Y. Ando, Topological insulator materials. J. Phys. Soc. Jpn. **82**, 102001 (2013). https://doi.org/10.7566/JPSJ.82.102001
3. N. Goldman, J.C. Budich, P. Zoller, Topological quantum matter with ultracold gases in optical lattices. Nat. Phys. **12**, 639 (2016). https://doi.org/10.1038/nphys3803
4. T. Ozawa, H.M. Price, A. Amo, N. Goldman, M. Hafezi, L. Lu, M.C. Rechtsman, D. Schuster, J. Simon, O. Zilberberg, I. Carusotto, Topological photonics. Rev. Mod. Phys. **91**, 015006 (2019). https://link.aps.org/doi/10.1103/RevModPhys.91.015006
5. D. Pesin, A.H. MacDonald, Spintronics and pseudospintronics in graphene and topological insulators. Nat. Mater. **11**, 409 (2012). https://doi.org/10.1038/nmat3305
6. C. Nayak, S.H. Simon, A. Stern, M. Freedman, S. Das Sarma, Non-Abelian anyons and topological quantum computation. Rev. Mod. Phys. **80**, 1083–1159 (2008). https://link.aps.org/doi/10.1103/RevModPhys.80.1083
7. V.L. Berezinskii, Destruction of long-range order in one-dimensional and two-dimensional systems possessing a continuous symmetry group. II. Quantum systems. J. Exp. Theor. Phys. **34**, 610 (1972)
8. J.M. Kosterlitz, D.J. Thouless, Ordering, metastability and phase transitions in two-dimensional systems. J. Phys. C: Solid State Phys. **6**, 1181–1203 (1973). https://doi.org/10.1088%2F0022-3719%2F6%2F7%2F010
9. N.D. Mermin, H. Wagner, Absence of ferromagnetism or antiferromagnetism in one- or two-dimensional isotropic Heisenberg models. Phys. Rev. Lett. **17**, 1133–1136 (1966). https://link.aps.org/doi/10.1103/PhysRevLett.17.1133
10. K. von Klitzing, G. Dorda, M. Pepper, New method for high-accuracy determination of the fine-structure constant based on quantized Hall resistance. Phys. Rev. Lett. **45**, 494–497 (1980). https://link.aps.org/doi/10.1103/PhysRevLett.45.494
11. D.J. Thouless, M. Kohmoto, M.P. Nightingale, M. den Nijs, Quantized Hall conductance in a two-dimensional periodic potential. Phys. Rev. Lett. **49**, 405–408 (1982). https://link.aps.org/doi/10.1103/PhysRevLett.49.405
12. B.I. Halperin, Quantized hall conductance, current-carrying edge states, and the existence of extended states in a two-dimensional disordered potential. Phys. Rev. B **25**, 2185–2190 (1982). https://link.aps.org/doi/10.1103/PhysRevB.25.2185
13. Y. Hatsugai, Chern number and edge states in the integer quantum Hall effect. Phys. Rev. Lett. **71**, 3697–3700 (1993). https://link.aps.org/doi/10.1103/PhysRevLett.71.3697
14. F.D.M. Haldane, Nonlinear field theory of large-spin Heisenberg antiferromagnets: semiclassically quantized solitons of the one-dimensional easy-axis néel state. Phys. Rev. Lett. **50**, 1153–1156 (1983). https://link.aps.org/doi/10.1103/PhysRevLett.50.1153
15. I. Affleck, T. Kennedy, E.H. Lieb, H. Tasaki, Rigorous results on valence-bond ground states in antiferromagnets. Phys. Rev. Lett. **59**, 799–802 (1987). https://link.aps.org/doi/10.1103/PhysRevLett.59.799

16. F.D.M. Haldane, Model for a quantum hall effect without landau levels: condensed-matter realization of the "parity anomaly". Phys. Rev. Lett. **61**, 2015–2018 (1988). https://link.aps.org/doi/10.1103/PhysRevLett.61.2015
17. C.L. Kane, E.J. Mele, Quantum spin Hall effect in graphene. Phys. Rev. Lett. **95**, 226801 (2005). https://link.aps.org/doi/10.1103/PhysRevLett.95.226801
18. B.A. Bernevig, T.L. Hughes, S.-C. Zhang, Quantum spin Hall effect and topological phase transition in HgTe quantum wells. Science **314**, 1757–1761 (2006). https://science.sciencemag.org/content/314/5806/1757
19. M. König, S. Wiedmann, C. Brüne, A. Roth, H. Buhmann, L.W. Molenkamp, X.-L. Qi, S.-C. Zhang, Quantum spin hall insulator state in HgTe quantum wells. Science **318**, 766–770 (2007). https://science.sciencemag.org/content/318/5851/766
20. A.R. Mellnik, J.S. Lee, A. Richardella, J.L. Grab, P.J. Mintun, M.H. Fischer, A. Vaezi, A. Manchon, E.-A. Kim, N. Samarth, D.C. Ralph, Spin-transfer torque generated by a topological insulator. Nature **511**, 449 (2014). https://doi.org/10.1038/nature13534
21. M. Diez, A.M.R.V.L. Monteiro, G. Mattoni, E. Cobanera, T. Hyart, E. Mulazimoglu, N. Bovenzi, C.W.J. Beenakker, A.D. Caviglia, Giant negative magnetoresistance driven by spin-orbit coupling at the laalo3/srtio3 interface. Phys. Rev. Lett. **115**, 016803 (2015). https://link.aps.org/doi/10.1103/PhysRevLett.115.016803
22. L. Fu, C.L. Kane, Topological insulators with inversion symmetry. Phys. Rev. B **76**, 045302 (2007). https://link.aps.org/doi/10.1103/PhysRevB.76.045302
23. D. Hsieh, D. Qian, L. Wray, Y. Xia, Y.S. Hor, R.J. Cava, M.Z. Hasan, A topological Dirac insulator in a quantum spin Hall phase. Nature **452**, 970 (2008). https://doi.org/10.1038/nature06843
24. M.Z. Hasan, C.L. Kane, Colloquium: topological insulators. Rev. Mod. Phys. **82**, 3045–3067 (2010). https://link.aps.org/doi/10.1103/RevModPhys.82.3045
25. X.-L. Qi, S.-C. Zhang, Topological insulators and superconductors. Rev. Mod. Phys. **83**, 1057–1110 (2011). https://link.aps.org/doi/10.1103/RevModPhys.83.1057
26. M. Sato, Y. Ando, Topological superconductors: a review. Rep. Prog. Phys. **80**, 076501 (2017). https://doi.org/10.1088%2F1361-6633%2Faa6ac7
27. E. Majorana, Teoria simmetrica dell'elettrone e del positrone. Il Nuovo Cimento (1924–1942) **14**, 171 (2008). https://doi.org/10.1007/BF02961314
28. A.Y. Kitaev, Unpaired majorana fermions in quantum wires. Phys.-Uspekhi **44**, 131–136 (2001). https://doi.org/10.1070%2F1063-7869%2F44%2F10s%2Fs29
29. A. Kitaev, Fault-tolerant quantum computation by anyons. Ann. Phys. **303**, 2–30 (2003). http://www.sciencedirect.com/science/article/pii/S0003491602000180
30. J.K. Pachos, *Introduction to Topological Quantum Computation* (Cambridge University Press, Cambridge, 2012)
31. N. Read, D. Green, Paired states of fermions in two dimensions with breaking of parity and time-reversal symmetries and the fractional quantum hall effect. Phys. Rev. B **61**, 10267–10297 (2000). https://link.aps.org/doi/10.1103/PhysRevB.61.10267
32. G.E. Volovik, *The Universe in a Helium Droplet*, vol. 117 (Oxford University Press, Oxford, 2003)
33. S. Deng, L. Viola, G. Ortiz, Majorana modes intime-reversal invariant s-wave topological superconductors. Phys. Rev. Lett. **108**, 036803 (2012). https://link.aps.org/doi/10.1103/PhysRevLett.108.036803
34. S. Deng, G. Ortiz, L. Viola, Multiband s-wave topological superconductors: role of dimensionality and magnetic field response. Phys. Rev. B **87**, 205414 (2013). https://link.aps.org/doi/10.1103/PhysRevB.87.205414
35. S. Ryu, Y. Hatsugai, Topological origin of zero-energy edge states in particle-hole symmetric systems. Phys. Rev. Lett. **89**, 077002 (2002). https://link.aps.org/doi/10.1103/PhysRevLett.89.077002
36. A. Altland, M.R. Zirnbauer, Nonstandard symmetry classes in mesoscopic normal-superconducting hybrid structures. Phys. Rev. B **55**, 1142–1161 (1997). https://link.aps.org/doi/10.1103/PhysRevB.55.1142

37. S. Ryu, A.P. Schnyder, A. Furusaki, A.W.W. Ludwig, Topological insulators and superconductors: tenfold way and dimensional hierarchy. New J. Phys. **12**, 065010 (2010). https://doi.org/10.1088/1367-2630/12/6/065010
38. L. Fu, Topological crystalline insulators. Phys. Rev. Lett. **106**, 106802 (2011). https://link.aps.org/doi/10.1103/PhysRevLett.106.106802
39. C.-K. Chiu, A.P. Schnyder, Classification of reflection-symmetry-protected topological semimetals and nodal superconductors. Phys. Rev. B **90**, 205136 (2014). https://link.aps.org/doi/10.1103/PhysRevB.90.205136
40. S. Deng, G. Ortiz, A. Poudel, L. Viola, Majorana flat bands in s-wave gapless topological superconductors. Phys. Rev. B **89**, 140507 (2014). https://link.aps.org/doi/10.1103/PhysRevB.89.140507
41. A. Damascelli, Z. Hussain, Z.-X. Shen, Angle-resolved photoemission studies of the cuprate superconductors. Rev. Mod. Phys. **75**, 473–541 (2003). https://link.aps.org/doi/10.1103/RevModPhys.75.473
42. J. Wiedenmann, E. Bocquillon, R.S. Deacon, S. Hartinger, O. Herrmann, T.M. Klapwijk, L. Maier, C. Ames, C. Brüne, C. Gould, A. Oiwa, K. Ishibashi, S. Tarucha, H. Buhmann, L.W. Molenkamp, 4π-periodic Josephson supercurrent in HgTe-based topological Josephson junctions. Nat. Commun. **7**, Article, 10303 (2016). https://doi.org/10.1038/ncomms10303
43. S.M. Albrecht, A.P. Higginbotham, M. Madsen, F. Kuemmeth, T.S. Jespersen, J. Nygrard, P. Krogstrup, C.M. Marcus, Exponential protection of zero modes in Majorana islands. Nature **531**, 206 EP (2016). https://doi.org/10.1038/nature17162
44. V. Mourik, K. Zuo, S.M. Frolov, S.R. Plissard, E.P.A.M. Bakkers, L.P. Kouwenhoven, Signatures of Majorana fermions in hybrid superconductor-semiconductor nanowire devices. Science **336**, 1003–1007 (2012). https://science.sciencemag.org/content/336/6084/1003
45. S. Nadj Perge, I.K. Drozdov, J. Li, H. Chen, S. Jeon, J. Seo, A.H. MacDonald, B.A. Bernevig, A. Yazdani, Observation of Majorana fermions in ferromagnetic atomic chains on a superconductor. Science **346**, 602–607 (2014). https://science.sciencemag.org/content/346/6209/602
46. E. Prodan, H. Schulz-Baldes, *Bulk and Boundary Invariants for Complex Topological Insulators: From k-Theory to Physics*, 1st edn., vol. 117 (Springer, Cham, 2016)
47. V. Mathai, G.C. Thiang, T-duality of topological insulators. J. Phys. A: Math. Theor. **48**, 42FT02 (2015). https://doi.org/10.1088%2F1751-8113%2F48%2F42%2F42ft02
48. K.C. Hannabuss, T-duality and the bulk-boundary correspondence. J. Geom. Phys. **124**, 421–435 (2018). http://www.sciencedirect.com/science/article/pii/S0393044017302966
49. J.C. Avila, H. Schulz-Baldes, C. Villegas-Blas, Topological invariants of edge states for periodic two-dimensional models. Math. Phys. Anal. Geom. **16**, 137–170 (2013). https://doi.org/10.1007/s11040-012-9123-9
50. C. Bourne, J. Kellendonk, A. Rennie, The k-theoretic bulk–edge correspondence for topological insulators. Ann. Henri Poincaré **18**, 1833–1866 (2017). https://doi.org/10.1007/s00023-016-0541-2
51. A.M. Essin, V. Gurarie, Bulk-boundary correspondence of topological insulators from their respective green's functions. Phys. Rev. B **84**, 125132 (2011). https://link.aps.org/doi/10.1103/PhysRevB.84.125132
52. A.M. Essin, V. Gurarie, Delocalization of boundary states in disordered topological insulators. J. Phys. A: Math. Theor. **48**, 11FT01 (2015). https://doi.org/10.1088%2F1751-8113%2F48%2F11%2F11ft01
53. R.S.K. Mong, V. Shivamoggi, Edge states and the bulk-boundary correspondence in Dirac Hamiltonians. Phys. Rev. B **83**, 125109 (2011). https://link.aps.org/doi/10.1103/PhysRevB.83.125109
54. C.W.J. Beenakker, D.I. Pikulin, T. Hyart, H. Schomerus, J.P. Dahlhaus, Fermion-parity anomaly of the critical supercurrent in the quantum spin-hall effect. Phys. Rev. Lett. **110**, 017003 (2013). https://link.aps.org/doi/10.1103/PhysRevLett.110.017003

55. A.R. Akhmerov, J.P. Dahlhaus, F. Hassler, M. Wimmer, C.W.J. Beenakker, Quantized conductance at the majorana phase transition in a disordered superconducting wire. Phys. Rev. Lett. **106**, 057001 (2011). https://link.aps.org/doi/10.1103/PhysRevLett.106.057001
56. X.-L. Qi, T.L. Hughes, S.-C. Zhang, Topological field theory of time-reversal invariant insulators. Phys. Rev. B **78**, 195424 (2008). https://link.aps.org/doi/10.1103/PhysRevB.78.195424
57. S.N. Kempkes, A. Quelle, C.M. Smith, Universalities of thermodynamic signatures in topological phases. Sci. Rep. **6**, Article, 38530 (2016). https://doi.org/10.1038/srep38530
58. N.W. Ashcroft, N.D. Mermin, *Solid State Physics*, 1st edn. (Holt, Rinehart and Winston, New York, 1976)
59. X.G. Wen, Q. Niu, Ground-state degeneracy of the fractional quantum Hall states in the presence of a random potential and on high-genus Riemann surfaces. Phys. Rev. B **41**, 9377–9396 (1990). https://link.aps.org/doi/10.1103/PhysRevB.41.9377
60. A. Kitaev, Anyons in an exactly solved model and beyond. Ann. Phys. **321**, January Special Issue, 2–111 (2006). http://www.sciencedirect.com/science/article/pii/S0003491605002381
61. A. Alase, E. Cobanera, G. Ortiz, L. Viola, Exact solution of quadratic fermionic hamiltonians for arbitrary boundary conditions. Phys. Rev. Lett. **117**, 076804 (2016). https://link.aps.org/doi/10.1103/PhysRevLett.117.076804
62. A. Alase, E. Cobanera, G. Ortiz, L. Viola, Generalization of Bloch's theorem for arbitrary boundary conditions: theory. Phys. Rev. B **96**, 195133 (2017). https://link.aps.org/doi/10.1103/PhysRevB.96.195133
63. E. Cobanera, A. Alase, G. Ortiz, L. Viola, Generalization of Bloch's theorem for arbitrary boundary conditions: interfaces and topological surface band structure. Phys. Rev. B **98**, 245423 (2018). https://link.aps.org/doi/10.1103/PhysRevB.98.245423
64. E. Cobanera, A. Alase, G. Ortiz, L. Viola, Exact solution of corner-modified banded block-Toeplitz eigensystems. J. Phys. A: Math. Theor. **50**, 195204 (2017). https://doi.org/10.1088/1751-8121/aa6046
65. A. Alase, E. Cobanera, G. Ortiz, L. Viola, Matrix factorization approach to bulk-boundary correspondence and stability of zero modes (in preparation)

Chapter 2
Generalization of Bloch's Theorem to Systems with Boundary

In this chapter, we generalize Bloch's theorem to systems of independent electrons subject to arbitrary BCs. Intuitively, one may expect that the structure of the energy eigenstates of a system, in which translational symmetry is only mildly broken by BCs, may still be characterized, since away from the boundary the eigenstates can only witness a clean system. In essence, the generalized Bloch theorem makes this idea precise, by providing an exact (often in fully closed-form) description of the eigenstates of the system's Hamiltonian in terms of generalized eigenstates of *non-unitary* representations of translational symmetry in infinite space, that is, with boundary at infinity and no torus topology. The main motivation behind formulating and deriving this generalization of Bloch's theorem is to eventually extend the band theory of solids to take into account the effect of boundary. Such a theory can possibly allow for a more transparent treatment of boundary physics and bulk-boundary correspondence in topological materials.

Our generalization of Bloch's theorem assumes tight-binding models of finite range as its starting point. Starting from Tamm's investigation in 1932, tight-binding models have been routinely used for the study of surface states, for topological and topologically trivial systems alike. Their predictions show excellent agreement with experiments. The use of finite range tight-binding models is justified particularly in topological systems, since the relevant topology is known to be a feature of low energy excitations of bounded Hamiltonians.[1] We will comment more about these aspects in Chap. 4.

Our generalization of Bloch's theorem leverages a separation of the time-independent Schrödinger equation into simultaneous bulk and boundary equations. The bulk equation is independent of the BCs, and can be solved algebraically. A unifying theme behind these results is an effective *analytic continuation to the complex plane of the standard Bloch's Hamiltonian* off the Brillouin zone. As a result, both exponentially decaying edge modes and more exotic modes with *power-*

[1]See Ref. [1] for more details on a continuum treatment of topological phases of matter.

A. Alase, *Boundary Physics and Bulk-Boundary Correspondence in Topological Phases of Matter*, Springer Theses, https://doi.org/10.1007/978-3-030-31960-1_2

law prefactors can emerge, provided the BCs allow them. Such exotic states were previously believed to arise only in systems with long-range couplings [2–6]. The analytic continuation of the Bloch Hamiltonian is remarkably useful because the original problem reduces to a matrix polynomial function. Interestingly, a recent study made use of similar polynomial structures for the purpose of topological classification [7].

The outline of this chapter is as follows. In Sect. 2.1, we provide a motivating example of a tight-binding Hamiltonian modeling an impurity at the edges, which illustrates the approach behind our general derivation. In Sect. 2.2, we reduce the eigenvalue problem of the many-electron finite-range quadratic fermionic Hamiltonian describing a system in one spatial dimension, subject to specified BCs, to the one of a single-particle BdG Hamiltonian that has the structure of a "corner-modified BBT matrix." We develop a structural characterization of the energy eigenstates for the many-electron systems under consideration, which culminates into our generalization of Bloch's theorem in Sect. 2.2.3. In Sect. 2.3, we discuss the implications of our generalization of Bloch's theorem. Like the usual Bloch's theorem, such a generalization is first and foremost a practical tool for calculations, granting direct access to exact energy eigenvalues and eigenstates. We provide two new procedures—one numerical and another algebraic—for carrying out the exact diagonalization of the single-particle Hamiltonian, based on the generalized Bloch theorem. The algebraic procedure, which may provide *closed-form* solutions to the problem, will be explicitly illustrated through a number of examples in Chap. 3. Crucially, our generalized Bloch theorem also allows derivation of an "indicator" for the bulk-boundary correspondence, which contains information from both the bulk *and* the BCs and is computationally more efficient than other indicators also applicable in the absence of translational symmetry [8]. We also establish some important connections between our generalized Bloch theorem and the widely employed "transfer matrix" approach [9]. Finally, we extend our generalized Bloch theorem in Sect. 2.4 by establishing procedures for the exact diagonalization of D-dimensional clean systems subject to arbitrary BCs (surface disorder included) on two parallel hyperplanes. We accomplish this by allowing for the BCs to be adjusted in order to conveniently describe surface relaxation, reconstruction, or disorder in terms of an appropriate boundary matrix. We also show in the same section how to diagonalize "multi-component" systems that host hyper-planar interfaces separating clean bulks, namely, "junctions."

2.1 An Impurity Problem as a Motivating Example

Consider the simple tight-binding Hamiltonian

$$\widehat{H}_{\text{tot}} = \widehat{H}_N = -\mathrm{t}\sum_{j=1}^{N-1}(\hat{\Phi}_j^\dagger \hat{\Phi}_{j+1} + \hat{\Phi}_{j+1}^\dagger \hat{\Phi}_j),$$

defined on an open chain of N (even) lattice sites with nearest-neighbor hopping strength t, where $\hat{\Phi}_j, \hat{\Phi}_j^\dagger$ denote annihilation and creation operator respectively corresponding to the fermionic orbital at site j, and satisfy the canonical anti-commutation relations. The corresponding single-particle Hamiltonian is [10]

$$H_N = -\mathsf{t}\sum_{j=1}^{N-1}(|j\rangle\langle j+1| + |j+1\rangle\langle j|),$$

where $\{|j\rangle,\ j = 1,\ldots,N\}$ is the orthonormal basis of single-particle states labeled by the lattice site. This Hamiltonian breaks translation-invariance due to the presence of the boundary, so that the crystal momentum is not a good quantum number. In fact, for any $k \in [-\pi,\pi)$, the state (not an eigenstate) $|k\rangle \equiv \frac{1}{\sqrt{N}}\sum_{j=1}^{N} e^{ikj}|j\rangle$ obeys

$$H_N|k\rangle = -2\mathsf{t}\cos k|k\rangle + \frac{\mathsf{t}}{\sqrt{N}}\Big(|1\rangle + e^{ik(N+1)}|N\rangle\Big), \tag{2.1}$$

with a similar relation holding for $-k$

$$H_N|-k\rangle = -2\mathsf{t}\cos k|-k\rangle + \frac{\mathsf{t}}{\sqrt{N}}\Big(|1\rangle + e^{-ik(N+1)}|N\rangle\Big). \tag{2.2}$$

The first term on the right-hand side of Eqs. (2.1)–(2.2) indicates that $|k\rangle$ and $|-k\rangle$ "almost" (for large N) satisfy the eigenvalue relation with energy $-2\mathsf{t}\cos k$, while the two terms in the brackets show that the eigenvalue relation is violated near the two edges of the chain. Under "periodic" BCs, $-2\mathsf{t}\cos k$ is the actual energy eigenvalue of the eigenstate $|k\rangle$ (and $|-k\rangle$), and k is the crystal momentum [11], given by $k = 2\pi q/N \in [-\pi,\pi)$, $q = -N/2,\ldots,N/2-1$.

Because of the identical first term $-2\mathsf{t}\cos k$ in Eqs. (2.1) and (2.2), the states $|k\rangle$ and $|-k\rangle$ can be linearly combined in order to cancel off the similar-looking boundary contributions. For $\alpha,\beta \in \mathbb{C}$, the eigenvalue relation

$$H_N\Big(\alpha|k\rangle + \beta|-k\rangle\Big) = -2\mathsf{t}\cos k\Big(\alpha|k\rangle + \beta|-k\rangle\Big)$$

is recovered provided that the constraint

$$\frac{\mathsf{t}}{\sqrt{N}}(\alpha+\beta)|1\rangle + \frac{\mathsf{t}}{\sqrt{N}}(\alpha e^{ik(N+1)} + \beta e^{-ik(N+1)})|N\rangle = 0$$

is satisfied. For this to hold, the coefficients of both $|1\rangle$ and $|N\rangle$ must vanish, which leads to the kernel equation

$$\mathsf{t}\begin{bmatrix} 1 & 1 \\ e^{ik(N+1)} & e^{-ik(N+1)} \end{bmatrix}\begin{bmatrix} \alpha \\ \beta \end{bmatrix} \equiv B\begin{bmatrix} \alpha \\ \beta \end{bmatrix} = 0. \tag{2.3}$$

The determinant of the above "boundary matrix" B must vanish, which happens if the condition $e^{i2k(N+1)} = 1$ is satisfied, that is, when $k = \pi q/(N+1)$, $q = 1, \ldots, N$. For each of these values of k, $\alpha = -\beta = 1/\sqrt{2}$ provides the required kernel vector of the boundary matrix, with the resulting N eigenvectors

$$|\epsilon_k\rangle \equiv \frac{|k\rangle - |-k\rangle}{\sqrt{2}} = i\sqrt{\frac{2}{N}}\sum_{j=1}^{N}\sin(kj)|j\rangle,$$

of energy $\epsilon_k = -2\mathsf{t}\cos k$. Notice that the allowed values of k differ from the case of periodic BCs [12].

Encouraged by these results, we change the Hamiltonian by adding an on-site potential at the edges,

$$W = \mathsf{w}(|1\rangle\langle 1| + |N\rangle\langle N|), \quad \mathsf{w} \in \mathbb{R},$$

so that the total single-particle Hamiltonian becomes $H_{\text{tot}} = H_N + W$. The boundary matrix B changes to

$$B \equiv \begin{bmatrix} \mathsf{t} + \mathsf{w}e^{ik} & \mathsf{t} + \mathsf{w}e^{-ik} \\ \mathsf{t}e^{ik(N+1)} + \mathsf{w}e^{ikN} & \mathsf{t}e^{-ik(N+1)} + \mathsf{w}e^{-ikN} \end{bmatrix}.$$

While it is harder to predict analytically the values of k for which it has a non-trivial kernel, it is interesting to examine the limit $\mathsf{w} \gg \mathsf{t}$. In this limit, we can approximate the relevant kernel condition as

$$B\begin{bmatrix} \alpha \\ \beta \end{bmatrix} \approx \mathsf{w}\begin{bmatrix} e^{ik} & e^{-ik} \\ e^{ikN} & e^{-ikN} \end{bmatrix}\begin{bmatrix} \alpha \\ \beta \end{bmatrix} = 0,$$

showing non-trivial solutions if and only if $e^{i2k(N-1)} = 1$. There are now $(N-2)$ k-values yielding stationary eigenstates as before. The two missing eigenstates are localized at the edges, and can be taken to be $|1\rangle$ and $|N\rangle$, to leading order in $\mathsf{t}/\mathsf{w} \ll 1$. These localized states are reminiscent of Tamm–Shockley modes [13, 14].

In hindsight, it is natural to ask whether this approach to diagonalization may be improved and extended to more general Hamiltonians. The answer is Yes, and the next sections provide the appropriate tools. We will begin by generalizing this approach to lattice systems in 1D.

2.2 Generalization of Bloch's Theorem for 1D Systems

2.2.1 Model Hamiltonians

Consider a non-interacting fermionic system defined on a 1D lattice, consisting of $j = 1, \ldots, N$ identical cells, each containing $m = 1, \ldots, d_{\text{int}}$ internal degrees of freedom, associated for instance with spin and orbital motion. Let the creation (annihilation) operator for the orbital labeled by (j, m) be denoted by $\hat{\Phi}^{\dagger}_{jm}$ ($\hat{\Phi}_{jm}$), and let

$$\hat{\Psi}^{\dagger}_{j} \equiv \left[\hat{\Phi}^{\dagger}_{j1} \cdots \hat{\Phi}^{\dagger}_{jd_{\text{int}}}\ \hat{\Phi}_{j1} \cdots \hat{\Phi}_{jd_{\text{int}}}\right]$$

be the corresponding $(2d_{\text{int}})$-dimensional fermionic operator basis. If the system under consideration is clean (disorder-free), except near the boundaries, its Hamiltonian is quadratic in fermionic operators, and can be expressed as

$$\widehat{H}_{\text{tot}} = \frac{1}{2}\sum_{j=1}^{N} \hat{\Psi}^{\dagger}_{j} h_0 \hat{\Psi}_{j} + \frac{1}{2}\sum_{r=1}^{R}\left(\sum_{j-1}^{N-r} \hat{\Psi}^{\dagger}_{j} h_r \hat{\Psi}_{j+r} + \text{H.c.}\right) + \frac{1}{2}\sum_{b,b'} \hat{\Psi}^{\dagger}_{b} W_{bb'} \hat{\Psi}_{b'}, \tag{2.4}$$

where the matrices h_r describe hopping and pairing among fermions situated r cells apart in the bulk, R is the *finite* range of hopping and pairing in the bulk with $R \ll N$, the indices $b, b' \in \{1, \ldots, R,\ N-R+1, \ldots, N\}$, and the matrices $W_{bb'}$ describe hopping and pairing among the fermions at the boundary. For arrays, such as $\hat{\Psi}^{\dagger}_{j}$ and $\hat{\Psi}_{j}$, we follow the convention that those appearing on the left (right) of a matrix are row (column) arrays. The particle-hole symmetry of the Hamiltonian imposes additional block structure

$$h_r = \begin{bmatrix} K_r & \Delta_r \\ -\Delta^*_r & -K^*_r \end{bmatrix}, \quad W_{bb'} = \begin{bmatrix} W^{(K)}_{bb'} & W^{(\Delta)}_{bb'} \\ -W^{(\Delta)*}_{bb'} & -W^{(K)*}_{bb'} \end{bmatrix}$$

on the hopping and pairing matrices, with the blocks satisfying the relations

$$K_r = K^{\dagger}_{-r}, \quad \Delta_r = -\Delta^{\mathsf{T}}_{-r}$$
$$W^{(K)}_{bb'} = W^{(K)\dagger}_{b'b}, \quad W^{(\Delta)}_{bb'} = -W^{(\Delta)\mathsf{T}}_{b'b}.$$

The translation-invariance (in the bulk) of the Hamiltonian in Eq. (2.4) is evident from the terms of the form $\sum_j \hat{\Psi}^{\dagger} h_r \hat{\Psi}_{j+r}$, which are invariant under the shift $j \mapsto j + \Delta j$, with Δj some integer. The finite extent of lattice sites j due to the presence of edges, as well as the terms involving $W_{bb'}$ are responsible for the loss of translation-invariance in $\widehat{H}_{\text{tot}}$. Hamiltonians of the form in Eq. (2.4) arise ubiquitously in mean-field descriptions of fermionic systems as realized in both solid-state and cold-atom platforms [10, 15, 16].

Analyzing the single-particle sector of $\widehat{H}_{\text{tot}}$ suffices to study its many-body energy eigenvalues and eigenvectors [10]. In the Bogoliubov de-Gennes formalism, the single-particle space $\mathcal{H}$ is a vector space that is isomorphic to the space spanned by linear fermionic operators $\{\hat{\Phi}_{jm}, \hat{\Phi}_{jm}^\dagger,\ j = 1, \ldots, N,\ m = 1, \ldots, d_{\text{int}}\}$. Let this isomorphism map the operators $\hat{\Phi}_{jm}$ and $\hat{\Phi}^\dagger_{jm}$ to the vectors $|j\rangle|m\rangle$ and $|j\rangle|m + d_{\text{int}}\rangle$ in $\mathcal{H}$, respectively. Notice that we have conveniently factorized $\mathcal{H}$ into the tensor product of two subsystems, $\mathcal{H} \simeq \mathbb{C}^N \otimes \mathbb{C}^{2d_{\text{int}}} \equiv \mathcal{H}_L \otimes \mathcal{H}_I$, associated with lattice and internal factors respectively. In order to bring the role of translation-invariance to the fore, we will use an unconventional ordering of the single-particle basis. The standard practice [10] is to arrange all creation operators ($\{|j\rangle|m\rangle,\ j = 1, \ldots, N,\ m = 1, \ldots, d_{\text{int}}\}$ in our notation), followed by all annihilation operators ($\{|j\rangle|m\rangle,\ j = 1, \ldots, N,\ m = d_{\text{int}} + 1, \ldots, 2d_{\text{int}}\}$) to form a basis. This ordering leads to the conventional structure

$$H_{\text{tot}} = \begin{bmatrix} K_{\text{tot}} & \Delta_{\text{tot}} \\ -\Delta^*_{\text{tot}} & -K^*_{\text{tot}} \end{bmatrix}$$

of the BdG Hamiltonian. We will use the order $\{|j\rangle|m\rangle,\ j = 1, \ldots, N,\ m = 1, \ldots, 2d_{\text{int}}\}$ of the single-particle basis, so that the particle-hole label absorbed in the internal degrees of freedom appears *after* the lattice label. In this order of the basis, the BdG Hamiltonian is given by $H_{\text{tot}} = H_N + W$, where

$$H_N = h_0 + \sum_{r=1}^{R} (T^r h_r + \text{H.c.}), \qquad W = \sum_{b,b'} |b\rangle\langle b'| W_{bb'}, \tag{2.5}$$

T is the left-shift operator $T|j\rangle = |j-1\rangle,\ \forall j \neq 1,\ T|1\rangle = 0$, and $T^\dagger$ implements the corresponding right shift. In the basis $\{|j\rangle|m\rangle,\ j = 1, \ldots, N,\ m = 1, \ldots, 2d_{\text{int}}\}$, we have

$$H_N = \begin{bmatrix} h_0 \;\ldots\; h_R & & 0 & \cdots & 0 \\ \vdots \;\;\ddots & \ddots & & \ddots & \vdots \\ h_R^\dagger & \ddots & \ddots & & 0 \\ & \ddots & & & \\ & & & \ddots & \\ 0 & \ddots & \ddots & & h_R \\ \vdots \;\;\ddots & & \ddots & \ddots & \vdots \\ 0 \;\cdots\; 0 & & h_R^\dagger & \cdots & h_0 \end{bmatrix},$$

$$W = \begin{bmatrix} w^{(l)}_{11} & \cdots & w^{(l)}_{1R} & 0 & w_{11} & \cdots & w_{1R} \\ \vdots & \ddots & \vdots & \vdots & \vdots & \ddots & \vdots \\ w^{(l)}_{R1} & \cdots & w^{(l)}_{RR} & \vdots & w_{R1} & \cdots & w_{RR} \\ 0 & \cdots & \cdots & 0 & \cdots & \cdots & 0 \\ w^{\dagger}_{11} & \cdots & w^{\dagger}_{1R} & \vdots & w^{(r)}_{11} & \cdots & w^{(r)}_{1R} \\ \vdots & \ddots & \vdots & \vdots & \vdots & \ddots & \vdots \\ w^{\dagger}_{R1} & \cdots & w^{\dagger}_{RR} & 0 & w^{(r)}_{R1} & \cdots & w^{(r)}_{RR} \end{bmatrix},$$

where we have used the notation

$$w^{(l)}_{bb'} \equiv W_{bb'}, \quad w^{(r)}_{bb'} \equiv W_{N-R+b,N-R+b'}, \qquad (2.6)$$
$$w_{bb'} \equiv W_{b,N-R+b'}.$$

Thus, H_{tot} is a "corner-modified" BBT matrix with $2R+1$ bands (see Sect. 5.1.1 for a rigorous definition). Namely, the r-th off-diagonal bands above and below the diagonal have blocks given by bulk hopping/pairing matrices h_r and $h_r^{\dagger}$, respectively, whereas boundary terms appear in the corner of the matrix. We emphasize that the BBT structure of H_N reflects the translation-invariance in the bulk with finite range of hopping and pairing.

If the Hamiltonian in Eq. (2.4) is particle number-conserving ($\Delta_r = 0 = W_r^{(\Delta)}$), then it suffices to diagonalize the single-particle Hamiltonian $H_{\text{tot}} = K + W^{(K)}$, where

$$K = K_0 + \sum_{r=1}^{R} (T^r K_r + \text{H.c.}), \quad W^{(K)} = \sum_{b,b'} |b\rangle\langle b'| W^{(K)}_{bb'}.$$

In this case too, H_{tot} is a corner-modified, BBT matrix, but the blocks $h_r = K_r$ are of size $d_{\text{int}} \times d_{\text{int}}$. In order to have a uniform notation, we shall use

$$d \equiv \begin{cases} d_{\text{int}} & \text{if } \Delta = 0 \text{ (number-conserving)} \\ 2d_{\text{int}} & \text{if } \Delta \neq 0 \text{ (number-non-conserving)} \end{cases}.$$

In this notation, the subsystem decomposition of the single-particle state space is always

$$\mathcal{H} \cong \mathbb{C}^N \otimes \mathbb{C}^d \equiv \mathcal{H}_L \otimes \mathcal{H}_I,$$

irrespective of whether the system is number-conserving or not. Correspondingly, we will denote the orthonormal bases of lattice and internal spaces by $\{|j\rangle,\ j = 1, \ldots, N\}$ and $\{|m\rangle,\ m = 1, \ldots, d\}$ respectively in both the cases.

Before analyzing Hamiltonians with arbitrary BCs, let us revisit the periodic (Born–von Karman) BCs, which are standardly used in calculations of band structure and bulk topological invariants [17].[2] For periodic BCs, the single-particle Hamiltonian H_{tot} in Eq. (2.5) is a circulant block-Toeplitz matrix, which may be expressed as

$$H_{\text{tot}} = H_p \equiv h_0 + \sum_{r=1}^{R} (V_N{}^r h_r + \text{H.c.}),$$

in terms of the cyclic shift operators

$$V_N \equiv T + (T^\dagger)^{N-1}, \quad V_N^\dagger = V_N^{-1} = T^\dagger + T^{N-1}.$$

Crucially, translational symmetry implies that H_p, V_N, and $V_N^\dagger$ form a commutative set, allowing for the eigenspectrum of H_p to be determined via standard discrete Fourier transform from the lattice to the momentum basis on $\mathcal{H}_L$. The Bloch Hamiltonian H_k is the matrix-valued symbol [18] of the circulant block-Toeplitz matrix without boundary terms,

$$H_k \equiv h_0 + \sum_{r=1}^{R} (e^{ikr} h_r + e^{-ikr} h_r^\dagger). \tag{2.7}$$

Here the crystal momentum k takes values from the first Brillouin zone. The Bloch's theorem in this setting dictates that the eigenvectors of H_p may be expressed as $|\epsilon\rangle \equiv |e^{ik}\rangle |u(\epsilon, e^{ik})\rangle$, where

$$|e^{ik}\rangle \equiv \frac{1}{\sqrt{N}} \sum_{j=1}^{N} e^{ikj} |j\rangle$$

are plane wave states in the lattice space $\mathcal{H}_L$, and $|u(\epsilon, e^{ik})\rangle$ is the eigenvector of the Bloch Hamiltonian H_k with eigenvalue ϵ. The reason for choosing e^{ik} to be the argument of the internal vector $|u(\epsilon, e^{ik})\rangle$ will become clear in the later sections.

As lattice translation ceases to be a symmetry of the Hamiltonian, the discrete Fourier transform fails to diagonalize H_{tot}. In particular, the left and right shift operators T and $T^\dagger$ *do not* share a common eigenbasis, calling for a different diagonalization approach. In the following sections, we investigate the structure

[2] It is illuminating to review the footnote 6 in Ref. [11] (p. 135), where the authors discuss the proof of the standard Bloch's theorem.

of the energy eigenstates of the single-particle Hamiltonian H_{tot} using a suitable "bulk-boundary separation" of the time-independent Schrödinger equation, which will culminate in the desired generalization of Bloch's theorem to systems described by such model Hamiltonians. Several steps in the following discussion require more rigorous mathematical justification. For mathematically oriented readers, we have included a detailed and more general discussion in Chap. 5, with pointers to the appropriate sections where required.

2.2.2 *Bulk-Boundary Separation of the Schrödinger Equation*

The Bulk-Boundary System of Equations

The motivating example of Sect. 2.1 suggests that it may be possible to isolate the extent to which boundary effects prevent bulk eigenstates from becoming eigenstates of the actual Hamiltonian H_{tot}. Consider Eqs. (2.1) and (2.2) in particular. We may condense them into a single "relative eigenvalue Equation," $P_B H_N|\pm k\rangle = (-2t\cos k)P_B|\pm k\rangle$, in terms of the projector $P_B \equiv \sum_{j=2}^{N-1}|j\rangle\langle j|$. The extension of this observation to the general class of Hamiltonians $H_{\text{tot}} = H_N + W$ requires only the knowledge of the range R in Eq. (2.4). Define two orthogonal "bulk and boundary projectors" as follows:

$$P_B \equiv \sum_{j=R+1}^{N-R} |j\rangle\langle j|, \quad P_\partial \equiv I_N - P_B,$$

with I_N the identity matrix on $\mathcal{H}$ (see Fig. 2.1). The defining property of the bulk projector is that it annihilates any boundary contribution W, that is, $P_B W = 0$. Because $P_B + P_\partial = I_N$, the "bulk-boundary system of equations,"

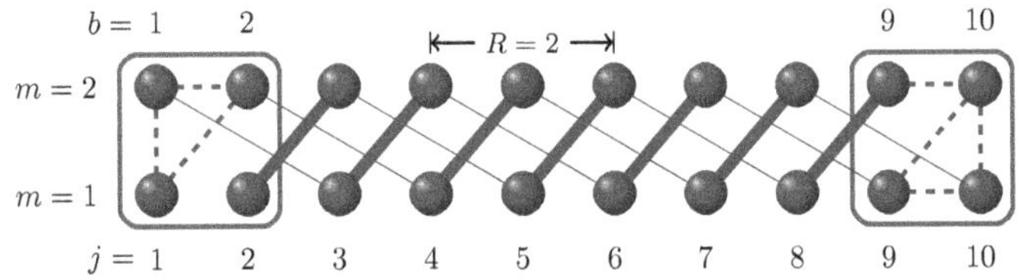

Fig. 2.1 Bulk-boundary separation for a system with two fermionic modes per unit cell, $d = 2$, and next-nearest-neighbor hopping, $R = 2$. Each blue circle stands for a fermionic mode. Thick and thin solid lines indicate two different hopping strengths in the bulk. Since the size of the boundary depends on the range R, the boundary comprises the first and last two unit cells of the chain. Dotted lines stand for arbitrary hopping strengths at the boundary. Figure adapted with permission from Ref. [19]. Copyrighted by the American Physical Society

$$\begin{cases} P_B H_N|\epsilon\rangle = \epsilon P_B|\epsilon\rangle, \\ (P_\partial H_N + W)|\epsilon\rangle = \epsilon P_\partial|\epsilon\rangle, \end{cases} \tag{2.8}$$

may be seen to be completely equivalent to the standard eigenvalue equation, $H_{\text{tot}}|\epsilon\rangle = \epsilon|\epsilon\rangle$ (see Sect. 5.2.1).

The bulk-boundary separation of the eigensystem problem is advantageous, because the bulk equation is, in a well-defined sense, translation-invariant. By extending the shift operator T infinitely in both directions, we obtain a translation-invariant *auxiliary Hamiltonian*,

$$H \equiv h_0 + \sum_{r=1}^{R}(V^r h_r + V^{-r} h_r^\dagger), \tag{2.9}$$

where $V \equiv \sum_{j\in\mathbb{Z}} |j\rangle\langle j+1|$ now denotes the generator of discrete translations on the (infinite-dimensional) lattice vector space spanned by $\{|j\rangle\}_{j\in\mathbb{Z}}$. The subtle difference between the Hamiltonians H_N and H is that while T is not invertible, V *is*, and in fact $V^{-1} = V^\dagger$. This difference is decisive in solving the corresponding eigenvalue problems. On the one hand, the eigenvalue equation $H|\psi\rangle = \epsilon|\psi\rangle$ is equivalent to the infinite system of linear equations

$$h_0|\psi_j\rangle + \sum_{r=1}^{R}\left(h_r|\psi_{j+r}\rangle + h_r^\dagger|\psi_{j-r}\rangle\right) = \epsilon|\psi_j\rangle, \quad j \in \mathbb{Z}, \quad |\psi\rangle \equiv \sum_{j\in\mathbb{Z}} |j\rangle|\psi_j\rangle. \tag{2.10}$$

On the other hand, the bulk equation $P_B H_N|\epsilon\rangle = \epsilon P_B|\epsilon\rangle$, with $|\epsilon\rangle \equiv \sum_{j=1}^{N} |j\rangle|\psi_j\rangle$ is equivalent to Eq. (2.10) but restricted to the finite domain $R < j \leq N - R$. Hence, the bulk equation is underdetermined (there are $2R$ more vector variables than constraints). In particular, if $|\psi\rangle$ is an eigenstate of the infinite Hamiltonian as above, then

$$|\epsilon\rangle = \sum_{j=1}^{N} |j\rangle\langle j|\psi\rangle \equiv P_{1,N}|\psi\rangle$$

is a solution of the bulk equation. It is in this sense of shared solutions with H that the bulk equation is, as anticipated, translation-invariant.

Algebraic Solution of the Bulk Equation

Let us revisit the energy eigenvalue equation Eq. (2.10) for the infinite Hamiltonian. Had the goal been to diagonalize H, one would have focused on finding energy eigenvectors associated with *normalized* states in Hilbert space. However, our model systems are of *finite*[3] extent, and we are only interested in using H as an auxiliary operator for finding all the translation-invariant solutions of the bulk equation. Hence, we will allow H to act on arbitrary vector sequences of the form $\Psi = \sum_{j\in\mathbb{Z}} |j\rangle|\psi_j\rangle$, possibly "well outside" the Hilbert space (possibly not square-summable), and so we will drop Dirac's ket notation. We will denote this space of vector sequences by $\mathcal{V}$, and the linear operators acting on this space in bold font, such as $\boldsymbol{H}$ and $\boldsymbol{V}$. From the standpoint of solving the bulk equation, every sequence that satisfies $\boldsymbol{H}\Psi = \epsilon\Psi$ is acceptable, so one must find them all. In the space of all sequences, the translational symmetry $\boldsymbol{V}$ remains invertible but is no longer unitary, because the notion of adjoint operator (and inner product) is not defined. This is important, because it means that translations need not have their eigenvalues on the unit circle, or be diagonalizable. Nonetheless, $[\boldsymbol{V}, \boldsymbol{H}] = 0$, and so, both these features have interesting physical consequences for finite systems.

We will refer to the space of solutions of the bulk equation as the "bulk solution space" and denote it by

$$\mathcal{M}_{1,N}(\epsilon) \equiv \text{Ker}\, P_B(H_N - \epsilon I_N),$$

for any fixed energy ϵ. Let $\mathcal{M}_{-\infty,\infty}(\epsilon) \equiv \text{Ker}\,(\boldsymbol{H} - \epsilon\boldsymbol{I})$ denote the space of eigenvectors of $\boldsymbol{H}$ of eigenvalue ϵ within the space $\mathcal{V}$, where $\boldsymbol{I}$ stands for identity operator on $\mathcal{V}$. In terms of these spaces, our analysis in Sect. 2.2.2 establishes the relation

$$\boldsymbol{P}_{1,N}\mathcal{M}_{-\infty,\infty} \subseteq \mathcal{M}_{1,N}, \tag{2.11}$$

where we dropped the argument ϵ. Translation invariance is equivalent to the properties $\boldsymbol{V}\mathcal{M}_{-\infty,\infty} \subseteq \mathcal{M}_{-\infty,\infty}$ and $\boldsymbol{V}^{-1}\mathcal{M}_{-\infty,\infty} \subseteq \mathcal{M}_{-\infty,\infty}$.[4] If the matrix h_R is invertible, Eq. (2.11) becomes $\boldsymbol{P}_{1,N}\mathcal{M}_{-\infty,\infty} = \mathcal{M}_{1,N}$ (see Theorem 5.7).

Since $\boldsymbol{V}$ commutes with $\boldsymbol{V}^{-1}$, these two symmetries share eigenvectors of the form

$$\Phi_{z,1}|u\rangle \equiv \left(\sum_{j\in\mathbb{Z}} z^j|j\rangle\right)|u\rangle, \quad z \in \mathbb{C}, \quad z \neq 0,$$

with $|u\rangle$ any internal state: there are d linearly independent eigenvectors of translations for each $z \neq 0$. As a simple but important consequence of the identities

$$\boldsymbol{V}\Phi_{z,1}|u\rangle = z\Phi_{z,1}|u\rangle, \quad \boldsymbol{V}^{-1}\Phi_{z,1}|u\rangle = z^{-1}\Phi_{z,1}|u\rangle,$$

[3]We will discuss systems in large-N limit in Sect. 2.3.2.

[4]For the vector space of semi-infinite systems extending to infinity in the positive direction, only the former relation, that is, $\boldsymbol{T}\mathcal{M}_{0,\infty} \subseteq \mathcal{M}_{0,\infty}$ holds true.

one finds that

$$\boldsymbol{H}\Phi_{z,1}\,|u\rangle = \Phi_{z,1}H(z)|u\rangle, \tag{2.12}$$

where

$$H(z) = h_0 + \sum_{r=1}^{R}(z^r h_r + z^{-r} h_r^\dagger) \tag{2.13}$$

acts on the internal space $\mathcal{H}_I$ only. We refer to $H(z)$ as the "reduced bulk Hamiltonian." Since $H_k = H(z = e^{ik})$ is the usual Bloch Hamiltonian of a 1D system with periodic BCs, $H(z)$ is the analytic continuation of H_k off the Brillouin zone.

One can similarly continue the energy dispersion relation off the Brillouin zone, by relating ϵ to z via

$$\det(H(z) - \epsilon \mathbb{1}_d) = 0, \tag{2.14}$$

where $\mathbb{1}_d$ is the identity on $\mathcal{H}_I$. In practice, it is advantageous to use the polynomial

$$p(\epsilon, z) \equiv z^{Rd} \det(H(z) - \epsilon \mathbb{1}_d). \tag{2.15}$$

We say that ϵ is "regular" if $p(\epsilon, z)$ is not the zero polynomial, and "singular" otherwise. That is, $p(\epsilon, z) \equiv 0$ for all z if ϵ is singular. Such a (slight) abuse of language (see Sect. 5.1.3) is permitted as we are interested in varying ϵ for a fixed Hamiltonian. For any given Hamiltonian of finite range R, as we assumed, there are at most Rd singular energies. Physically, singular energies correspond to flat bands, as one can see by restriction to the Brillouin zone. We can now state the first useful result (see Theorem 5.8 for formal proof):

Theorem 2.1 *For any system size $N > 2R$ and ϵ is regular, the number of independent solutions of the bulk equation is* $\dim \mathcal{M}_{1,N}(\epsilon) = 2Rd$.

This result ties well with the physical meaning of the number $2Rd = \dim(\text{Range } P_\partial)$ as counting the total number of degrees of freedom on the boundary, which is equal to the dimension of the boundary subspace. The condition $N > 2R$ implies that the system is big enough to contain at least one site in the bulk.

Extended-Support Bulk Solutions at Regular Energies

The solutions of the bulk equation that are inherited from $\boldsymbol{H}$ have non-vanishing support on the full lattice space $\mathcal{H}_L$, and are labeled by the eigenvalues of $\boldsymbol{V}$, possibly together with a second "quantum number" that appears because $\boldsymbol{V}$ is not unitary on the space of all sequences. For any $z \neq 0$, if $|u\rangle$ satisfies the eigenvalue equation $H(z)|u\rangle = \epsilon|u\rangle$, then Eq. (2.12) implies that $\Phi_{z,1}|u\rangle$ is an eigenvector of $\boldsymbol{H}$ with eigenvalue ϵ. In order to be more systematic, let $\{z_\ell\}_{\ell=1}^{n}$ denote the n *distinct* non-zero roots of Eq. (2.15), and $\{s_\ell\}_{\ell=1}^{n}$ their respective multiplicities. For generic values of ϵ, $H(z_\ell)$ has exactly s_ℓ eigenvectors $\{|u_{\ell s}\rangle\}_{s=1}^{s_\ell}$ in $\mathcal{H}_I$, satisfying $H(z_\ell)|u_{\ell s}\rangle = \epsilon|u_{\ell s}\rangle, \quad s = 1, \ldots, s_\ell$. Since $\boldsymbol{H}\Phi_{z_\ell,1}|u_{\ell s}\rangle = \epsilon\Phi_{z_\ell,1}|u_{\ell s}\rangle$, the states

$$\boldsymbol{P}_{1,N}\Phi_{z_\ell,1}|u_{\ell s}\rangle = \sum_{j=1}^{N} z_\ell^j|j\rangle|u_{\ell s}\rangle \equiv |z_\ell, 1\rangle|u_{\ell s}\rangle \tag{2.16}$$

are solutions of the bulk equation. Intuitively, these states are "eigenstates of the Hamiltonian up to BCs."

For a few isolated values of ϵ, $H(z_\ell)$ can have less than s_ℓ eigenvectors. However, the number of eigenvectors of $\boldsymbol{H}$ is still s_ℓ (see Theorem 5.13), as we illustrate here by example. Suppose for concreteness that

$$\boldsymbol{H} - \epsilon\boldsymbol{I} = -\frac{\mathsf{t}}{2}(\boldsymbol{V} + \boldsymbol{V}^{-1}) - \epsilon\boldsymbol{I} = -\frac{\mathsf{t}}{2}\boldsymbol{V}^{-1}\prod_{\ell=1,2}(\boldsymbol{V} - z_\ell\boldsymbol{I}), \quad \mathsf{t} \in \mathbb{R}.$$

Since $R = 1$ and $d = 1$, we expect two eigenvectors for each value of ϵ. One concludes that the eigenspace of energy ϵ is spanned by the sequences $\Phi_{z_\ell,1}$, $\ell = 1, 2$, if $z_1 \neq z_2$. But, if $\epsilon = \pm\mathsf{t}$, then $z_1 = z_2 = \mp 1$, and

$$\boldsymbol{H} \mp \mathsf{t}\boldsymbol{I} = -\frac{\mathsf{t}}{2}\boldsymbol{V}^{-1}(\boldsymbol{V} - z_1\boldsymbol{I})^2.$$

How can one get two independent solutions in this case? It turns out that, in addition to $\Phi_{z_1,1}$, $(\boldsymbol{V} - z_1\boldsymbol{I})^2$ contributes another sequence to the kernel of $\boldsymbol{H} - \epsilon\boldsymbol{I}$, namely, $\Phi_{z_1,2} = \sum_{j\in\mathbb{Z}} j z_1^{j-1}|j\rangle$. There are two eigenvectors in total, even though there is only one root.

Returning to the general case, the sequences

$$\Phi_{z,v} = \frac{1}{(v-1)!}\partial_z^{v-1}\Phi_{z,1} = \sum_{j\in\mathbb{Z}} \frac{j^{(v-1)}}{(v-1)!} z^{j-v+1}|j\rangle, \tag{2.17}$$
$$j^{(v)} \equiv j(j-1)\ldots(j-v+1), \quad j^{(0)} \equiv 1,$$

span the kernel of $(\boldsymbol{V} - z\boldsymbol{I})^s$ for $v = 1, \ldots, s$. In other words, $\Phi_{z,v}$ is a "generalized eigenvector" of the translational symmetry $\boldsymbol{V}$ of rank v with eigenvalue z. We refer to eigenvectors with $v > 1$ as the *power-law solutions* of the bulk equation (solutions with a power-law prefactor). They exist because translations are not diagonalizable in the full space of sequences (as opposed to the Hilbert space of square-summable sequences), leading to the new quantum number v.

The power-law solutions of the bulk equation may be found from the action of $\boldsymbol{H}$ on the generalized eigenvectors of $\boldsymbol{V}$. For arbitrary internal state $|u_x\rangle$, we have

$$\boldsymbol{H}\Phi_{z,x}|u_x\rangle = \frac{1}{(x-1)!}\partial_z^{x-1}\Phi_{z,1}H(z)|u_x\rangle. \tag{2.18}$$

Then one can show from Eqs. (2.17) and (2.18) that the action of $\boldsymbol{H}$ on the vector sequence $\Psi = \sum_{x=1}^{v}\Phi_{z,x}|u_x\rangle$, where $\{|u_x\rangle\}$ are arbitrary internal states, is given by

$$\boldsymbol{H}\Psi = \sum_{x=1}^{v}\sum_{x'=1}^{v}\Phi_{z,x}[H_v(z)]_{xx'}|u_{x'}\rangle. \tag{2.19}$$

Here, $H_v(z)$ is an *upper-triangular block-Toeplitz* matrix with non-trivial blocks

$$[H_v(z)]_{xx'} \equiv \frac{1}{(x'-x)!}\partial_z^{x'-x}H(z), \quad 1 \le x \le x' \le v. \tag{2.20}$$

In matrix form, by letting $H^{(x)} \equiv \partial_z^x H(z)$, we have

$$H_v(z) = \begin{bmatrix} H^{(0)} & H^{(1)} & \frac{1}{2}H^{(2)} & \cdots & \frac{1}{(v-1)!}H^{(v-1)} \\ 0 & \ddots & \ddots & \ddots & \vdots \\ \vdots & \ddots & \ddots & \ddots & \frac{1}{2}H^{(2)} \\ \vdots & & \ddots & \ddots & H^{(1)} \\ 0 & \cdots & \cdots & 0 & H^{(0)} \end{bmatrix}.$$

We refer to $H_v(z)$ as the "generalized reduced bulk Hamiltonian of order v." Notice that $H_1(z) = H(z)$. In the partial basis

$$\Phi_z = \left[\Phi_{z,1} \ \dots \ \Phi_{z,v}\right], \tag{2.21}$$

organized as a row vector, the entries of $|u\rangle = \left[|u_1\rangle \ \dots \ |u_v\rangle\right]^{\mathsf{T}}$ are the vector-valued coordinates of Ψ, $\Psi = \Phi_z|u\rangle = \sum_{x=1}^{v}\Phi_{z,x}|u_x\rangle$. Then, Eq. (2.19) can be rewritten as

$$\boldsymbol{H}\Phi_z|u\rangle = \Phi_z H_v(z)|u\rangle.$$

Now it becomes clear that for Ψ to be an eigenvector of $\boldsymbol{H}$, the required condition is $H_v(z)|u\rangle = \epsilon|u\rangle$, which is analogous to the condition derived for the generic case $v = 1$. If a root z_ℓ of Eq. (2.14) has multiplicity s_ℓ, then $\boldsymbol{H}$ has precisely s_ℓ linearly independent eigenvectors corresponding to z_ℓ. This provides a characterization of the eigenstates of $\boldsymbol{H}$, which may be regarded as extending Bloch's theorem to $\boldsymbol{H}$ viewed as a linear transformation on the space of all vector-valued sequences (see Theorem 5.13):

Theorem 2.2 *For fixed, regular* ϵ, *let* $\{z_\ell\}_{\ell=1}^n$ *denote the distinct non-zero roots of Eq.* (2.14), *with respective multiplicities* $\{s_\ell\}_{\ell=1}^n$. *Then, the eigenspace of* $\boldsymbol{H}$ *of energy* ϵ *is a direct sum of n vector spaces spanned by generalized eigenstates of* $\boldsymbol{V}$ *of the form*

$$\Psi_{\ell s} = \Phi_{z_\ell}|u_{\ell s}\rangle = \sum_{v=1}^{s_\ell} \Phi_{z_\ell, v}|u_{\ell s v}\rangle, \quad s = 1, \ldots, s_\ell,$$

where the linearly independent vectors $\{|u_{\ell s}\rangle\}_{s=1}^{s_\ell}$ *are chosen in such a way that* $H_{s_\ell}(z_\ell)|u_{\ell s}\rangle = \epsilon|u_{\ell s}\rangle$, *and* $|u_{\ell s}\rangle = \left[|u_{\ell s 1}\rangle \; \ldots \; |u_{\ell s s_\ell}\rangle\right]^{\mathsf{T}}$.

Once the eigenvectors of $\boldsymbol{H}$ are calculated, the bulk solutions of extended support are readily obtained by projection. Let, for $v \geq 1$,

$$|z, v\rangle \equiv \boldsymbol{P}_{1,N}\Phi_{z,v} = \sum_{j=1}^{N} \frac{j^{(v-1)}}{(v-1)!} z^{j-v+1}|j\rangle$$

be the projections of generalized eigenvectors of $\boldsymbol{V}$. Then

$$\mathcal{B}_{\text{ext}} \equiv \{|\psi_{\ell s}\rangle, \; s = 1, \ldots, s_\ell, \; \ell = 1, \ldots, n\}$$

describes a basis of the extended solutions of the bulk equation, where

$$|\psi_{\ell s}\rangle = \sum_{v=1}^{s_\ell} |z_\ell, v\rangle|u_{\ell s v}\rangle \quad \forall \ell, s. \tag{2.22}$$

It is worth noting that an energy value ϵ lies inside an energy band if and only if at least one of the roots $\{z_\ell\}_{\ell=1}^n$ is of unit norm ($|z_\ell| = 1$). If none of the roots lie on the unit circle, then ϵ necessarily lies in a band gap (or above or below all energy bands). However, it may happen that some, though not all, of the roots $\{z_\ell\}$ lie on the unit circle, as also evidenced by use of the transfer matrix method [20]. Such energy values, in fact, describe the observed phenomenon of "surface resonance" [21]. The bulk eigenstates at such energies can have contributions from exponentially decaying states (corresponding to $|z_\ell| \neq 1$) with large amplitude near the surface, and Bloch-wave like states (corresponding to $|z_\ell| = 1$) that penetrate deep into the bulk. Whether such states are physical depends on compatibility with the BCs.

Remark 2.3 The bulk equation bears power-law solutions only at a few isolated values of ϵ [22]. However, *linear combinations* of $v = 1$ solutions show power-law-like behavior, as soon as two or more of the roots of Eq. (2.14) are sufficiently close to each other. Suppose, for instance, that for some ϵ, two of the roots of Eq. (2.14) coincide at z_*. For energy differing from ϵ by a small amount $\delta\epsilon$, the double root

z_* bifurcates into two roots slightly away from each other, with values $z_* \pm \delta z$. The relevant bulk solution space is spanned by

$$|z_* + \delta z, 1\rangle + |z_* + \delta z, 1\rangle \approx 2|z_*, 1\rangle,$$
$$|z_* + \delta z, 1\rangle - |z_* + \delta z, 1\rangle \approx 2(\delta z/z_*)|z_*, 2\rangle,$$

showing that the second vector has indeed a close resemblance to the power-law solution $|z_*, 2\rangle$. Similar considerations apply if $d > 1$, as it is typically the case in physical applications. Assuming that the relevant bulk solutions at energy $\epsilon + \delta\epsilon$ are described by analytic vector functions $|\psi(z_* + \delta z)\rangle$ and $|\psi(z_* - \delta z)\rangle$, then, from the above analysis, it is clear that for energy ϵ, the power-law bulk solution will be proportional to

$$\lim_{\delta z \to 0} (|\psi(z_* + \delta z)\rangle - |\psi(z_* - \delta z)\rangle) \propto \partial_z |\psi(z_*)\rangle. \tag{2.23}$$

We will make use of this observation for the calculation of power-law solutions in several applications in Chap. 3.

Emergent Solutions at Regular Energies

While the extended solutions of the bulk equation correspond to the non-zero roots of Eq. (2.14), the polynomial $p(\epsilon, z)$ defined in Eq. (2.15) may also include $z_0 = 0$ as a root of multiplicity s_0, that is, we may generally write

$$p(\epsilon, z) = z^{Rd} \det(H(z) - \epsilon \mathbb{1}_d) \equiv c \prod_{\ell=0}^{n} (z - z_\ell)^{s_\ell}, \quad c \neq 0.$$

However, $|z = 0\rangle|u\rangle = 0$ does not describe any state of the system. This observation suggests that the extended solutions of the bulk equation may fail to account for all $2Rd$ solutions we expect for regular ϵ. That this is indeed the case follows from a known result in the theory of matrix polynomials [23], implying that $2Rd = 2s_0 + \sum_{\ell=1}^{n} s_\ell$ for matrix polynomials associated with Hermitian Toeplitz matrices. Hence, the number of solutions of the bulk equation of the form given in Eq. (2.22) is

$$\sum_{\ell=1}^{n} s_\ell = 2Rd - 2s_0. \tag{2.24}$$

We call the missing $2s_0$ solutions of the bulk equation "emergent," because they are no longer controlled by $\boldsymbol{H}$ and (non-unitary) translational symmetry, but rather they appear only because of the truncation of the infinite lattice down to a finite one, and only if $\det h_R = 0$ (see Theorem 5.7). Emergent solutions are a direct, albeit non-generic, manifestation of translation-symmetry-breaking; nonetheless, remarkably, they can also be determined by the analytic continuation of the Bloch Hamiltonian, in a precise sense.

The key to computing the emergent solutions is to relate the problem of solving the bulk equation to a *half-infinite* Hamiltonian, rather than the doubly-infinite $\boldsymbol{H}$ we have exploited thus far. We will denote the space of vector sequences of the form $\sum_{j=1}^{\infty} |j\rangle|\upsilon_j\rangle$ by $\mathcal{V}_\infty$. Let us define the "unilateral shifts"

$$\boldsymbol{S} = \sum_{j=1}^{\infty} |j\rangle\langle j+1|, \quad \boldsymbol{S}^\star = \sum_{j=1}^{\infty} |j+1\rangle\langle j|$$

on the semi-infinite lattice whose cells are labeled by $j = 1, \ldots, \infty$. The Hamiltonian

$$\boldsymbol{H}_\infty \equiv h_0 + \sum_{r=1}^{R} (\boldsymbol{S}^r h_r + (\boldsymbol{S}^\star)^r h_r^\dagger) \tag{2.25}$$

is then the half-infinite counterpart of $\boldsymbol{H}$. Let us also define the projector

$$\boldsymbol{P}_{R+1,\infty} \equiv \sum_{j=R+1}^{\infty} |j\rangle\langle j| = (\boldsymbol{S}^\star)^R \boldsymbol{S}^R.$$

Suppose there is a state $\Upsilon \in \mathcal{V}_\infty$ that solves the equation $\boldsymbol{P}_{R+1,\infty}(\boldsymbol{H}_\infty - \epsilon \boldsymbol{I}_\infty)\Upsilon = 0$. Then one can check that $|\psi\rangle = \boldsymbol{P}_{1,N}\Upsilon$ is a solution of the bulk equation, Eq. (2.8). Clearly, some of the bulk solutions we arrive at in this way using $\boldsymbol{H}_\infty$ will coincide with those obtained from $\boldsymbol{H}$. These are precisely the extended solutions we already computed in Sect. 2.2.2. In contrast, the emergent solutions are obtained *only* from $\boldsymbol{H}_\infty$.

Left-Localized Emergent Bulk Solutions In analogy to the sequences $\Phi_{z,v}$ associated with $\boldsymbol{T}$ in Eq. (2.17), let us define states

$$\Upsilon_{z,1}^+ \equiv \sum_{j=0}^{\infty} z^j |j+1\rangle,$$

$$\Upsilon_{z,v}^+ \equiv \frac{1}{(v-1)!} \frac{d^{v-1}}{dz^{v-1}} \Upsilon_{z,1}^+, \quad v = 2, 3, \ldots. \tag{2.26}$$

in such a way that $\Upsilon_{0,v}^+ = |j = v\rangle$ and, also,

$$\Upsilon_z^+|u\rangle \equiv \sum_{x=1}^{v} \Upsilon_{z,x}^+|u_x\rangle = \begin{bmatrix}\Upsilon_{z,1}^- & \cdots & \Upsilon_{z,v}^+\end{bmatrix}\begin{bmatrix}|u_1\rangle \\ \vdots \\ |u_v\rangle\end{bmatrix}.$$

It is then immediate to verify that

$$K_+(\epsilon, S)\Upsilon_{z,1}^+|u_1\rangle = \Upsilon_{z,1}^+ K_+(\epsilon, z)|u_1\rangle.$$

Moreover, using Eq. (2.26), one also obtains

$$K_+(\epsilon, S)\Upsilon_z^+|u\rangle = \begin{bmatrix}\Upsilon_{z,1}^+ & \cdots & \Upsilon_{z,v}^+\end{bmatrix} K_{+,v}(\epsilon, z)\begin{bmatrix}|u_1\rangle \\ \vdots \\ |u_v\rangle\end{bmatrix}, \tag{2.27}$$

in terms of the upper-triangular $v \times v$ block matrix

$$[K_{+,v}(\epsilon, z)]_{xx'} = \frac{1}{(x'-x)!}\frac{d^{x'-x}K^-(\epsilon, z)}{dz^{x'-x}}, \quad 1 \le x \le x' \le v.$$

It will be crucial for later use to notice that $K_{+,v}(\epsilon, z)$ is a block-Toeplitz matrix. Explicitly, such a matrix takes the form

$$K_+(\epsilon) \equiv \begin{bmatrix}
h_R^\dagger & \cdots & h_0 - \epsilon \mathbb{1}_d & \cdots & h_R & 0 & \cdots & 0 \\
 & \ddots & & \ddots & & \ddots & \ddots & \vdots \\
 & & & & & & \ddots & 0 \\
 & & \ddots & & \ddots & & \ddots & \\
 & & & & & & & h_R \\
 & & & \ddots & & \ddots & & \vdots \\
 & & & & & & & h_0 - \epsilon \mathbb{1}_d \\
0 & & & & \ddots & & & \vdots \\
\vdots & \ddots & & & & & & \\
0 & \cdots & 0 & & & & & h_R^\dagger
\end{bmatrix}, \tag{2.28}$$

for systems with fairly large $s_0 > 2R + 1$.

Both $K_{+,v}(\epsilon, z)$ and $H_v(z)$ are defined by the same formula, recall Eq. (2.55). The key difference between the two is that $K_{+,v}(\epsilon, z)$ is well-defined also at $z = 0$. So suppose that $z_0 = 0$ is a root of $p(\epsilon, z)$ of multiplicity $s_0 > 0$. Then, there are precisely s_0 independent solutions of the equation (see Theorem 5.14) $K_{+,s_0}(\epsilon, z_0 = 0)|u_s^+\rangle = 0$, $s = 1, \ldots, s_0$. The corresponding emergent bulk solutions are

$$|\psi_s^+\rangle = \boldsymbol{P}_{1,N}\Upsilon_0^+|u_s^+\rangle = \sum_{j=1}^{s_0}|j\rangle|u_{sj}^+\rangle. \qquad (2.29)$$

They are localized on the left edge over the first s_0 sites. For Hermitian Hamiltonians, $s_0 \leq dR$ necessarily.

Right-Localized Emergent Bulk Solutions Left-localized emergent bulk solutions cannot appear alone; they can only appear in conjunction with a set of right-localized emergent bulk solutions. The reason is as follows. Consider the unitary, Hermitian operator

$$U = U^\dagger \equiv \sum_{j=1}^{N}|N-j+1\rangle\langle j| \otimes \mathbb{1}_d, \quad U^2 = \mathbb{1}_{dN},$$

which implements a mirror transformation of the lattice, by acting trivially on internal states. The transformed Hamiltonian is the Hermitian block-Toeplitz matrix

$$\widetilde{H}_N = UH_NU = \mathbb{1}_N \otimes h_0 + \sum_{r=1}^{R}(T^r \otimes h_r^\dagger + T^{r\,\dagger} \otimes h_r),$$

in which the hopping matrices have been exchanged as $h_r \leftrightarrow h_r^\dagger$. Therefore, the left-localized emergent bulk solutions for $\widetilde{H}_N$ are dictated by the matrix $\widetilde{K}_+(\epsilon)$ with entries $[\widetilde{K}_+(\epsilon)]_{ij} = [K_+(\epsilon)]_{ij}^\dagger$. If $|\widetilde{\psi}^+\rangle$ denotes a left-localized emergent solution for $\widetilde{H}_N$, then

$$0 = P_B(\widetilde{H}_N - \epsilon)|\widetilde{\psi}_s^+\rangle = UP_B(H-\epsilon)U|\widetilde{\psi}_s^+\rangle$$

(recall that P_B is the bulk projector), implying that the state $U|\widetilde{\psi}_s^+\rangle = \sum_{j=1}^{s_0}|N-j+1\rangle|\tilde{u}_{sj}^+\rangle$ is an emergent bulk solution for H_N, localized on the *right* edge. Similarly, the left-localized emergent bulk solutions of H_N are in one-to-one correspondence with the right-localized emergent solutions of $\widetilde{H}_N$. This conclusion relies heavily on the commutation relation $P_BU = UP_B$, which is always necessarily true for "closed" systems (Hermitian Hamiltonians), as we considered here.

But how can we compute the right-localized emergent bulk solutions directly in terms of H_N? Let $|\widetilde{\psi}_s^+\rangle = \sum_{j=1}^{s_0}|j\rangle|\tilde{u}_{sj}^+\rangle$, $s = 1,\ldots,s_0$, denote the left-localized emergent solutions associated with $\widetilde{H}_N$, and let

$$|\psi_s^-\rangle \equiv \sum_{j=1}^{s_0}|N-s_0+j\rangle|u_{sj}^-\rangle = U|\widetilde{\psi}_s^+\rangle, \quad s = 1,\ldots,s_0, \qquad (2.30)$$

denote the corresponding right-localized emergent solutions of H_N, so that $|u^-_{sj}\rangle \equiv |\tilde{u}^+_{s,s_0-j+1}\rangle$. Because $|u^-_s\rangle = \tilde{U}|\tilde{u}^+_s\rangle$, with $\tilde{U} = \tilde{U}^\dagger = \sum_{j=1}^{s_0} |j\rangle\langle s_0 - j + 1|$, we conclude that $K_-(\epsilon)$ is related to $\widetilde{K}_+(\epsilon)$ via $K_-(\epsilon) = \tilde{U}\widetilde{K}_+(\epsilon)\tilde{U}$. This leads to the entries

$$[K_-(\epsilon)]_{ij} = [\widetilde{K}_+(\epsilon)]_{s_0-i+1,s_0-j+1} = [K_+(\epsilon)]^\dagger_{ji},$$

thanks to the fact that $K_+(\epsilon)$ is a block-Toeplitz matrix. Hence, $K_-(\epsilon) = [K_+(\epsilon)]^\dagger$, as desired. In what follows, we shall denote the bases of left- and right-localized emergent bulk solutions by $\mathcal{B}_+ \equiv \{|\psi_{0s}\rangle\}_{s=1}^{s_0}$ and $\mathcal{B}_- \equiv \{|\psi_{(n+1)s}\rangle\}_{s=1}^{s_0}$, respectively.

Bulk-Localized States at Singular Energies

If h_R is *not* invertible, there can be at most a *finite* number of singular energy values (usually referred to as flat bands), leading to bulk-localized solutions: these solutions are finitely supported and appear everywhere in the bulk. Hence, a singular energy *cannot* be excluded from the physical spectrum of a finite system by way of BCs. In contrast, emergent solutions are also finitely supported but necessarily "anchored" to the edges (and only appearing for regular values of ϵ).

Recall that if ϵ is singular, then $\det(H(z) - \epsilon\mathbb{1}_d) = 0$ holds for any value of z. Thus, there exists an analytic vector function,

$$|v(z)\rangle \equiv \sum_{\delta=0}^{\delta_0} z^{-\delta}|v_\delta\rangle, \quad \delta_0 = (d-1)2Rd, \tag{2.31}$$

satisfying $H(z)|v(z)\rangle = \epsilon|v(z)\rangle$ for all z. To obtain $|v(z)\rangle$, one can construct the "adjugate matrix" of $(H(z) - \epsilon\mathbb{1}_d)$.[5] Hence,

$$(H(z) - \epsilon\mathbb{1}_d)\mathrm{adj}(H(z) - \epsilon\mathbb{1}_d) = \det(H(z) - \epsilon\mathbb{1}_d)\mathbb{1}_d = 0,$$

and so one can use *any* of the non-zero columns of $\mathrm{adj}(H(z) - \epsilon\mathbb{1}_d)$, suitably premultiplied by a power of z, for the vector polynomial $|v(z)\rangle$. By matching powers of z, this becomes

[5] Recall that the adjugate matrix $\mathrm{adj}(A)$ associated with a square matrix A is constructed out of the signed minors of A and satisfies $\mathrm{adj}(A)A = \det(A)\mathbb{1}$.

$$\begin{bmatrix} h_R & 0 & \cdots & 0 \\ h_{R-1} & h_R & \ddots & \vdots \\ \vdots & \ddots & \ddots & 0 \\ \vdots & \ddots & \ddots & \ddots \\ h_R^\dagger & \ddots & \ddots & h_R \\ & \ddots & \ddots & \vdots \\ 0 & & \ddots & \vdots \\ \vdots & \ddots & \ddots & h_{R-1}^\dagger \\ 0 & \cdots & 0 & h_R^\dagger \end{bmatrix} \begin{bmatrix} |v_0\rangle \\ |v_1\rangle \\ \vdots \\ |v_{\delta_0}\rangle \end{bmatrix} = 0. \tag{2.32}$$

The idea now is to use the linearly independent solutions of Eq. (2.32) to construct finite-support solutions of the bulk equation. Let us denote such solutions by $|v_\mu\rangle \equiv \left[|v_{\mu 0}\rangle\ |v_{\mu 1}\rangle \ldots |v_{\mu\delta_0}\rangle\right]^{\mathsf{T}}$, for $\mu = 1, \ldots, \mu_0$. One can check directly that the finitely supported sequences

$$\Psi_{j\mu} \equiv \sum_{\delta=0}^{\delta_0} |j+\delta\rangle |v_{\mu\delta}\rangle, \quad j \in \mathbb{Z}, \quad \mu = 1, \ldots, \mu_0$$

all satisfy $(\boldsymbol{H} - \epsilon \boldsymbol{I})\Psi_{js} = 0$ because $|v_\mu\rangle$ obeys Eq. (2.32). Hence, the states $\boldsymbol{P}_{1,N}\Psi_{j\mu}$ provide finitely supported solutions of the bulk equation. In addition, as long as $2R < j < N - 2R - \delta_0$, the boundary equation is also satisfied trivially, and so *all* such states become eigenvectors of $H_N + W$ with the singular energy ϵ. This is why singular energies, if present for the infinite system, are necessarily also part of the spectrum of the finite system and display macroscopic degeneracy of order $\mathcal{O}(N)$.

Let us further remark that the sequences $\Psi_{j\mu}$ and associated solutions of the bulk equation need *not* be linearly independent. To obtain a complete (rather than over-complete), set of solutions for flat bands, one would require a technical tool, the "Smith normal form" [18], which is discussed in Sect. 5.1.3. We refer the reader to Sect. 5.2.2 for more details, and to Ref. [24] for additional related discussion on flat bands.

The Boundary Matrix

For regular energies, the bulk solutions determine a subspace of the full Hilbert space (Theorem 2.1), whose dimension $2Rd \ll dN$ for typical applications. While not all bulk solutions are eigenstates of $H_{\text{tot}} = H_N + W$, the actual eigenstates must necessarily appear as bulk solutions. Hence, the bulk-boundary separation in

Eq. (2.8), and, in particular, the bulk equation, identifies by way of a translational symmetry analysis a *small* search subspace. In order to find the energy eigenstates efficiently, one must solve the boundary equation on this search subspace. Since the boundary equation is linear, its restriction to the space of bulk solutions can be represented by a matrix, the "boundary matrix." The latter is a square matrix that combines our basis of bulk solutions with the relevant BCs.

Let $\mathcal{B} \equiv \mathcal{B}^{+} \cup \mathcal{B}_{\text{ext}} \cup \mathcal{B}^{-}$ be a basis for $\mathcal{M}_{1,N}$. Then, building on the previous section, and setting $s_{(n+1)} = s_0$, the ansatz state

$$|\epsilon, \boldsymbol{\alpha}\rangle \equiv |\Psi_{\mathcal{B}}\rangle\boldsymbol{\alpha} = \sum_{\ell=0}^{n+1}\sum_{s=1}^{s_\ell} \alpha_{\ell s}|\psi_{\ell s}\rangle$$

represents the solutions of the bulk equation parameterized by the $2Rd$ amplitudes $\boldsymbol{\alpha}$, where

$$\boldsymbol{\alpha} \equiv \begin{bmatrix} \alpha_{01} & \cdots & \alpha_{(n+1)s_{(n+1)}} \end{bmatrix}^{\mathsf{T}}, \qquad |\Psi_{\mathcal{B}}\rangle \equiv \begin{bmatrix} |\psi_{01}\rangle & \cdots & |\psi_{(n+1)s_{(n+1)}}\rangle \end{bmatrix}. \tag{2.33}$$

Moreover, let as before $b = 1, \ldots, R, N-R+1, \ldots, N$ label the boundary sites. Then,

$$\begin{aligned} P_B(H_{\text{tot}} - \epsilon I_N)|\epsilon, \boldsymbol{\alpha}\rangle &= 0 \quad \text{and} \\ P_\partial(H_{\text{tot}} - \epsilon I_N)|\epsilon, \boldsymbol{\alpha}\rangle &= \sum_b |b\rangle\langle b|(H_N + W - \epsilon I_N)|\Psi_{\mathcal{B}}\rangle\boldsymbol{\alpha}. \end{aligned} \tag{2.34}$$

In particular, the boundary equation is equivalent to the requirement that $\langle b|(H_N + W - \epsilon I_N)|\Psi_{\mathcal{B}}\rangle\boldsymbol{\alpha} = 0$ for all boundary sites. Since $\langle b|(H_N + W - \epsilon I_N)|\Psi_{\mathcal{B}}\rangle \equiv \langle b|H_\epsilon|\Psi_{\mathcal{B}}\rangle$ denotes a row array of internal states, it is possible to organize these arrays into the boundary matrix

$$B(\epsilon) \equiv \begin{bmatrix} \langle 1|H_\epsilon|\psi_{01}\rangle & \cdots & \langle 1|H_\epsilon|\psi_{(n+1)s_{(n+1)}}\rangle \\ \vdots & & \vdots \\ \langle R|H_\epsilon|\psi_{01}\rangle & \cdots & \langle R|H_\epsilon|\psi_{(n+1)s_{(n+1)}}\rangle \\ \langle N-R+1|H_\epsilon|\psi_{01}\rangle & \cdots & \langle N-R+1|H_\epsilon|\psi_{(n+1)s_{(n+1)}}\rangle \\ \vdots & & \vdots \\ \langle N|H_\epsilon|\psi_{01}\rangle & \cdots & \langle N|H_\epsilon|\psi_{(n+1)s_{(n+1)}}\rangle \end{bmatrix}. \tag{2.35}$$

By construction, the boundary matrix B is a block matrix of block-size $d \times 1$. In terms of this matrix, Eq. (2.34) provides the useful identity

$$H_{\text{tot}}|\epsilon, \boldsymbol{\alpha}\rangle = \epsilon|\epsilon, \boldsymbol{\alpha}\rangle + \sum_{b,s} |b\rangle B_{bs}(\epsilon)\boldsymbol{\alpha}_s, \quad \epsilon \in \mathbb{R}. \tag{2.36}$$

One may write an analogous equation in Fock space by defining an array

$$\eta^{\dagger}_{\epsilon,\boldsymbol{\alpha}} \equiv \sum_{j=1}^{N} \langle j|\epsilon, \boldsymbol{\alpha}\rangle \hat{\Psi}^{\dagger}_{j}.$$

Then Eq. (2.36) translates into

$$[\widehat{H}_{\text{tot}}, \eta^{\dagger}_{\epsilon,\boldsymbol{\alpha}}] = \epsilon\, \eta^{\dagger}_{\epsilon,\boldsymbol{\alpha}} + \sum_{b,s} \hat{\Psi}^{\dagger}_{b} B_{bs}(\epsilon)\boldsymbol{\alpha}_{s}. \tag{2.37}$$

It is interesting to notice that this (many-body) relation remains true even if ϵ is allowed to be a complex number.

2.2.3 Formulation of the Generalized Bloch Theorem

The bulk-boundary separation of the energy eigenvalue equation shows that actual energy eigenstates are necessarily linear combinations of solutions of the bulk equation. This observation leads to a generalization of Bloch's theorem for independent fermions under arbitrary BCs:

Theorem 2.3 (Generalized Bloch Theorem in 1D) *Let $H_{tot} = H_N + W$ denote the single-particle Hamiltonian of a clean system subject to BCs described by $W = P_\partial W$. If ϵ is a regular energy eigenvalue of H_{tot} of degeneracy $\mathscr{K}_\epsilon$, the associated eigenstates can be taken to be of the form $|\epsilon, \boldsymbol{\alpha}_{\mathrm{k}}\rangle = |\Psi_{\mathcal{B}}\rangle\boldsymbol{\alpha}_{\mathrm{k}}$, $\mathrm{k} = 1, \ldots, \mathscr{K}_\epsilon$, where $\{\boldsymbol{\alpha}_{\mathrm{k}},\ \mathrm{k} = 1, \ldots, \mathscr{K}_\epsilon\}$ is a basis of the kernel of the boundary matrix $B(\epsilon)$ at energy ϵ.*

In short, $(H_N + W)|\epsilon, \boldsymbol{\alpha}\rangle = \epsilon|\epsilon, \boldsymbol{\alpha}\rangle$ if and only if $B(\epsilon)\boldsymbol{\alpha} = 0$, in which case it also follows from Eq. (2.37) that $\eta^{\dagger}_{\epsilon,\boldsymbol{\alpha}}$ is a normal fermionic mode of the many-body Hamiltonian $\widehat{H}$. From now on, we will refer to energy eigenstates of the form $|\Psi_{\mathcal{B}}\rangle\boldsymbol{\alpha}_{\mathrm{k}}$ as "generalized Bloch states." Recall that H acts on $\mathcal{H} \cong \mathbb{C}^N \otimes \mathbb{C}^d$, with couplings of finite range R. A lower bound on N should be obeyed, in order for the above theorem to apply. If $\det h_R \neq 0$, since there are no emergent solutions or flat bands, generalized Bloch states describe the allowed energy eigenstates as soon as $N > 2R$, independently of d. If h_R fails to be invertible, we should require that $N > 2\max(s_0, R)$ to ensure that emergent solutions on opposite edges do not overlap, and are thus independent. Since $s_0 \leq Rd$, this condition is satisfied for any $N > 2Rd$. In general, $N > 2R(d + 1)$ always suffices for generalized Bloch states to describe *generic* energy eigenstates (see Theorem 5.15).

We further note that if ϵ is *not* an energy eigenvalue, the kernel of $B(\epsilon)$ is trivial. Thus, the degeneracy of a single-particle energy level coincides with the dimension of the kernel of $B(\epsilon)$. Let $\rho(\omega)$ denote the single-particle density of states. Combining its definition with the generalized Bloch theorem, we then see that

$$\rho(\omega) = \sum_{\det B(\epsilon)=0} [\dim \operatorname{Ker} B(\epsilon)]\, \delta(\hbar\omega - \epsilon),$$

an alternative formula to the usual

$$\rho(\omega) = -\frac{1}{\pi} \operatorname{Im} \operatorname{Tr} (H_N + W - \hbar\omega + i0^+)^{-1},$$

from the theory of Green's functions [25]. Another interesting, closely related formula is

$$\mathcal{Z}_W = \operatorname{Tr} e^{-\beta(H_N+W)} = \sum_{\det B(\epsilon)=0} \dim \operatorname{Ker} B(\epsilon)\, e^{-\beta\epsilon},$$

for the partition function of the single-particle Hamiltonian, with the dependence on BCs highlighted [26].

The generalized Bloch theorem relies on the complete solution of the bulk equation, Eq. (2.53). As the latter describes an unconventional *relative* eigenvalue problem for the (generally) *non-Hermitian* operator $P_B H_N$, the standard symmetry analysis of quantum mechanics does not apply. It is nonetheless possible to decompose the solution space of the bulk equation into symmetry sectors, if the Hamiltonian obeys unitary symmetries that also commute with the bulk projector P_B. Assume that a unitary operator $\mathcal{S}$ commutes with *both* $H = H_N + W$ and P_B. Then any vector in the bulk solution space satisfies

$$P_B(H_N + W - \epsilon \mathbb{1})|\psi\rangle = 0 \Rightarrow \mathcal{S}^\dagger P_B(H_N + W - \epsilon I_N)\mathcal{S}|\psi\rangle = 0.$$

This implies that the bulk solution space is invariant under the action of $\mathcal{S}$. Therefore, there exists a basis of the bulk solution space in which the action of $\mathcal{S}$ is block-diagonal. This leads to multiple eigenstate ansätze, each labeled by an eigenvalue of $\mathcal{S}$. Further, $\mathcal{S}^\dagger P_B \mathcal{S} = 0$ implies that the boundary subspace (i.e., the kernel of P_B) is also invariant under $\mathcal{S}$. After finding a basis of the boundary subspace in which $\mathcal{S}$ is block-diagonal, the boundary matrix itself splits into several matrices, each labeled by an eigenvalue of $\mathcal{S}$. We will use this strategy in some of the applications in Sects. 2.4.2 and 3.2.

Recovering the Standard Bloch's Theorem

We now show that one consistently recovers the conventional Bloch's theorem for periodic BCs. In this case, the appropriate matrix W reads $W = W_p \equiv \sum_{r=1}^{R} (T^{N-r} h_r^\dagger + \text{H.c.})$, since then one can check that

$$H_p = H_N + W_p = h_0 + \sum_{r=1}^{R}(V_N^r h_r + \text{H.c.}),$$

in terms of the fundamental circulant matrix V_N. The Bloch states are the states that diagonalize H_p and V_N simultaneously. Theorem 2.3 guarantees that we can choose the eigenstates of H_p to be linear combinations of translation-invariant and emergent solutions. Thus, we only need to check if these linear combinations include eigenstates of V_N. There is no hope of retaining the emergent solutions, because they are localized and too few in number (at most $2Rd$) to be rearranged into eigenstates of V_N. The same holds for translation-invariant solutions with a power-law prefactor. Hence, the search subspace that is compatible with the translational symmetry V_N is described by the simplified ansatz

$$|\epsilon, \boldsymbol{\alpha}\rangle = \sum_{\ell=1}^{n} \alpha_{\ell 1}|\psi_{\ell 1}\rangle,$$

n being the number of distinct roots $\{z_\ell\}$ as before. Now, $V_N|\psi_{\ell 1}\rangle = z_\ell|\psi_{\ell 1}\rangle - z_\ell(1 - z_\ell^N)|N\rangle|u_{\ell s_\ell 1}\rangle$, and so the generalized Bloch states can only be eigenstates of V_N if $e^{ik_\ell N} = 1$ with $z_\ell = e^{ik_\ell}$, and all but one entry in $\boldsymbol{\alpha}$ vanish. That is, $|\epsilon, \boldsymbol{\alpha}\rangle = |\epsilon, k_\ell\rangle \equiv |z_\ell, 1\rangle|u_{\ell 1,1}\rangle$. As one may verify, $H_p|\epsilon, k_\ell\rangle = |z_\ell, 1\rangle H(z_\ell)|u_{\ell 1,1}\rangle = \epsilon|\epsilon, k_\ell\rangle$, showing that $|\epsilon, k_\ell\rangle$ is indeed compatible with the boundary matrix. Manifestly, $|\epsilon, k_\ell\rangle$ is an eigenstate of H_p in the standard Bloch form—thereby recovering the conventional Bloch's theorem for periodic BCs, as desired.

A Criterion for the Absence of Localized Eigenstates

Symmetry conditions paired with suitable BCs can exclude completely edge modes, topological or otherwise. Here we identify one particularly useful sufficient condition that guarantees the absence of edge modes. It relies on the analytic diagonalization of an operator from the family

$$T_\theta + T_\theta^\dagger \equiv e^{-i\theta}T + e^{i\theta}T^\dagger, \quad \theta \in [0, 2\pi).$$

Physically, one may think of θ as an applied electric field, but it may also arise due to transverse momentum. The combination $T + T^\dagger$ is singled out by a symmetry argument. The $\mathbb{Z}_2$ mirror symmetry,

$$U_{\mathcal{M}} \equiv \sum_{j=1}^{N} |N - j + 1\rangle\langle j|, \quad U_{\mathcal{M}}^\dagger = U_{\mathcal{M}}^{-1} = U_{\mathcal{M}},$$

exchanges the two shift operators, $U_{\mathcal{M}} T U_{\mathcal{M}} = T^\dagger$, so that $U_{\mathcal{M}}(T + T^\dagger)U_{\mathcal{M}} = T^\dagger + T$. The eigenstates and eigenvalues of $T + T^\dagger$ are known [27], and were recomputed in (see Theorem 5.14)

$$(T + T^\dagger)|k_q\rangle = 2\cos\left(\frac{\pi q}{N+1}\right)|k_q\rangle, \quad q = 1, \ldots, N,$$

with unnormalized eigenvectors

$$|k_q\rangle = \sum_{j=1}^{N} \sin\left(\frac{\pi q j}{N+1}\right)|j\rangle.$$

Let $\mathrm{X} \equiv \sum_{j=1}^{N} j\,|j\rangle\langle j|$ denote the position operator. Then it is easy to check that $[X, T] = -T$, and thus $e^{i\theta \mathrm{X}} T e^{-i\theta \mathrm{X}} = e^{-i\theta} T$. In particular, $e^{i\theta \mathrm{X}}(T + T^\dagger)e^{-i\theta \mathrm{X}} = e^{-i\theta} T + e^{i\theta} T^\dagger$. It follows that the eigenstates of $T_\theta + T_\theta^\dagger$ are given by

$$|k_q, \theta\rangle = \sum_{j=1}^{N} \sin\left(\frac{\pi q j}{N+1}\right) e^{i\theta j}|j\rangle, \quad q = 1, \ldots, N.$$

Assume now that *all* the matrices h_r entering the single-particle Hamiltonian of interest satisfy the relation $h_r^\dagger = e^{i2r\theta} h_r$, for some choice of θ, that is,

$$H_{\text{tot}} = h_0 + \sum_{r=1}^{R} (T_\theta^r + T_\theta^{\dagger\, r}) e^{ir\theta} h_r + W.$$

Then, it is easy to see that H_{tot} can be rewritten as

$$H_{\text{tot}} = h_0 + \sum_{r=1}^{R} (T_\theta + T_\theta^\dagger)^r\, \tilde{h}_r + W' + W,$$

in terms of new hopping matrices $\tilde{h}_r$ and boundary contribution W' with the *same* finite range R (for example, $(T + T^\dagger)^3 = T^3 + 3T - |1\rangle\langle 2| - |N-1\rangle\langle N| + \text{H.c.}$). If the original BCs are such that $W = -W'$, then H_{tot} can be expressed as a function of $T_\theta + T_\theta^\dagger$. It follows that no localized eigenstate can exist in this case. We will encounter such a situation in Sect. 3.2.2.

2.3 Implications of the Generalized Bloch Theorem

2.3.1 *Algorithms for Computing Energy Eigenvalues and Eigenstates*

The results of Sect. 2.2 can be used to develop diagonalization algorithms for the relevant class of single-particle Hamiltonians. We will describe two such algorithms. The first treats ϵ as a parameter for numerical search. The second is inspired by the algebraic Bethe ansatz, as suggested by comparing our Eq. (2.36) to Eq. (28) of Ref. [28].

Numerical "Scan-in-Energy" Diagonalization

The procedure described in this section is a special instance of the *Eigensystem Algorithm* described in Ref. [29], specialized to Hermitian matrices. It employs a search for energy eigenvalues along the real line, and takes advantage of the results of Sect. 2.2 to determine whether a given number is an eigenvalue. The overall procedure is schematically depicted in Fig. 2.2. The first part of the algorithm finds

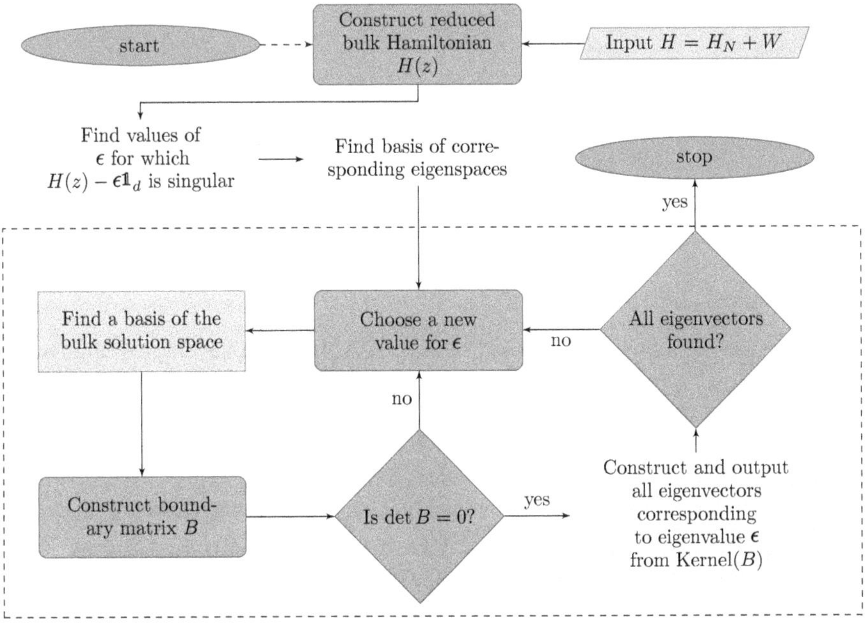

Fig. 2.2 Flowchart of the numerical diagonalization algorithm. The steps inside the dashed rectangle form the loop for scanning over ϵ. The crucial step is solving the bulk equation, which encompasses steps (4)–(8) as described in the text. Figure adapted with permission from Ref. [19]. Copyrighted by the American Physical Society

all eigenvectors of H that correspond to the flat (dispersionless) energy band, if any exists. Two steps are entailed:

1. Find all real values of ϵ for which $\det(H(z) - \epsilon \mathbb{1}_d)$ vanishes for any z. Output these as singular eigenvalues of H_{tot}.
2. For each of the eigenvalues found in step (1), find and output a basis of the corresponding eigenspace of H_{tot} using any conventional algorithm.

In implementing step (2) above, one can leverage the analysis of Sect. 2.2.2. The following part of the algorithm, which repeats until all eigenvectors of H_{tot} are found, proceeds according to the following steps:

3. Choose a seed value of ϵ, different from those eigenvalues found already.
4. Find all n distinct non-zero roots of the equation $\det(H(z) - \epsilon \mathbb{1}_d) = 0$. Let these roots be $\{z_\ell,\ \ell = 1, \ldots, n\}$, and their respective multiplicities $\{s_\ell,\ \ell = 1, \ldots, n\}$.
5. For each such roots, construct the generalized reduced bulk Hamiltonian $H_{s_\ell}(z_\ell)$ (Eq. (2.20)).
6. Find a basis of the eigenspace of $H_{s_\ell}(z_\ell)$ with eigenvalue ϵ. Let the basis vectors be $\{|u_{\ell s}\rangle,\ s = 1, \ldots, s_\ell\}$. The bulk solution corresponding to (ℓ, s) is $|\psi_{\ell s}\rangle = |z_\ell, 1\rangle|u_{\ell s}\rangle$, with Φ_{z_ℓ} defined in Eq. (2.21).
7. If h_R is non-invertible, find $s_0 = Rd - \sum_{\ell=1}^{n} s_\ell/2$. Construct matrix $K_+(\epsilon)$ as described in Eq. (2.28), and $K_-(\epsilon) = [K_+(\epsilon)]^\dagger$.
8. Find bases of the kernels of $K_+(\epsilon)$ and $K_-(\epsilon)$. Let the basis vectors be $\{|u_s^+\rangle,\ s = 1, \ldots, s_0\}$ and $\{|u_s^-\rangle,\ s = 1, \ldots, s_0\}$, respectively. The emergent bulk solutions corresponding to each s are follow from Eqs. (2.29) and (2.30).
9. Construct the boundary matrix $B(\epsilon)$ (Eq. (2.35)).
10. If $\det B(\epsilon) = 0$, output ϵ as an eigenvalue. Find a basis $\{\boldsymbol{\alpha}_\text{k},\ \text{k} = 1, \ldots, \mathscr{K}_\epsilon\}$ of the kernel of $B(\epsilon)$. Then a basis of the eigenspace of H corresponding to energy ϵ is $\{|\epsilon_\text{k}\rangle = |\Psi_\mathcal{B}\rangle\boldsymbol{\alpha}_\text{k},\ \text{k} = 1, \ldots, \mathscr{K}_\epsilon\}$, with $|\Psi_\mathcal{B}\rangle$ being defined in Eq. (2.33). If all dN eigenvectors are not yet found, then go back to step (3).
11. If $\det B(\epsilon) \neq 0$, choose a new value of ϵ as dictated by the relevant root-finding algorithm.[6] Go back to step (4).

Some considerations are in order, in regard to the fact that the determinant of $B(\epsilon)$ plotted as a function of energy ϵ may display finite-precision inaccuracies that appear as fictitious roots. Such issues arise at those ϵ where two (or more) of the roots of Eq. (2.14) cross as a function of ϵ, due to the non-orthogonality of the basis $\mathcal{B}$ that results from the procedure described in Sect. 2.2.2. Let ϵ_* be a value of energy for which this happens, so that the bulk equation bears a power-law solution. For $\epsilon \approx \epsilon_*$ (except ϵ_* itself), Eq. (2.14) has two roots that are very close in value, so that the corresponding bulk solutions overlap almost completely. This results in

[6]One may use any conventional root-finding algorithm suited for continuous functions to implement this step. In practice, we find that the determinant of the boundary matrix is *analytic* near most values of ϵ, which can be further leveraged to improve the process.

a boundary matrix having two nearly identical columns, with determinant vanishing in the limit $\epsilon \to \epsilon_*$, *irrespective* of ϵ_* being an eigenvalue of H (hence, a physical solution). However, if we calculate $B(\epsilon)$ *exactly at* ϵ_*, then the basis $\mathcal{B}$ contains power-law solutions, and accurately indicates whether ϵ_* is an eigenvalue. This also means that the function $\det B(\epsilon)$ has a discontinuity at $\epsilon = \epsilon_*$.

A simple way to identify those fictitious roots is as follows. Rewrite the polynomial in Eq. (2.15) as

$$p(\epsilon, z) = \sum_{r=s_0}^{2Rd-s_0} p_r(\epsilon) z^r, \tag{2.38}$$

which is treated as a polynomial in z with coefficients depending on ϵ (if s_0 changes with ϵ, we use the smallest possible value of s_0 in Eq. (2.38). $p(\epsilon, z)$ has double roots at ϵ_* if and only if the "discriminant" $\mathscr{D}(P(\epsilon_*, z)) = 0$ [30]. The latter gives a polynomial expression in ϵ, of degree $\mathcal{O}(Rd)$. By finding the roots of this equation, one can obtain all the values of ϵ for which fictitious roots of $\det B(\epsilon)$ may appear. To check whether these roots are true eigenvalues, one then needs to construct $B(\epsilon)$ by including the power-law solutions in the ansatz.

We further note that, while the ansatz is not continuous at such values of ϵ, the fact that the bulk solution space is the kernel of the linear operator $P_B(H_N + W - \epsilon)$ implies that it must change *smoothly* with ϵ. A way to improve numerical accuracy would be to construct an orthonormal basis (e.g., via Gram-Schmidt orthogonalization) of $\mathcal{M}_{1,N}(\epsilon)$ at each ϵ, and use this basis to construct a modified boundary matrix $\tilde{B}(\epsilon)$. In practice, one may *directly* compute the new determinant by using

$$\det \tilde{B}(\epsilon) = \frac{\det B(\epsilon)}{\sqrt{\det \mathscr{G}(\epsilon)}},$$

where $\mathscr{G} \equiv \langle \Psi_{\mathcal{B}} | \Psi_{\mathcal{B}} \rangle$ is the Gramian matrix [31] of the basis of bulk solutions obtained in steps (4)–(8) of the algorithm, with entries $\mathscr{G}_{\ell s, \ell' s'} \equiv \langle \psi_{\ell s} | \psi_{\ell' s'} \rangle$. In fact, it can be checked that the bulk solutions

$$\left\{ |\tilde{\psi}_{\ell s}\rangle \equiv \sum_{\ell=0}^{n+1} \sum_{s=1}^{s_\ell} [\mathscr{G}^{-1/2}]_{\ell' s', \ell s} |\psi_{\ell' s'}\rangle, \quad s = 1, \ldots, s_\ell, \quad \ell = 0, \ldots, n+1 \right\}$$

form an orthonormal basis of the bulk solution space $\mathcal{M}_{1,N}$. The calculation of the entries of the Gramian is straightforward thanks to the analytic result

$$\langle z, 1 | z', 1 \rangle = \begin{cases} \frac{z^* z' - (z^* z')^{N+1}}{1 - z^* z'} & \text{if } z' \neq 1/z^* \\ N & \text{if } z' = 1/z^* \end{cases}.$$

In regard to the time and space complexity of the algorithm, the required resources depend entirely on those needed to compute the boundary matrix $B(\epsilon)$. For generic ϵ, regardless of the invertibility of h_R, the size of $B(\epsilon)$ is $2Rd \times 2Rd$, independently of N. Calculation of each of its entries is also simple from the point of view of complexity, thanks to the fact that $H = H_N + W$ is symmetrical (see Definition 5.2).[7] Accordingly, both the number of steps and the memory space used by this algorithm do not scale with the system size N, making this approach computationally more efficient than conventional methods of diagonalization of generic Hermitian matrices [32].

Algebraic Diagonalization

The scan-in-energy algorithm can be further developed into an algorithm that yields an analytic solution (often closed-form), in the same sense as the Bethe ansatz method does for a different class of (interacting) quantum integrable systems. The idea is to obtain, for generic values of ϵ, an analytic expression for $B(\epsilon)$, since its determinant will then provide a condition for ϵ to be an eigenvalue, and the corresponding eigenvectors can be obtained from its kernel. As mentioned, for generic ϵ, the extended bulk solutions do not include any power-law solutions. This property can be exploited to derive an analytic expression for $B(\epsilon)$ in such a generic setting. The values of ϵ for which power-law solutions appear, or the analytic expression fails for other reasons, can be dealt with on a case-by-case basis.

By the Abel–Ruffini theorem, a completely closed-form solution *by radicals* in terms of ϵ can be achieved if the degree in z of the characteristic polynomial of the reduced bulk Hamiltonian is at most four. If this is not the case, the roots $\{z_\ell\}$ do not possess an algebraic expression in terms of ϵ and entries of H. The workaround is then to consider $\{z_\ell\}$ as free variables, with the constraint that each of them satisfy the characteristic equation of $H(z)$. With these tools in hand, the following procedure can be used to find an analytical solution for generic values of ϵ:

1. Construct the polynomial $p(\epsilon, z)$ in Eq. (2.38), which is a bivariate polynomial in ϵ and z. Determine s_0 using $s_0 = 2Rd - \deg(p(\epsilon, z))$, where deg(.) denotes the degree of the polynomial in z.
2. Assuming that ϵ and z satisfy $p(\epsilon, z) = 0$, find an expression for the eigenvector $|u(\epsilon, z)\rangle$ of $H(z)$ with eigenvalue ϵ.
3. Consider variables $\{z_\ell,\ \ell = 1, \ldots, 2Rd - 2s_0\}$, each satisfying $P_\epsilon(z_\ell) = 0$. Each of these corresponds to a bulk solution $|z_\ell, 1\rangle|u(\epsilon, z_\ell)\rangle$.
4. If h_R is not invertible, construct matrices $K_+(\epsilon)$ and $K_-(\epsilon) = [K_+(\epsilon)^\dagger]$ (Eq. (2.28)).

[7] Note that Hermiticity of H necessarily results in a "symmetrical" corner-modified block-Toeplitz matrix.

5. Find bases for their kernels, each of which contains s_0 vectors. Let these be $\{|u_s^+(\epsilon)\rangle,\ s = 1, \ldots, s_0\}$ and $\{|u_s^-(\epsilon)\rangle,\ s = 1, \ldots, s_0\}$. These correspond to finite-support solutions of the bulk equation.
6. Construct the boundary matrix $B(\epsilon) \equiv B(\epsilon, \{z_\ell\})$ (Eq. (2.35)).
7. The condition for ϵ being an eigenvalue of H is $\det B(\epsilon, \{z_\ell\}) = 0$. Therefore, a complete characterization of eigenvalues is

$$\{p(\epsilon, z_\ell) = 0, \quad \ell = 1, \ldots, n\}, \quad \det B(\epsilon, \{z_\ell\}) = 0.$$

8. If $\deg(p(\epsilon, z)) \leq 4$, substitute for each z_ℓ the closed-form expression of the corresponding root $z_\ell(\epsilon)$. The eigenvalue condition in step (7) simplifies to a single equation, $\det B(\epsilon, \{z_\ell(\epsilon)\}) = 0$.
9. For every eigenvalue ϵ, the kernel vector $\boldsymbol{\alpha}(\epsilon, \{z_\ell\})$ of $B(\epsilon, \{z_\ell\})$ provides the corresponding eigenvector of H.

In steps (2), (5), and (9), we need to obtain an analytic expression for the basis of the kernel of a square symbolic matrix of fixed kernel dimension in terms of its entries. This can be done in many different ways, and often is possible by inspection. One possible way was described in Sect. 2.2.2 in connection to evaluating $\text{Ker}(H(z) - \epsilon \mathbb{1}_d)$ for singular values of ϵ. The above analysis does not hold when ϵ satisfies any of the following conditions:

1. $\det(H(z) - \epsilon \mathbb{1}_d) = 0$ has one or more double roots. This is equivalent to $\mathscr{D}(p(\epsilon, z)) = 0$, as discussed in Sect. 2.3.1. This is a polynomial equation in terms of ϵ, the roots of which yield all required values of ϵ.
2. The coefficient $p_{s_0}(\epsilon)$ of z^{s_0} in $p(\epsilon, z)$ vanishes. Equivalently, ϵ is a root of $p_{s_0}(\epsilon) = 0$.
3. Each entry of $|u(\epsilon, z)\rangle$ vanishes. Such points are identified by solving simultaneously the equations $\langle m|u(\epsilon, z)\rangle = 0,\ m = 1, \ldots, d$ and $p(\epsilon, z) = 0$, Since a necessary and sufficient condition for these polynomials (in z) to have a common root is that their resultant vanishes [30], we find the relevant values of ϵ by equating the pairwise resultants to zero.
4. $\{|u_s^+(\epsilon)\rangle,\ s = 1, \ldots, s_0\}$ or $\{|u_s^-(\epsilon)\rangle,\ s = 1, \ldots, s_0\}$ are linearly dependent. To find such values of ϵ, one may form the corresponding Gramian matrix and equate its determinant to zero.

For all the values of ϵ thus identified, $B(\epsilon)$ is calculated by following steps (4)–(10) in the scan-in-energy algorithm. To summarize, this algebraic procedure achieves diagonalization in analytic form: the upshot is a system of *polynomial equations*, whose simultaneous roots are the eigenvalues, and an analytic expression for the eigenvectors, with *parametric dependence* on the eigenvalue.

2.3.2 An Indicator of Bulk-Boundary Correspondence

A main motivation behind the development of the generalized Bloch theorem is to achieve a more rigorous understanding of the bulk-boundary correspondence. In this section, we take a first step by presenting an indicator of bulk-boundary correspondence based on the results from Sect. 2.2. The indicator is built out of the boundary matrix and, therefore, encodes information from the bulk *and* the BCs.

For a system of size N, the existence of localized modes at energy ϵ reflects into a non-trivial kernel of the boundary matrix, which we now denote by $B_N(\epsilon)$ in order to emphasize the dependence on N and ϵ. As we increase N without changing the BCs, the energy ϵ of the bound modes (that is, modes that remain asymptotically normalizable) attains a limiting value. For instance, in topologically non-trivial, particle-hole or chiral- symmetric systems under hard-wall BCs, the mid-gap bound modes attain zero energy in the large-N limit. This convergence of bound modes and their energies is nicely captured by a modified version of the boundary matrix in the limit $N \gg 1$, which we now construct.

Consider a system of N sites in a ring topology, as shown in Fig. 2.3a, so as to allow non-zero contribution from the matrix $w_{bb'}$ in the BCs described by W (see Eq. (2.6)). Let us assume that the system hosts one or more bound modes near the junction formed by the two ends, which converge in the large-N limit to energy ϵ. The resulting modes are the bound modes of a bridge configuration that extends to infinity on both sides, and where the boundary region is shown in Fig. 2.3b. For each N, we may express the bound eigenstate as in Eq. (5.35). Such bound states have contributions *only* from those bulk solutions that are normalizable for $N \gg 1$. The extended-support solutions corresponding to $|z_\ell| = 1$ are not normalizable, and therefore must drop out from the ansatz. Further, while the amplitude of those

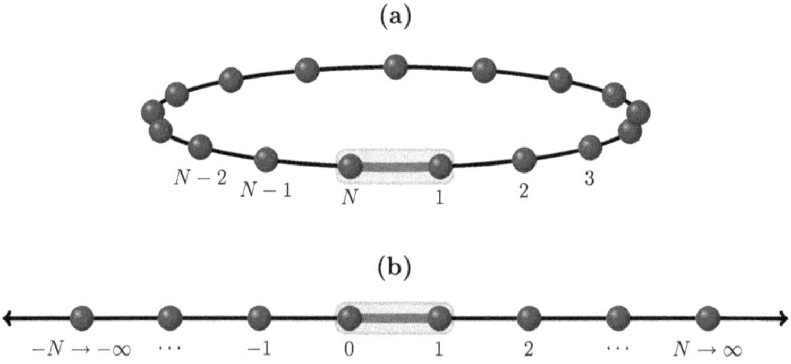

Fig. 2.3 Ring (**a**) vs bridge (**b**) configurations of a chain Hamiltonian, $d = 1 = R$. The solid (black) lines denote nearest-neighbor bulk hopping, whereas the thick (red) line indicates hopping between the left ($j = N$) and the right ($j = 1$) boundary (shaded gray rectangle). The bound states of (**a**) converge to the ones of (**b**) in the large-N limit. Figure adapted with permission from Ref. [19]. Copyrighted by the American Physical Society

corresponding to $|z_\ell| > 1$ blows up near $j = N$, they *remain* normalizable in the limit. This becomes apparent once we rescale such solutions by z_ℓ^{-N}. These rescaled solutions almost vanish at $j = 1$ for large N. Based on these considerations, we propose a modified ansatz for finite N,

$$|\epsilon, \boldsymbol{\alpha}\rangle_N \equiv \sum_{\substack{|z_\ell|<1, \\ \ell=0}} \sum_{s=1}^{s_\ell} \alpha_{\ell s}|\psi_{\ell s}\rangle + \sum_{\substack{|z_\ell|>1, \\ \ell=n+1}} \sum_{s=1}^{s_\ell} \alpha_{\ell s} z_\ell{}^{-N}|\psi_{\ell s}\rangle \tag{2.39}$$

expressed in terms of at most $2Rd$ amplitudes.

The above ansatz may be used to compute a corresponding boundary matrix $B_N(\epsilon)$ in the same way as described in Sect. 2.2.2. Note that $B_N(\epsilon)$ may not capture the bound modes appearing at finite N since, by construction, it does not incorporate contributions from extended-support solutions corresponding to $|z_\ell| = 1$. However, $B_\infty(\epsilon) \equiv \lim_{N\to\infty} B_N(\epsilon)$ is now well-defined, and describes accurately the presence and exact form of bound modes in the limit. The condition for a non-trivial kernel becomes $\det[B_N^\dagger(\epsilon)B_N(\epsilon)] = 0$. Based on this condition, we define the quantity

$$\mathscr{I}_\epsilon \equiv \log\{\det[B_\infty(\epsilon)^\dagger B_\infty(\epsilon)]\}, \tag{2.40}$$

as an "indicator of bulk-boundary correspondence." This captures precisely the interplay between the bulk properties and the BCs that may lead to the emergence of bound modes, in the sense that, as we parametrically change either or both of the reduced bulk Hamiltonian and the BCs, $\mathscr{I}_\epsilon$ shows a singularity at (and only at) the parameter value for which the system hosts bound modes at energy ϵ. Unlike most other topological indicators that are derived from bulk properties (i.e., in a torus topology), our indicator is constructed from a boundary matrix, that incorporates the relevant properties of the bulk. In cases where the bound modes are protected by a symmetry, this allows for the indicator to be computed for arbitrary BCs that respect the symmetry, paving the way to characterizing the robustness of the bound modes against classes of boundary perturbations.

An interesting situation is that of $w_{bb'} = 0$, in which case the large-N limit consists of two disjoint semi-infinite chains. Then $B_\infty(\epsilon)$ is block-diagonal,

$$B_\infty(\epsilon) = \begin{bmatrix} B_\infty^+(\epsilon) & 0 \\ 0 & B_\infty^-(\epsilon) \end{bmatrix},$$

where B_∞^+ (B_∞^-) may be interpreted as the boundary matrix of a semi-infinite chain, describing the edge modes at the left (right) edge, respectively.

While the indicator $\mathscr{I}_\epsilon$ of Eq. (2.40) signals the presence of bound states, it does not convey information about the degeneracy of that energy level, which is nevertheless contained in the boundary matrix. Therefore, it is often useful to also study the behavior of the "degeneracy indicator" as a function of ϵ:

$$\mathscr{K}_\epsilon \equiv \dim \text{Ker}[B_\infty(\epsilon)].$$

In practice, this is obtained by counting the number of zero singular values of $B_\infty(\epsilon)$.

Remark With reference to the discussion in Sect. 2.3.1, recall that in numerical computations, $B_\infty(\epsilon)$ signals fictitious roots whenever the bulk equation has a power-law solution. In such cases, we once again remedy the issue by resorting to the Gramian. Then the corrected value of the indicator is given by

$$\mathscr{I}_\epsilon = \log \left\{ \frac{\det[B_\infty(\epsilon)^\dagger B_\infty(\epsilon)]}{\det \mathscr{G}(\epsilon)} \right\}.$$

Thus, the correct degeneracy of the energy is obtained by counting zero (within numerical accuracy) singular values of the matrix $\tilde{B}_\infty(\epsilon) = B_\infty(\epsilon)\mathscr{G}(\epsilon)^{-1/2}$.

2.3.3 *Transfer Matrix in Light of the Generalized Bloch Theorem*

Starting with the work in Refs. [33, 34], the transfer matrix has remained the tool of choice for computing wavefunctions of localized eigenstates [20, 35–37] including, as mentioned, recent studies of Majorana wavefunctions in both clean and disordered Kitaev wires [38]. In this section, we revisit the transfer matrix approach to band-structure determination in the light of our generalized Bloch theorem. In particular, we show how, in situations where the transfer matrix fails to be diagonalizable, our analysis makes it possible to give physical meaning to the generalized eigenvectors by relating them to the power-law solutions discussed in Sect. 2.2.2.

Basics of the Standard Transfer Matrix Method

While our conclusions apply more generally to arbitrary finite-range clean models, for concreteness we refer in our discussion to the simplest setting where both approaches are applicable, namely, a 1D chain with nearest-neighbor hopping. We further focus on open (hard-wall) BCs, as most commonly employed in transfer matrix studies. The relevant single-particle-Hamiltonian H_N is then a tridiagonal block-Toeplitz matrix, with entries $h_1^\dagger$, h_0 and h_1 along the three diagonals. Generically, h_1 is assumed to be invertible. The starting point of the method entails obtaining the recurrence relation between eigenvector components. Specifically, if

$|\epsilon\rangle = \sum_{j=1}^{N} |j\rangle|\psi_j\rangle$ is an eigenvector of H with energy eigenvalue ϵ relative to the usual Hilbert space factorization $\mathcal{H} = \mathcal{H}_L \otimes \mathcal{H}_I$, the components $|\psi_j\rangle$ satisfy the recurrence relation

$$h_1^\dagger|\psi_{j-1}\rangle + (h_0 - \epsilon \mathbb{1}_d)|\psi_j\rangle + h_1|\psi_{j+1}\rangle = 0, \quad 2 \leq j \leq N-1. \tag{2.41}$$

In terms of the $2d \times 2d$ "transfer matrix"

$$t(\epsilon) \equiv \begin{bmatrix} 0 & \mathbb{1}_d \\ -h_1^{-1}h_1^\dagger & -h_1^{-1}(h_0 - \epsilon\mathbb{1}_d) \end{bmatrix}, \tag{2.42}$$

the above recurrence relation may be reformulated as

$$P_{j,j+1}|\epsilon\rangle = t(\epsilon)P_{j-1,j}|\epsilon\rangle, \quad 2 \leq j \leq N-1, \tag{2.43}$$

where we have written $P_{j,j+1}|\epsilon\rangle \equiv \left[|\psi_j\rangle \ |\psi_{j+1}\rangle\right]^{\mathsf{T}}$. Thus,

$$P_{j+1,j+2}|\epsilon\rangle = t(\epsilon)^j P_{1,2}|\epsilon\rangle, \quad 0 \leq j \leq N-2, \tag{2.44}$$

which can be leveraged for obtaining the complete set of eigenvectors of H_N. We can define $|\psi_0\rangle, |\psi_{N+1}\rangle$ by using the relations

$$P_{1,2}|\epsilon\rangle = t(\epsilon)P_{0,1}|\epsilon\rangle, \quad P_{N,N+1}|\epsilon\rangle = t(\epsilon)P_{N-1,N}|\epsilon\rangle,$$

so that $P_{N,N+1}|\epsilon\rangle = T(\epsilon)P_{0,1}|\epsilon\rangle$ in terms of the matrix $T(\epsilon) \equiv t(\epsilon)^N$.[8] Open BCs enforce $|\psi_0\rangle = 0 = |\psi_{N+1}\rangle$. Substituting these boundary values leads to

$$\begin{bmatrix} |\psi_N\rangle \\ 0 \end{bmatrix} = \begin{bmatrix} T_{11}(\epsilon) & T_{12}(\epsilon) \\ T_{21}(\epsilon) & T_{22}(\epsilon) \end{bmatrix} \begin{bmatrix} 0 \\ |\psi_1\rangle \end{bmatrix},$$

which has a non-trivial solution if and only if

$$\det T_{22}(\epsilon) = 0. \tag{2.45}$$

Therefore, all values of ϵ that obey the above condition are eigenvalues of H_N. For each eigenvalue, the corresponding $|\psi_1\rangle$ is obtained as the kernel of $T_{22}(\epsilon)$. In practice, $T(\epsilon)$ is calculated by first diagonalizing $t(\epsilon)$ by a similarity transformation, and then exponentiating the eigenvalues along its diagonal [9].

As can be appreciated from this example, the standard version of the transfer matrix method relies on invertibility of certain matrices, although "inversion-free"

[8] We distinguish the cumulative transfer matrix $T(\epsilon)$ from the shift operator T by always writing it with the argument ϵ.

[39, 40] or partially inversion-free [20] modifications have also been suggested. In the standard case, the only prerequisite for constructing $t(\epsilon)$ at each step is the banded structure of the single-particle Hamiltonian and, most importantly, $T(\epsilon)$ is assumed to be diagonalizable.

Connections to the Generalized Bloch Theorem

In order to relate the above analysis to the generalized Bloch formalism, the key observation is to note that the set of equations in Eq. (2.41) constitute the complete bulk equation, as described in Sect. 2.2.2. Consequently, Eq. (2.43) is satisfied by any bulk solution $|\psi\rangle \in \mathcal{M}_{1,N}$, where $\mathcal{M}_{1,N}$ denotes the bulk solution space as usual. It is insightful to recast Eq. (2.44) in the form $t(\epsilon)^j P_{1,2}|\psi\rangle = P_{1,2}T^j|\psi\rangle$, $0 \le j \le N-2$, suggesting that the action of the transfer matrix in the bulk solution space is closely related to the one of the left shift T. When restricted to $\mathcal{M}_{1,N}$, the above yields the following operator identity:

$$(t(\epsilon) - z\mathbb{1}_{2d})^j P_{1,2}\Big|_{\mathcal{M}_{1,N}} = P_{1,2}(T - zI_N)^j\Big|_{\mathcal{M}_{1,N}}, \tag{2.46}$$

with $z \in \mathbb{C}$. This relation may be used to establish a direct connection between the basis of the bulk solution space described in the generalized Bloch theorem, and the Jordan structure of the transfer matrix. In the absence of power-law solutions, each bulk solution $|\psi_{\ell s}\rangle$ is annihilated by $P_{1,2}(T - z_\ell I_N) = P_{1,2}[P_B(T - z_\ell I_N)]$. In such cases, Eq. (2.46) reads

$$(t(\epsilon) - z_\ell \mathbb{1}_{2d})P_{1,2}|\psi_{\ell s}\rangle = P_{1,2}(T - z_\ell I_N)|\psi_{\ell s}\rangle = 0,$$

implying that $P_{1,2}|\psi_{\ell s}\rangle$ is an eigenvector of $t(\epsilon)$ with eigenvalue z_ℓ. Naturally, a Bloch wave-like bulk solution corresponds to an eigenvalue on the unit circle, whereas an exponential solution corresponds to one inside or outside the unit circle, in agreement with the literature [9].

While, as remarked, the transfer matrix is typically assumed to be diagonalizable, we now show that generalized eigenvectors of $t(\epsilon)$ *are* physically meaningful, and in fact related to the power-law solutions of the bulk equation. Let ϵ be a value of energy for which power-law solutions are present. We can then generalize our earlier calculation for the eigenvectors of the transfer matrix by noting that each $|\psi_{\ell s}\rangle$ is annihilated by $P_{1,2}(T - z_\ell I_N)^{s_\ell}$, where s_ℓ is the multiplicity of the root z_ℓ as usual. Then, a similar calculation reveals that $P_{1,2}|\psi_{\ell s}\rangle$ is a generalized eigenvector of $t(\epsilon)$, satisfying

$$(t(\epsilon) - z_\ell \mathbb{1}_{2d})^{s_\ell} P_{1,2}|\psi_{\ell s}\rangle = 0.$$

Thus, *generalized eigenvectors of the transfer matrix are projections of solutions with a power-law prefactor*. In some non-generic scenarios, they indeed contribute to the energy eigenstates, as we discussed.[9]

Returning to the general case, a number of additional remarks are worth making, in regard to points of contact and differences between the transfer matrix approach and our generalized Bloch theorem. First, the eigenstate ansatz obtained from the analytic continuation of the Bloch Hamiltonian provides a *global* characterization of energy eigenvectors (and generalized eigenvectors), as opposed to the local characterization afforded within the transfer matrix approach, whereby each eigenvector is reconstructed "iteratively" for any given eigenvalue. Further to that, the generalized Bloch theorem unveils the role of non-unitary representations of translational symmetry for finite systems. Perhaps most importantly, the two methods differ in the way BCs are handled. Clearly, in both approaches it is necessary to match BCs in order to obtain the physical energy spectrum. While open BCs are most commonly used in transfer matrix calculations, the method has also been applied to relaxed surfaces [9] and generalized periodic BCs [41], all of which belong to the class of BCs considered in our setting. In this sense, it is tempting to compare Eq. (2.45) with the condition on the determinant of the boundary matrix, $\det B(\epsilon) = 0$. However, the class of BCs to which the transfer matrix approach can be successfully applied is not a priori clear, thus whether such a condition can be established for as general a class of BCs as our theorem covers has not been investigated to the best of our knowledge.

From a numerical standpoint, the computational complexity of the standard transfer matrix method for clean systems (when applicable) is independent of the system size N, as is the case of our scan-in-energy algorithm in Sect. 2.3.1. In those cases where inversion of certain matrices is a difficulty and inversion-free approaches are used [39, 40], the latter also have a comparable computational complexity to our method. Interestingly, all approaches so far that are truly inversion-free rely at some point or another on the solution of a non-linear eigenvalue problem. Thanks to the fact that, as noted, the construction of $t(\epsilon)$ in the generic case relies only on the banded structure of H_N, bulk disorder can be handled efficiently within transfer matrix approaches, albeit for a limited class of BCs. For general BCs as we consider, it is thus natural to combine the transfer matrix approach with the bulk-boundary separation we have introduced, in order to still find solutions efficiently: the transfer matrix can be employed to find all possible solutions of the bulk equation in the presence of bulk disorder, and the latter can then be used as input for the boundary matrix that provides a condition for energy eigenstates.

---

[9]Mathematically, the connections between the bulk solutions and the generalized eigenvectors of the transfer matrix may be seen as a result of the fact that the transfer matrix is a "linearization" of the non-linear eigenvalue problem associated with the reduced bulk Hamiltonian, see e.g. Ref. [18]. In the current example, the non-linear eigenvalue equation of $H(z)$, $[z^{-1}h_1^\dagger + (h_0 - \epsilon) + zh_1]\,|u\rangle = 0$, is equivalent to the standard eigenvalue equation of $T(\epsilon)$, namely, $T(\epsilon)(P_{1,2}|z, 1\rangle|u\rangle) = z(P_{1,2}|z, 1\rangle|u\rangle)$. In this sense, the solutions with power-law prefactor can be thought of as the generalized eigenvectors of the non-linear eigenvalue problem of the reduced bulk Hamiltonian.

2.4 Extensions of the Generalized Bloch Theorem

In this section, we will further extend the generalization of Bloch's theorem to two scenarios. First, we will formulate the theorem for D-dimensional systems with arbitrary BCs imposed on two parallel hyperplanes. We will then show how interfaces can be handled using the same approach.

2.4.1 *Higher-Dimensional Systems*

Quadratic Hamiltonians with Arbitrary Boundary Conditions in Higher Dimensions

Consider first a D-dimensional, translation-invariant *infinite* system of independent fermions. Such a system is described in full generality by a quadratic, not necessarily particle-number-conserving, Hamiltonian in Fock space. In a lattice approximation, the vector position of a given fermion in the regular crystal lattice can be written as the sum of a Bravais lattice vector and a basis vector [11]. We will include these basis vectors as part of the internal labels, and denote Bravais lattice vectors as $\mathbf{j} \equiv \sum_{\mu=1}^{D} j_\mu \mathbf{a}_\mu$, with $\mathbf{a}_1, \ldots, \mathbf{a}_D$ primitive vectors of the Bravais lattice Λ_D, and $j_\mu \in \mathbb{Z}$. An orthonormal basis of the Hilbert space of single-particle states is thus labeled by Bravais lattice vectors $\mathbf{j}$, and a finite number of internal labels, $m = 1, \ldots, d_{\text{int}}$. We denote by $\hat{\Phi}_{\mathbf{j}m}$ ($\hat{\Phi}_{\mathbf{j}m}^\dagger$) the fermionic annihilation (creation) operator corresponding to lattice vector $\mathbf{j}$ and internal state m. The Hamiltonian of an infinite translation-invariant system can then be written as

$$\widehat{H} = \sum_{\mathbf{r},\, \mathbf{j} \in \Lambda_D} \left[\hat{\Phi}_{\mathbf{j}}^\dagger K_{\mathbf{r}} \hat{\Phi}_{\mathbf{j}+\mathbf{r}} + \frac{1}{2}(\hat{\Phi}_{\mathbf{j}}^\dagger \Delta_{\mathbf{r}} \hat{\Phi}_{\mathbf{j}+\mathbf{r}}^\dagger + \text{H.c.})\right], \tag{2.47}$$

with $\hat{\Phi}_{\mathbf{j}}^\dagger \equiv \left[\hat{\Phi}_{\mathbf{j}1}^\dagger \cdots \hat{\Phi}_{\mathbf{j}d_{\text{int}}}^\dagger\right]$, $\mathbf{r}, \mathbf{j}$ Bravais lattice vectors, and the $d_{\text{int}} \times d_{\text{int}}$ hopping and pairing matrices $K_{\mathbf{r}}$, $\Delta_{\mathbf{r}}$ satisfying $K_{-\mathbf{r}} = K_{\mathbf{r}}^\dagger$, $\Delta_{-\mathbf{r}} = -\Delta_{\mathbf{r}}^{\mathrm{T}}$.

As the infinite system is translation-invariant in all D directions, it is customary to introduce the volume containing the electrons by imposing Born–von Karman (periodic) BCs over a macroscopic volume commensurate with the primitive cell of Λ_D. If the allowed $\mathbf{j}$'s correspond to $j_\mu = 1, \ldots, M_\mu$, and the total number of primitive cells is $M \equiv M_1 M_2 \ldots M_D$, then $\hat{\Phi}_{\mathbf{k}}^\dagger \equiv (1/\sqrt{M}) \sum_{\mathbf{j} \in \Lambda_D} e^{i\mathbf{k}\cdot\mathbf{j}} \hat{\Phi}_{\mathbf{j}}^\dagger$ defines the Fourier-transformed array of creation operators of *real* Bloch wave-vector (or crystal momentum), $\mathbf{k} \equiv \sum_{\mu=1}^{D} \frac{k_\mu}{M_\mu} \mathbf{b}_\mu$, with k_μ integers such that k lies inside the Brillouin zone (BZ). , and $\mathbf{b}_\mu$ defines the reciprocal lattice vectors satisfying $\mathbf{a}_\mu \cdot \mathbf{b}_\nu = 2\pi \delta_{\mu\nu}$, with $\delta_{\mu\nu}$ being the Kronecker's delta [11]. By letting $*$ denote complex conjugation, one can express the Hamiltonian of Eq. (2.48) in momentum space as

$$\widehat{H} = \frac{1}{2} \sum_{\mathbf{k}\in\mathrm{BZ}} [\hat{\Phi}^\dagger_{\mathbf{k}} K_{\mathbf{k}} \hat{\Phi}_{\mathbf{k}} + \hat{\Phi}^\dagger_{-\mathbf{k}} K^*_{-\mathbf{k}} \hat{\Phi}_{-\mathbf{k}} + \hat{\Phi}^\dagger_{\mathbf{k}} \Delta_{\mathbf{k}} \hat{\Phi}^\dagger_{-\mathbf{k}} + \hat{\Phi}_{\mathbf{k}} \Delta^*_{-\mathbf{k}} \hat{\Phi}_{-\mathbf{k}}],$$

which has a block structure in terms of the matrices

$$K_{\mathbf{k}} \equiv \sum_{\mathbf{r}\in\Lambda_D} e^{i\mathbf{k}\cdot\mathbf{r}} K_{\mathbf{r}}, \quad \Delta_{\mathbf{k}} \equiv \sum_{\mathbf{r}\in\Lambda_D} e^{i\mathbf{k}\cdot\mathbf{r}} \Delta_{\mathbf{r}}.$$

Now let us terminate this system along two parallel lattice hyperplanes, or "hypersurfaces" henceforth—resulting in open BCs. The terminated system is translation-invariant along $D-1$ lattice vectors parallel to the hypersurfaces, so that we can associate with it a Bravais lattice Λ_{D-1} of spatial dimension $D-1$, known as the "surface mesh" [21]. If $\mathbf{m}_1, \ldots, \mathbf{m}_{D-1}$ denote the primitive vectors of Λ_{D-1}, then any point $\mathbf{j}_\parallel \in \Lambda_{D-1}$ can be expressed as $\mathbf{j}_\parallel = \sum_{\mu=1}^{D-1} j_\mu \mathbf{m}_\mu$, where j_μ are integers. Let us choose a lattice vector s of Λ_D that is not in the surface mesh (and therefore, not parallel to the two hypersurfaces). We will call s the "stacking vector." Since $\{\mathbf{m}_1, \ldots, \mathbf{m}_{D-1}, \mathsf{s}\}$ are not the primitive vectors of Λ_D in general, the Bravais lattice $\bar{\Lambda}_D$ generated by them may cover only a subset of points in Λ_D. Therefore, in general, each primitive cell of $\bar{\Lambda}_D$ may enclose a number $J > 1$ of points of Λ_D. As a result, there are a total of $\bar{d}_{\text{int}} = J d_{\text{int}}$ fermionic degrees of freedom attached to each point $\mathbf{j}_\parallel + j\mathsf{s}$ of $\bar{\Lambda}_D$ with j an integer (see Fig. 2.4). Let us denote the corresponding creation (annihilation) operators by $\hat{\Phi}^\dagger_{\mathbf{j}_\parallel j1}, \ldots, \hat{\Phi}^\dagger_{\mathbf{j}_\parallel j\bar{d}_{\text{int}}}$ ($\hat{\Phi}_{\mathbf{j}_\parallel j1}, \ldots, \hat{\Phi}_{\mathbf{j}_\parallel j\bar{d}_{\text{int}}}$). For each $\mathbf{j}_\parallel$ in the surface mesh, we define the array of the basis of fermionic operators by

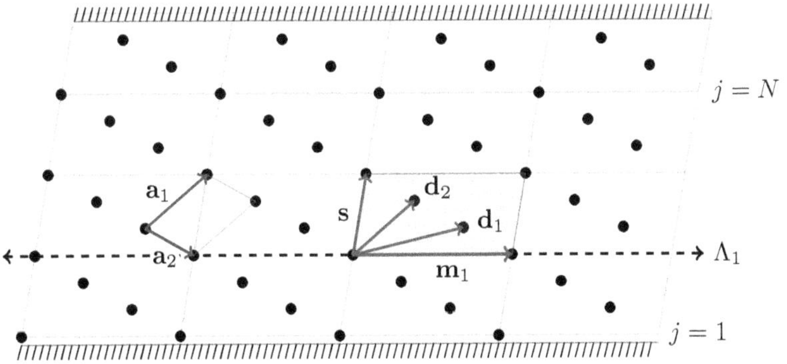

Fig. 2.4 Example of a 2D Bravais lattice Λ_2 terminated along two parallel lines (bordered with pattern). $\mathbf{a}_1$ and $\mathbf{a}_2$ denote the primitive vectors of Λ_2, and its primitive cell is shaded (in blue). The dotted (black) line connects the points of the surface mesh (Λ_1) generated by the primitive vector $\mathbf{m}_1$. The primitive cell of the Bravais lattice $\bar{\Lambda}_2$, generated by $\mathbf{m}_1$ and the stacking vector s, is also shown (shaded in brown). The original lattice Λ_2 is obtained by attaching the basis vectors $\mathbf{d}_1$ and $\mathbf{d}_2$ to each point of $\bar{\Lambda}_2$. Figure adapted with permission from Ref. [42]. Copyrighted by the American Physical Society

$$\hat{\Phi}^\dagger_{\mathbf{j}_\parallel} \equiv \left[\hat{\Phi}^\dagger_{\mathbf{j}_\parallel,1} \cdots \hat{\Phi}^\dagger_{\mathbf{j}_\parallel,N}\right], \quad \hat{\Phi}^\dagger_{\mathbf{j}_\parallel,j} \equiv \left[\hat{\Phi}^\dagger_{\mathbf{j}_\parallel j1} \cdots \hat{\Phi}^\dagger_{\mathbf{j}_\parallel j\bar{d}_{\text{int}}}\right],$$

where the integer N is proportional to the separation between the two hypersurfaces. For arrays, such as $\hat{\Phi}^\dagger_{\mathbf{j}_\parallel}$ and $\hat{\Phi}_{\mathbf{j}_\parallel}$, we shall follow the convention that the arrays appearing on the left (right) of a matrix are row (column) arrays. In the above basis, the many body Hamiltonian of the system, subject to open BCs on the hypersurfaces, can be expressed as

$$\widehat{H}_N = \sum_{\mathbf{j}_\parallel,\mathbf{r}_\parallel \in \Lambda_{D-1}} \left[\hat{\Phi}^\dagger_{\mathbf{j}_\parallel} K_{\mathbf{r}_\parallel} \hat{\Phi}_{\mathbf{j}_\parallel+\mathbf{r}_\parallel} + \frac{1}{2}(\hat{\Phi}^\dagger_{\mathbf{j}_\parallel} \Delta_{\mathbf{r}_\parallel} \hat{\Phi}^\dagger_{\mathbf{j}_\parallel+\mathbf{r}_\parallel} + \text{H.c.})\right],$$

where $\mathbf{j}_\parallel$, $\mathbf{r}_\parallel$ are vectors in the surface mesh, and $K_{\mathbf{r}_\parallel}$, $\Delta_{\mathbf{r}_\parallel}$ are $N\bar{d}_{\text{int}} \times N\bar{d}_{\text{int}}$ hopping and pairing matrices that satisfy $K_{-\mathbf{r}_\parallel} = K^\dagger_{\mathbf{r}_\parallel}$, $\Delta_{-\mathbf{r}_\parallel} = -\Delta^{\mathsf{T}}_{\mathbf{r}_\parallel}$ by virtue of fermionic statistics. Thanks to the assumptions of clean, finite-range system, these are BBT matrices: explicitly, if $R \geq 1$ is the range of hopping and pairing, we may write $[S_{\mathbf{r}_\parallel}]_{jj'} \equiv S_{\mathbf{r}_\parallel,j'-j} \equiv S_{\mathbf{r}_\parallel,r}$, with

$$S_{\mathbf{r}_\parallel,r} = 0 \qquad \text{if} \qquad |r| > R, \;\; \forall \mathbf{r}_\parallel, \qquad \text{where } S = K, \Delta.$$

Next, we enforce periodic BCs along the directions $\mathbf{m}_1, \ldots, \mathbf{m}_{D-1}$ in which translation- invariance is retained, by restricting to those lattice points $\mathbf{j}_\parallel = \sum_{\mu=1}^{D-1} j_\mu \mathbf{m}_\mu$ where for each μ, j_μ takes values from $\{1, \ldots, N_\mu\}$, N_μ being a positive integer. Let $\mathbf{n}_1, \ldots, \mathbf{n}_{D-1}$ denote the primitive vectors of the surface reciprocal lattice, which is the $(D-1)$-dimensional lattice reciprocal to the surface mesh Λ_{D-1}, satisfying $\mathbf{m}_\mu \cdot \mathbf{n}_\nu = 2\pi\delta_{\mu\nu}$ for $\mu, \nu = 1, \ldots, D-1$. The Wigner–Seitz cell of the surface reciprocal lattice is the "surface Brillouin zone," denoted by SBZ. In the Fourier-transformed basis defined by

$$\hat{\Phi}^\dagger_{\mathbf{k}_\parallel} \equiv \sum_{\mathbf{j}_\parallel}^{\Lambda_{D-1}} \frac{e^{i\mathbf{k}_\parallel\cdot\mathbf{j}_\parallel}}{\sqrt{N_S}} \hat{\Phi}^\dagger_{\mathbf{j}_\parallel}, \qquad N_S = N_1 \ldots N_{D-1}, \tag{2.48}$$

where $\mathbf{k}_\parallel = \sum_{\mu=1}^{D-1} \frac{k_\mu}{N_\mu} \mathrm{n}_\mu$ and the integers k_μ are crystal momenta in the SBZ, we can then express the relevant many-body Hamiltonian in terms of "virtual wires" labeled by $\mathbf{k}_\parallel$. That is, $\widehat{H}_N \equiv \sum_{\mathbf{k}_\parallel \in \text{SBZ}} \widehat{H}_{\mathbf{k}_\parallel,N}$, where

$$\widehat{H}_{\mathbf{k}_\parallel,N} = \frac{1}{2}(\hat{\Phi}^\dagger_{\mathbf{k}_\parallel} K_{\mathbf{k}_\parallel} \hat{\Phi}_{\mathbf{k}_\parallel} - \hat{\Phi}_{-\mathbf{k}_\parallel} K^*_{-\mathbf{k}_\parallel} \hat{\Phi}^\dagger_{-\mathbf{k}_\parallel} + \hat{\Phi}^\dagger_{\mathbf{k}_\parallel} \Delta_{\mathbf{k}_\parallel} \hat{\Phi}^\dagger_{-\mathbf{k}_\parallel} - \hat{\Phi}_{-\mathbf{k}_\parallel} \Delta^*_{-\mathbf{k}_\parallel} \hat{\Phi}_{\mathbf{k}_\parallel})$$
$$+ \frac{1}{2}\text{Tr}\, K_{\mathbf{k}_\parallel}. \tag{2.49}$$

Here, Tr denotes trace and the $N\bar{d}_{\text{int}} \times N\bar{d}_{\text{int}}$ matrices $S_{\mathbf{k}_\parallel}$, for $S = K, \Delta$, have entries

$$[S_{\mathbf{k}_\parallel}]_{jj'} \equiv S_{\mathbf{k}_\parallel, j'-j} \equiv S_{\mathbf{k}_\parallel, r} \equiv \sum_{\mathbf{r}_\parallel} e^{i\mathbf{k}_\parallel \cdot \mathbf{r}_\parallel} S_{\mathbf{r}_\parallel, r},$$

and the finite-range assumption requires that

$$S_{\mathbf{k}_\parallel, r} = 0 \text{ if } |r| > R, \quad \forall \mathbf{k}_\parallel \in \text{SBZ}, \quad \text{where } S = K, \Delta. \tag{2.50}$$

Physically, non-ideal surfaces may result from processes such as surface relaxation or reconstruction, as well as from the presence of surface disorder (see Fig. 2.5). In our setting, these may be described as effective BCs, modeled by a Hermitian operator of the form

$$\widehat{W} \equiv \sum_{\mathbf{j}_\parallel, \mathbf{j}'_\parallel} \Big[\hat{\Phi}^\dagger_{\mathbf{j}_\parallel} W^{(K)}_{\mathbf{j}_\parallel, \mathbf{j}'_\parallel} \hat{\Phi}_{\mathbf{j}_\parallel} + \frac{1}{2} (\hat{\Phi}^\dagger_{\mathbf{j}_\parallel} W^{(\Delta)}_{\mathbf{j}_\parallel, \mathbf{j}'_\parallel} \hat{\Phi}^\dagger_{\mathbf{j}'_\parallel} + \text{H.c.}) \Big],$$

subject to the constraints from fermionic statistics,

$$W^{(K)}_{\mathbf{j}'_\parallel, \mathbf{j}_\parallel} = \big[W^{(K)}_{\mathbf{j}_\parallel, \mathbf{j}'_\parallel} \big]^\dagger, \qquad W^{(\Delta)}_{\mathbf{j}'_\parallel, \mathbf{j}_\parallel} = -\big[W^{(\Delta)}_{\mathbf{j}_\parallel, \mathbf{j}'_\parallel} \big]^{\mathsf{T}}.$$

Since such non-idealities at the surface are known to influence only the first few atomic layers near the surfaces, we assume that $\widehat{W}$ affects only the first R boundary slabs of the lattice, so that (see also Fig. 2.4)

$$\big[W^{(S)}_{\mathbf{j}_\parallel, \mathbf{j}'_\parallel} \big]_{jj'} = 0 \quad \forall \mathbf{j}_\parallel, \mathbf{j}'_\parallel, \quad \text{if} \quad j \text{ or } j' \in \{R+1, \ldots, N-R\}, \qquad S = K, \Delta.$$

The total Hamiltonian subject to arbitrary BCs is

$$\widehat{H}_{\text{tot}} \equiv \widehat{H}_N + \widehat{W}.$$

Let $j \equiv b = 1, \ldots, R,\ N-R+1, \ldots, N$ label boundary sites. In general, only $\widehat{H}_N$ will be able to be decoupled by Fourier transform, whereas $\widehat{W}$ retains cross-terms of the form

$$[W^{(S)}_{\mathbf{q}_\parallel, \mathbf{k}_\parallel}]_{bb'} = \sum_{\mathbf{j}_\parallel, \mathbf{j}'_\parallel} e^{i(\mathbf{k}_\parallel \cdot \mathbf{j}'_\parallel - \mathbf{q}_\parallel \cdot \mathbf{j}_\parallel)} \big[W^{(S)}_{\mathbf{j}_\parallel, \mathbf{j}'_\parallel} \big]_{bb'}, \qquad S = K, \Delta.$$

If the system is not particle-conserving, then we reorder the fermionic operator basis according to

$$\hat{\Psi}^\dagger_{\mathbf{k}_\parallel} \equiv \left[\hat{\Psi}^\dagger_{\mathbf{k}_\parallel,1} \cdots \hat{\Psi}^\dagger_{\mathbf{k}_\parallel,N}\right], \qquad \hat{\Psi}^\dagger_{\mathbf{k}_\parallel,j} \equiv \left[\hat{\Phi}^\dagger_{\mathbf{k}_\parallel,j}\ \hat{\Phi}_{-\mathbf{k}_\parallel,j}\right].$$

The single-particle Hamiltonian can then be expressed as

$$H_{\text{tot}} = H_N + W = \sum_{\mathbf{k}_\parallel} |\mathbf{k}_\parallel\rangle\langle\mathbf{k}_\parallel| H_{\mathbf{k}_\parallel,N} + \sum_{\mathbf{q}_\parallel,\mathbf{k}_\parallel} |\mathbf{q}_\parallel\rangle\langle\mathbf{k}_\parallel| W_{\mathbf{q}_\parallel,\mathbf{k}_\parallel}, \tag{2.51}$$

where $H_{\mathbf{k}_\parallel,N}$ is the single-particle (BdG) Hamiltonian corresponding to Eq. (2.49). The matrix

$$T \equiv \sum_{\mathbf{k}_\parallel} |\mathbf{k}_\parallel\rangle\langle\mathbf{k}_\parallel| \sum_{j=1}^{N-1} |j\rangle\langle j+1| = \sum_{j=1}^{N-1} |j\rangle\langle j+1|$$

implements a shift along the direction $\mathbf{s}$, where we have used the completeness relation $\mathbf{1} = \sum_{\mathbf{k}_\parallel\in\text{SBZ}} |\mathbf{k}_\parallel\rangle\langle\mathbf{k}_\parallel|$. Letting $r = j' - j$ as before, we have

$$H_{\mathbf{k}_\parallel,N} = h_{\mathbf{k}_\parallel,0} + \sum_{r=1}^{R} [T^r h_{\mathbf{k}_\parallel,r} + \text{H.c.}], \tag{2.52}$$

$$h_{\mathbf{k}_\parallel,r} = \sum_{\mathbf{r}_\parallel} e^{i\mathbf{k}_\parallel\cdot\mathbf{r}_\parallel} h_{\mathbf{r}_\parallel,r}, \qquad h_{\mathbf{r}_\parallel,r} = \begin{bmatrix} K_{\mathbf{r}_\parallel,r} & \Delta_{\mathbf{r}_\parallel,r} \\ -\Delta^*_{\mathbf{r}_\parallel,r} & -K^*_{\mathbf{r}_\parallel,r} \end{bmatrix},$$

whereas the single-particle boundary modification $W_{\mathbf{q}_\parallel,\mathbf{k}_\parallel}$ in Eq. (2.51) is given by

$$W_{\mathbf{q}_\parallel,\mathbf{k}_\parallel} = \begin{bmatrix} W^{(K)}_{\mathbf{q}_\parallel,\mathbf{k}_\parallel} & W^{(\Delta)}_{\mathbf{q}_\parallel,\mathbf{k}_\parallel} \\ -[W^{(\Delta)}_{-\mathbf{q}_\parallel,-\mathbf{k}_\parallel}]^* & -[W^{(K)}_{-\mathbf{q}_\parallel,-\mathbf{k}_\parallel}]^* \end{bmatrix}.$$

In the simpler case where the system is particle-conserving, we have $h_{\mathbf{r}_\parallel,r} = K_{\mathbf{r}_\parallel,r}$ and $W_{\mathbf{q}_\parallel,\mathbf{k}_\parallel} = W^{(K)}_{\mathbf{q}_\parallel,\mathbf{k}_\parallel}$. Reflecting the different ways in which a surface may deviate from its ideal structure (Fig. 2.5), we may consider BCs as belonging to three different categories of increasing complexity:

- *Relaxed BCs*—In the process of surface relaxation, the atoms in the surface slab displace from their ideal position in such a way that the surface (and the bulk) layers remain translation-invariant along $\mathbf{m}_1, \ldots, \mathbf{m}_{D-1}$. Therefore, $\mathbf{k}_\parallel$ remains a good quantum number, and $W_{\mathbf{q}_\parallel,\mathbf{k}_\parallel} = \delta_{\mathbf{q}_\parallel,\mathbf{k}_\parallel} W_{\mathbf{k}_\parallel,\mathbf{k}_\parallel}$. In particular, $W_{\mathbf{q}_\parallel,\mathbf{k}_\parallel} = 0$ for each $\mathbf{q}_\parallel$, $\mathbf{k}_\parallel$ for open BCs, which falls in this category.
- *Reconstructed BCs*—If the surfaces undergo reconstruction, then the total system can have lower periodicity than the one with ideal surfaces. This scenario is also referred to as "commensurate" surface reconstruction [21]. In this case, W may retain some cross-terms of the form $W_{\mathbf{q}_\parallel,\mathbf{k}_\parallel}$. However, not all values $\mathbf{k}_\parallel$

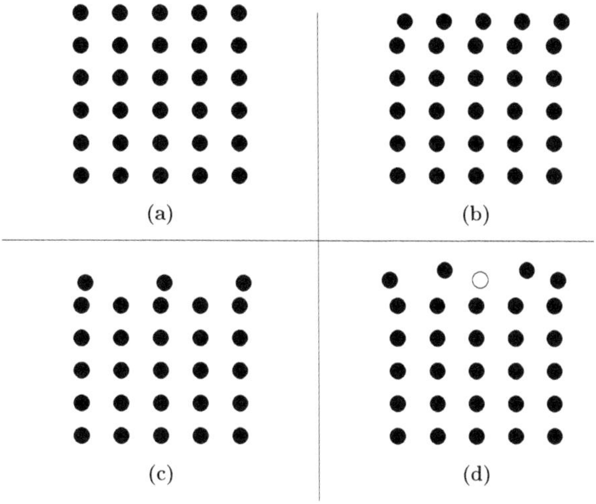

Fig. 2.5 (**a**) Sketch of a $D = 2$ crystal with ideal surface. The remaining panels show the same crystal with (**b**) relaxed, (**c**) reconstructed, and (**d**) disordered surface. The unfilled circle in panel (**d**) shows a surface impurity atom. Figure adapted with permission from Ref. [42]. Copyrighted by the American Physical Society

are expected to have cross-terms in this way, and the system can still be block-diagonalized. For example, for 2×1 reconstruction of the (111) surface of Silicon crystals, each block of the Hamiltonian will consist of only $2 \times 1 = 2$ values of $\mathbf{k}_{\|}$, whereas for its 7×7 reconstruction, each block includes 49 values of $\mathbf{k}_{\|}$ [21].

- *Disordered BCs*—If the surface reconstruction is "non-commensurate," or if the surface suffers from disorder, then the Hamiltonian cannot be block-diagonalized any further in general. Non-commensurate reconstruction of a surface is likely to happen in the case of adsorption.

Our setting is general enough to model adsorption as well as thin layer deposition up to a few atomic layers. For the rest of this section, we will assume that the system is subject to the most general type of disordered BCs.

Bulk-Boundary Separation and the Statement of the Theorem

The first needed ingredient towards formulating the generalized Bloch theorem is now a description of the eigenstates of the single-particle Hamiltonian $H_{\mathbf{k}_{\|},N}$ of the virtual wire labeled by $\mathbf{k}_{\|}$, given in Eq. (2.52). Let

$$d \equiv \begin{cases} \bar{d}_{\text{int}} & \text{if } \Delta = 0 = W^{(\Delta)}, \\ 2\bar{d}_{\text{int}} & \text{if } \Delta \neq 0 \text{ or } W^{(\Delta)} \neq 0. \end{cases}$$

Then, the projector

$$P_B = \sum_{j=R+1}^{N-R} |j\rangle\langle j|$$

determined by the range R of the virtual chains is the bulk projector. By definition, the matrix W describing BCs satisfies $P_B W = 0$, whereby it follows that $P_B H_{\text{tot}} = P_B(H_N + W) = P_B H_N$. Accordingly, building on the exact bulk-boundary separation also used in Sect. 2.2, the "bulk equation" to be solved reads

$$P_B H_N |\psi\rangle = \epsilon P_B |\psi\rangle, \qquad \epsilon \in \mathbb{R}. \tag{2.53}$$

The $d \times d$ analytic continuation of the Bloch Hamiltonian now takes the form

$$H_{\mathbf{k}_\parallel}(z) \equiv h_0 + \sum_{r=1}^{R} (z^r h_{\mathbf{k}_\parallel,r} + z^{-r} h^\dagger_{\mathbf{k}_\parallel,r}), \qquad z \in \mathbb{C}, \tag{2.54}$$

acting on the d-dimensional internal space spanned by states $\{|m\rangle,\ m = 1, \ldots, d\}$. The bulk solutions with power-law prefactor can be deduced from the $dv \times dv$ generalized Bloch Hamiltonians with block-entries

$$[H_{\mathbf{k}_\parallel,v}(z)]_{xx'} \equiv \frac{\partial_z^{x'-x} H_{\mathbf{k}_\parallel}(z)}{(x'-x)!} = \frac{H^{(x'-x)}_{\mathbf{k}_\parallel}(z)}{(x'-x)!}, \qquad 1 \le x \le x' \le v, \tag{2.55}$$

with $H^{(0)}_{\mathbf{k}_\parallel}(z) = H_{\mathbf{k}_\parallel}(z)$ given in Eq. (2.54). In array form,

$$H_{\mathbf{k}_\parallel,v}(z) = \begin{bmatrix} H^{(0)} & H^{(1)} & \frac{1}{2}H^{(2)} & \cdots & \frac{1}{(v-1)!}H^{(v-1)} \\ 0 & \ddots & \ddots & \ddots & \vdots \\ \vdots & \ddots & \ddots & \ddots & \frac{1}{2}H^{(2)} \\ \vdots & & \ddots & \ddots & H^{(1)} \\ 0 & \cdots & \cdots & 0 & H^{(0)} \end{bmatrix},$$

where the label (z) and the subscript $\mathbf{k}_\parallel$ were dropped for brevity. If the matrix $h_{\mathbf{k}_\parallel,R}$ is not invertible, then the related matrix polynomial

$$K_{\mathbf{k}_\parallel,+}(\epsilon, z) \equiv z^R (H_{\mathbf{k}_\parallel}(z) - \epsilon \mathbb{1}_d) \tag{2.56}$$

will be required for computation of emergent solutions. The $dv \times dv$ block matrix $K_{\mathbf{k}_\parallel v,+}(\epsilon, z)$ is defined by the same formula as in Eq. (2.55). The important difference between these two matrices is that $K_{\mathbf{k}_\parallel v,+}(\epsilon, z)$ is well-defined at $z = 0$, whereas $H_{\mathbf{k}_\parallel,v}(z)$ is not. These block matrices act on column arrays of v internal states, which can be expressed in the form $|u\rangle = \left[|u_1\rangle \ldots |u_v\rangle\right]^{\mathsf{T}}$, where each of the entries is an internal state.

For fixed but arbitrary ϵ, the expression

$$p_{\mathbf{k}_\parallel}(\epsilon, z) \equiv \det K_{\mathbf{k}_\parallel,+}(\epsilon, z) \tag{2.57}$$

defines a family of polynomials in z. We call a given value of ϵ "singular" if $p_{\mathbf{k}_\parallel}(\epsilon, z)$ vanishes identically for all z for some value of $\mathbf{k}_\parallel$. Otherwise, ϵ is "regular." At a singular value of the energy, z becomes independent of ϵ for some $\mathbf{k}_\parallel$. Physically, singular energies correspond to "flat bands," at fixed $\mathbf{k}_\parallel$. As explained in Part I, flat bands are not covered by the generalized Bloch theorem and require separate treatment.[10] In the following, we will concentrate on the *generic case where* ϵ *is regular.*

For regular energies, $p_{\mathbf{k}_\parallel}(\epsilon, z)$ can be factorized in terms of its *distinct* roots as

$$p_{\mathbf{k}_\parallel}(\epsilon, z) = c \prod_{\ell=0}^{n} (z - z_\ell)^{s_\ell}, \qquad c \in \mathbb{C},$$

with c a non-vanishing constant and $z_0 = 0$ by convention. If zero is not a root, then $s_0 = 0$. The z_ℓ, $\ell = 1, \ldots, s_\ell$, are the distinct non-zero roots of multiplicity $s_\ell \geq 1$. The number of solutions of the kernel equation

$$(H_{\mathbf{k}_\parallel, s_\ell}(z_\ell) - \epsilon \mathbb{1}_{d s_\ell})|u\rangle = 0 \tag{2.58}$$

coincides with the multiplicity s_ℓ of z_ℓ (see Theorem 5.13). We will denote a complete set of independent solutions of Eq. (2.58) by $|u_{\ell s}\rangle$, $s = 1, \ldots, s_\ell$, where each $|u_{\ell s}\rangle$ has $d \times 1$ block-entries

$$|u_{\ell s}\rangle = \left[|u_{\ell s 1}\rangle \ldots |u_{\ell s s_\ell}\rangle\right]^{\mathsf{T}}.$$

Moreover, if we define

$$K_{\mathbf{k}_\parallel,+}(\epsilon) \equiv K_{\mathbf{k}_\parallel, s_0,+}(\epsilon, z_0 = 0) \equiv K_{\mathbf{k}_\parallel,-}(\epsilon)^\dagger,$$

then it is also the case that the kernel equations

$$K_{\mathbf{k}_\parallel,+}(\epsilon)|u\rangle = 0, \qquad K_{\mathbf{k}_\parallel,-}(\epsilon)|u\rangle = 0$$

have each s_0 solutions. We will denote a basis of solutions of these kernel equations by $\{|u_s^+\rangle,\ s = 1, \ldots, s_0\}$ and $\{|u_s^-\rangle,\ s = 1, \ldots, s_0\}$ each with block-entries

---

[10] Recall that, if $h_{k_\parallel, R}$ is *not* invertible, flat bands may exist for at most a *finite* number of singular energy values, for each $k_\parallel$. Since the corresponding eigenstates can be chosen to be bulk-localized, they always enter the physical spectrum of the finite-size Hamiltonian $H = H_N + W$, as they are completely insensitive to the BCs. In this sense, they are properly outside the scope of the generalized Bloch theorem. Nonetheless, we showed in Sect. 2.2.2 how to compute the Compactly supported eigenstates of flat bands directly from the analytic continuation of the Bloch Hamiltonian.

$$|u^{\pm}_{\ell s}\rangle = \left[|u^{\pm}_{\ell s 1}\rangle \ \dots \ |u^{\pm}_{\ell s s_0}\rangle \right]^{\mathsf{T}}, \qquad \ell = 0, n+1.$$

In order to make the connection to the lattice degrees of freedom, let us introduce the lattice states

$$|z, v\rangle \equiv \sum_{j=1}^{N} \frac{j^{(v-1)}}{(v-1)!} z^{j-v+1} |j\rangle - \frac{1}{(v-1)!} \partial_z^{v-1} |z, 1\rangle, \tag{2.59}$$

with $j^{(0)} = 1$ and $j^{(v)} = (j - v + 1)(j - v + 2) \dots j$ for v a positive integer. The states

$$|\mathbf{k}_{\parallel}\rangle|\psi_{\mathbf{k}_{\parallel}\ell s}\rangle \equiv \sum_{v=1}^{s_\ell} |\mathbf{k}_{\parallel}\rangle|z_\ell, v\rangle|u_{\ell s v}\rangle, \qquad s = 1, \dots, s_\ell,$$

$$|\mathbf{k}_{\parallel}\rangle|\psi_{\mathbf{k}_{\parallel}0 s}\rangle \equiv \sum_{j=1}^{s_0} |\mathbf{k}_{\parallel}\rangle|j\rangle|u^{+}_{sj}\rangle, \qquad s = 1, \dots, s_0,$$

$$|\mathbf{k}_{\parallel}\rangle|\psi_{\mathbf{k}_{\parallel}(n+1)s}\rangle \equiv \sum_{j=1}^{s_0} |\mathbf{k}_{\parallel}\rangle|N - j + s_0\rangle|u^{-}_{sj}\rangle, \qquad s = 1, \dots, s_0, \tag{2.60}$$

form a complete set of independent solutions of the bulk equation, Eq. (2.53). Intuitively speaking, these states are eigenstates of the Hamiltonian "up to BCs." For regular energies as we assumed, there are exactly $2Rd = 2s_0 + \sum_{\ell=1}^{n} s_\ell$ solutions of the bulk equation for each value of $\mathbf{k}_{\parallel}$. The solutions associated with the non-zero roots are "extended bulk solutions," and the ones associated with $z_0 = 0$ are "emergent." Emergent bulk solutions are perfectly localized around the edges of the system in the direction perpendicular to the hypersurfaces.

By letting $s_{n+1} \equiv s_0$, the "ansatz"

$$|\epsilon, \boldsymbol{\alpha}\rangle \equiv \sum_{\mathbf{k}_{\parallel} \in \mathrm{SBZ}} \sum_{\ell=0}^{n+1} \sum_{s=1}^{s_\ell} \alpha_{\mathbf{k}_{\parallel}\ell s} |\mathbf{k}_{\parallel}\rangle|\psi_{\mathbf{k}_{\parallel}\ell s}\rangle$$

describes the most general solution of the bulk equation in terms of $2Rd$ amplitudes $\boldsymbol{\alpha}$ for each value of $\mathbf{k}_{\parallel}$. We call it an ansatz because the states $|\epsilon, \boldsymbol{\alpha}\rangle$ provide the appropriate search space for determining the energy eigenstate of the full Hamiltonian $H_{\text{tot}} = H_N + W$.

As a direct by-product of the above analysis, it is interesting to note that a *necessary* condition for H to admit an eigenstate of exponential behavior localized on the left (right) edge is that some of the roots $\{z_\ell\}$ of the equation $\det K_{\mathbf{k}_{\parallel},+}(\epsilon, z) = 0$ be inside (outside) the unit circle. Therefore, one simply needs to compute all roots of det $K_{\mathbf{k}_\parallel,+}(\epsilon, z)$ to know whether localized edge states may exist in principle.

We are finally in a position to impose arbitrary BCs. As before, we once again let $b = 1, \ldots, R; N - R + 1, \ldots, N$ to be a variable for the boundary sites. Then the boundary matrix is the block matrix

$$[B(\epsilon)]_{\mathbf{q}_\parallel b,\mathbf{k}_\parallel \ell s} = \delta_{\mathbf{q}_\parallel,\mathbf{k}_\parallel} \langle b|(H_{\mathbf{k}_\parallel,N} - \epsilon I_N|\psi_{\mathbf{k}_\parallel \ell s}\rangle + \langle b|W(\mathbf{q}_\parallel, \mathbf{k}_\parallel)|\psi_{\ell s}\rangle,$$

with non-square $d \times 1$ blocks (one block per boundary site b and crystal momentum $\mathbf{k}_\parallel$). By construction,

$$(H_{\text{tot}} - \epsilon I_N)|\epsilon, \boldsymbol{\alpha}\rangle = \sum_{\mathbf{q}_\parallel,b} \sum_{\mathbf{k}_\parallel,\ell,s} |\mathbf{q}_\parallel\rangle|b\rangle[B(\epsilon)]_{\mathbf{q}_\parallel b,\mathbf{k}_\parallel \ell s}\alpha_{\mathbf{k}_\parallel \ell s},$$

for *any* regular value of $\epsilon \in \mathbb{C}$. Hence, an ansatz state represents an energy eigenstate if and only if

$$\sum_{\mathbf{k}_\parallel,\ell,s} [B(\epsilon)]_{\mathbf{q}_\parallel b,\mathbf{k}_\parallel \ell s}\alpha_{\mathbf{k}_\parallel \ell s} = 0 \qquad \forall\, \mathbf{q}_\parallel, b,$$

for all boundary sites b and crystal momenta $\mathbf{q}_\parallel$, or, more compactly, $B(\epsilon)\boldsymbol{\alpha} = 0$. We will now state the extension of generalized Bloch theorem for clean systems subject to arbitrary BCs on two parallel hyperplanes:

Theorem 2.4 (Generalized Bloch Theorem) *Let $H_{tot} = H_N + W$ denote a single-particle Hamiltonian as specified above (Eq. (2.51)), for a slab of thickness $N > 2Rd$. Let $B(\epsilon)$ be the associated boundary matrix. If ϵ is an eigenvalue of H_{tot} and a regular energy of $H(z)$, the corresponding eigenstates of H_{tot} are of the form*

$$|\epsilon, \boldsymbol{\alpha}_{\mathrm{k}}\rangle = \sum_{\mathbf{k}_\parallel} \sum_{\ell=0}^{n+1} \sum_{s=1}^{s_\ell} \alpha^{(k)}_{\mathbf{k}_\parallel \ell s} |\mathbf{k}_\parallel\rangle|\psi_{\mathbf{k}_\parallel \ell s}\rangle, \qquad \mathrm{k} = 1, \ldots, \mathcal{K}_\epsilon,$$

where the amplitudes $\boldsymbol{\alpha}_{\mathrm{k}}$ are determined as a complete set of independent solutions of the kernel equation $B(\epsilon)\boldsymbol{\alpha}_{\mathrm{k}} = 0$, and the degeneracy $\mathcal{K}_\epsilon$ of the energy level ϵ coincides with the dimension of the kernel of the boundary matrix, $\mathcal{K}_\epsilon =$ dim Ker $B(\epsilon)$.

Similar to the 1D case, the lower bound $N > 2Rd$ on the thickness of the lattice is imposed in order to ensure that the emergent solutions on opposite edges of the system have zero overlap and are thus necessarily independent. It can be weakened to $N > 2R$ in the generic case where det $h_{\mathbf{k}_\parallel,R} \neq 0$, because in this case $s_0 = 0$ and there are no emergent solutions.

2.4.2 Interfaces

A second extension of our theoretical framework addresses the exact diagonalization of systems with internal boundaries, namely, interfaces between distinct bulks. In the spirit of keeping technicalities to a minimum, we focus on the simplest setting whereby two bulks with identical reduced Brillouin zones are separated by one interface. The extension to multi-component systems is straightforward, and can be pursued as needed by mimicking the procedure to be developed next.

Since the lattice vectors for the two bulks forming the interface are the same, the primitive vectors of the surface mesh $\{\mathbf{m}_\mu,\ \mu = 1, \ldots, D-1\}$ and the stacking vector s, are shared by both bulks. Further, the number J of unit cells of Λ_D in each unit cell of $\bar{\Lambda}_D$ is also equal for the two bulks. Let us assume that the two bulks are described by systems that are half-infinite in the directions $-\mathsf{s}$ and s, respectively. The bulk of system number one (left, $\mathrm{i} = 1$) occupies sites corresponding to $j = 0, -1, \ldots, -\infty\}$, whereas the bulk of system number two (right, $\mathrm{i} = 2$) occupies the remaining sites, corresponding to $j = 1, \ldots, \infty$ in the direction s. In analogy to the case of a single bulk, we may write single-particle Hamiltonians for the left and right bulks in terms of appropriate shift operators, namely,

$$T_1 \equiv \sum_{j=-\infty}^{-1} |j\rangle\langle j+1|, \quad T_2 \equiv \sum_{j=1}^{\infty} |j\rangle\langle j+1|.$$

Then $H_\mathrm{i} = \sum_{\mathbf{k}_\parallel} |\mathbf{k}_\parallel\rangle\langle\mathbf{k}_\parallel| H_{\mathrm{i}\mathbf{k}_\parallel}$, where

$$H_{\mathrm{i}\mathbf{k}_\parallel} = h_{\mathrm{i}\mathbf{k}_\parallel 0} + \sum_{r=1}^{R_\mathrm{i}} \left[T_\mathrm{i}^r h_{\mathrm{i}\mathbf{k}_\parallel r} + \mathrm{H.c.}\right],$$

with the corresponding bulk projectors given by

$$P_{B_1} \equiv \sum_{j=-\infty}^{-R_1} |j\rangle\langle j|, \quad P_{B_2} \equiv \sum_{j=R_2+1}^{\infty} |j\rangle\langle j|.$$

The projector onto the interface is $P_\partial = I - P_{B_1} - P_{B_2}$. The Hamiltonian for the total system is of the form

$$H_\mathrm{tot} = H_1 + W + H_2,$$

with $P_{B_\mathrm{i}} W = 0,\ \mathrm{i} = 1, 2$. In this context, W describes an *internal BC*, that is, physically, it accounts for the various possible ways of joining the two bulks. For simplicity, let us assume that W is translation-invariant in all directions parallel to the interface, so that we may write $W = \sum_{\mathbf{k}_\parallel} |\mathbf{k}_\parallel\rangle\langle\mathbf{k}_\parallel| W_{\mathbf{k}_\parallel}$. The next step is to split the Schrödinger equation $(H_\mathrm{tot} - \epsilon I)|\epsilon\rangle = 0$ into a bulk-boundary system of equations. This is possible by observing that an arbitrary state of the total system may be decomposed as $|\Psi\rangle = P_1|\Psi\rangle + P_2|\Psi\rangle$ in terms of the left and right projectors

$$P_1 \equiv \sum_{j=-\infty}^{0} |j\rangle\langle j| \otimes \mathbb{1}_d, \quad P_2 \equiv \sum_{j=1}^{\infty} |j\rangle\langle j| \otimes \mathbb{1}_d,$$

and that the following identities hold: $P_{B_1}(H_1 - \epsilon I)P_2 = 0 = P_{B_2}(H_2 - \epsilon I)P_1$. Hence, the bulk-boundary system of equations for the interface (or junction) takes the form

$$\begin{aligned} P_{B_1}(H_1 - \epsilon\mathbb{1})P_1|\epsilon\rangle &= 0, \\ P_\partial(H_1 + W + H_2 - \epsilon\mathbb{1})|\epsilon\rangle &= 0, \\ P_{B_2}(H_2 - \epsilon\mathbb{1})P_2|\epsilon\rangle &= 0. \end{aligned}$$

We may now solve for fixed but arbitrary ϵ the bottom and top bulk equations just as in the previous section. The resulting simultaneous solutions of the two bulk equations are expressible as

$$|\epsilon, \boldsymbol{\alpha}_{\mathbf{k}_\parallel}\rangle = |\epsilon, \boldsymbol{\alpha}_{1\mathbf{k}_\parallel}\rangle + |\epsilon, \boldsymbol{\alpha}_{2\mathbf{k}_\parallel}\rangle = \sum_{\mathrm{i}=1,2}\sum_{\mathbf{k}_\parallel} |\mathbf{k}_\parallel\rangle \otimes \Big(\sum_{\ell=0}^{n_\mathrm{i}}\sum_{v=1}^{s_{\mathrm{i}\ell}} \alpha_{\mathrm{i}\ell s}|\psi_{\mathrm{i}\mathbf{k}_\parallel \ell s}\rangle\Big), \tag{2.61}$$

where $\{|\psi_{\mathrm{i}\mathbf{k}_\parallel \ell s}\rangle\}$ are solutions of the bulk equation for the ith bulk. In such situations, we extend the definition of the lattice state $|z, v\rangle$ to a bi-infinite lattice by allowing the index j in Eq. (2.59) to take all integer values. We refer to $|\epsilon, \boldsymbol{\alpha}_{\mathrm{i}\mathbf{k}_\parallel}\rangle$, $\mathrm{i} = 1, 2$, as the eigenstate ansatz for the i-th bulk. For $|\epsilon, \boldsymbol{\alpha}_{\mathbf{k}_\parallel}\rangle$ to be an eigenstate of the full system, the column array of complex amplitudes $\boldsymbol{\alpha}_{\mathbf{k}_\parallel} = \begin{bmatrix}\boldsymbol{\alpha}_{1\mathbf{k}_\parallel} & \boldsymbol{\alpha}_{2\mathbf{k}_\parallel}\end{bmatrix}^{\mathsf{T}}$ must satisfy the boundary equation $B(\epsilon)\boldsymbol{\alpha}_{\mathbf{k}_\parallel} = 0$, in terms of the interface boundary matrix,

$$[B_{\mathbf{k}_\parallel}(\epsilon)]_{b,\mathrm{i}\ell s} = \langle b|(H_{1\mathbf{k}_\parallel} + W + H_{2\mathbf{k}_\parallel} - \epsilon I)|\psi_{\mathrm{i}\mathbf{k}_\parallel \ell s}\rangle,$$

where the boundary index $b \equiv -R_1 + 1, \ldots, 0; 1, \ldots, R_2$.

We conclude this chapter with some remarks on the diagonalization algorithms and the indicator. Both the diagonalization algorithms described in Sect. 2.3.1 can be extended to higher-dimensional systems for the most general case of disordered BCs, as well as for interfaces of finite extent. However, since the boundary matrix can have cross-terms between the virtual wires, we correspondingly have to deal with a single (non-decoupled) boundary matrix of size $2RdN^{D-1} \times 2RdN^{D-1}$. Finding the kernel of this boundary matrix has time complexity $\mathcal{O}(N^{3D-3})$ in the case of numerical algorithm, which will be reflected in the performance of the overall algorithm. Both the indicator $\mathcal{I}_\epsilon$ and degeneracy $\mathcal{K}_\epsilon$ retain their utility in higher-dimensional cases. For relaxed boundary conditions, one can compute these quantities for virtual wires, as we will encounter in the next chapter.

References

1. A.M. Tanhayi, G. Ortiz, B. Seradjeh, On the role of self-adjointness in the continuum formulation of topological quantum phases. Amer. J. Phy. **84**, 858 (2016).
2. S. Nadj-Perge, I.K. Drozdov, J. Li, H. Chen, S. Jeon, J. Seo, A.H. MacDonald, B.A. Bernevig, A. Yazdani, Observation of Majorana fermions in ferromagnetic atomic chains on a superconductor. Science **346**, 602–607 (2014). https://science.sciencemag.org/content/346/6209/602
3. W. DeGottardi, M. Thakurathi, S. Vishveshwara, D. Sen, Majorana fermions in superconducting wires: effects of long-range hopping, broken time-reversal symmetry, and potential land-scapes. Phys. Rev. B **88**, 165111 (2013). https://link.aps.org/doi/10.1103/PhysRevB.88.165111
4. G. Ortiz, J. Dukelsky, E. Cobanera, C. Esebbag, C. Beenakker, Many-body characterization of particle-conserving topological superfluids. Phys. Rev. Lett. **113**, 267002 (2014). https://link.aps.org/doi/10.1103/PhysRevLett.113.267002
5. D. Vodola, L. Lepori, E. Ercolessi, A.V. Gorshkov, G. Pupillo, Kitaev chains with long-range pairing. Phys. Rev. Lett. **113**, 156402 (2014). https://link.aps.org/doi/10.1103/PhysRevLett.113.156402
6. F. Pientka, L.I. Glazman, F. von Oppen, Topological superconducting phase in helical Shiba chains. Phys. Rev. B **88**, 155420 (2013). https://link.aps.org/doi/10.1103/PhysRevB.88.155420
7. N. Read, Compactly supported Wannier functions and algebraic k-theory. Phys. Rev. B **95**, 115309 (2017). https://link.aps.org/doi/10.1103/PhysRevB.95.115309
8. A.Y. Kitaev, Unpaired Majorana fermions in quantum wires. Phys.-Uspekhi **44**, 131–136 (2001). https://doi.org/10.1070%2F1063-7869%2F44%2F10s%2Fs29
9. D.H. Lee, J.D. Joannopoulos, Simple scheme for surface-band calculations. I. Phys. Rev. B **23**, 4988–4996 (1981). https://link.aps.org/doi/10.1103/PhysRevB.23.4988
10. J.P. Blaizot, G. Ripka, *Quantum Theory of Finite Systems* (MIT Press, Cambridge, 1986)
11. N.W. Ashcroft, N.D. Mermin, *Solid State Physics*, 1st edn. (Holt, Rinehart and Winston, New York, 1976)
12. H.J. Mikeska, W. Pesch, Boundary effects on static spin correlation functions in the isotropicx—y chain at zero temperature. Zeitschrift für Physik B Condens. Matter **26**, 351–353 (1977). https://doi.org/10.1007/BF01570745
13. I.E. Tamm, On the possible bound states of electrons on a crystal surface. Physikalische Zeitschrift der Sowjetunion **1**, 733 (1932)
14. W. Shockley, On the surface states associated with a periodic potential. Phys. Rev. **56**, 317–323 (1939). https://link.aps.org/doi/10.1103/PhysRev.56.317
15. G. Seifert, Tight-binding density functional theory: an approximate KohnSham DFT scheme. J. Phys. Chem. A **111**, PMID: 17439198, 5609–5613 (2007). https://doi.org/10.1021/jp069056r
16. L. Jiang, T. Kitagawa, J. Alicea, A.R. Akhmerov, D. Pekker, G. Refael, J.I. Cirac, E. Demler, M.D. Lukin, P. Zoller, Majorana fermions in equilibrium and in driven cold-atom quantum wires. Phys. Rev. Lett. **106**, 220402 (2011). https://link.aps.org/doi/10.1103/PhysRevLett.106.220402
17. B.A. Bernevig, T.L. Hughes, *Topological Insulators and Topological Superconductors* (Princeton University Press, Princeton, 2013)
18. I. Gohberg, P. Lancaster, L. Rodman, *Matrix Polynomials* (Academic, New York, 1982)
19. A. Alase, E. Cobanera, G. Ortiz, L. Viola, Generalization of Bloch's theorem for arbitrary boundary conditions: theory. Phys. Rev. B **96**, 195133 (2017). https://link.aps.org/doi/10.1103/PhysRevB.96.195133
20. V. Dwivedi, V. Chua, Of bulk and boundaries: generalized transfer matrices for tight-binding models. Phys. Rev. B **93**, 134304 (2016). https://link.aps.org/doi/10.1103/PhysRevB.93.134304
21. F. Bechstedt, *Principles of Surface Physics*, 1st edn. (Springer, Berlin, 2012)

22. W.F. Trench, A note on computing eigenvalues of banded Hermitian Toeplitz matrices. SIAM J. Sci. Comput. **14**, 248 (1993). https://doi.org/10.1137/0914015
23. F. De Terán, F.M. Dopico, P. Van Dooren, Matrix polynomials with completely prescribed eigenstructure. SIAM J. Matrix Anal. Appl. **36**, 302 (2015). https://doi.org/10.1137/140964138
24. J.C. Avila, H. Schulz-Baldes, C. Villegas-Blas, Topological invariants of edge states for periodic two-dimensional models. Math. Phys. Anal. Geom. **16**, 137–170 (2013). https://doi.org/10.1007/s11040-012-9123-9
25. L.E. Ballentine, *Quantum Mechanics: A Modern Development*, 2nd edn. (World Scientific Publishing Company, Singapore, 2014)
26. A. Quelle, E. Cobanera, C.M. Smith, Thermodynamic signatures of edge states in topological insulators. Phys. Rev. B **94**, 075133 (2016). https://link.aps.org/doi/10.1103/PhysRevB.94.075133
27. M. Püschel, J.M. Moura, The algebraic approach to the discrete cosine and sine transforms and their fast algorithms. SIAM J. Comput. **32**, 1280–1316 (2003). https://doi.org/10.1137/S009753970139272X
28. G. Ortiz, R. Somma, J. Dukelsky, S. Rombouts, Exactly-solvable models derived from a generalized Gaudin algebra. Nucl. Phys. B **707**, 421–457 (2005). https://doi.org/10.1016/j.nuclphysb.2004.11.008
29. E. Cobanera, A. Alase, G. Ortiz, L. Viola, Exact solution of corner-modified banded block-Toeplitz eigensystems. J. Phys. A: Math. Theor. **50**, 195204 (2017). https://doi.org/10.1088/1751-8121/aa6046
30. I.M. Gelfand, M. Kapranov, A. Zelevinsky, *Discriminants, Resultants, and Multidimensional Determinants* (Springer, Berlin, 2008)
31. R.A. Horn, C.R. Johnson, *Matrix Analysis* (Cambridge University Press, Cambridge, 2012)
32. J. Demmel, I. Dumitriu, O. Holtz, Fast linear algebra is stable. Numer. Math. **108**, 59–91 (2007). https://doi.org/10.1007/s00211-007-0114-x
33. Y. Hatsugai, Chern number and edge states in the integer quantum Hall effect. Phys. Rev. Lett. **71**, 3697–3700 (1993). https://link.aps.org/doi/10.1103/PhysRevLett.71.3697
34. Y. Hatsugai, Edge states in the integer quantum Hall effect and the Riemann surface of the Bloch function. Phys. Rev. B **48**, 11851 (1993). https://link.aps.org/doi/10.1103/PhysRevB.48.11851
35. R.S.K. Mong, V. Shivamoggi, Edge states and the bulk-boundary correspondence in Dirac Hamiltonians. Phys. Rev. B **83**, 125109 (2011). https://link.aps.org/doi/10.1103/PhysRevB.83.125109
36. P. Delplace, D. Ullmo, G. Montambaux, Zak phase and the existence of edge states in graphene. Phys. Rev. B **84**, 195452 (2011). https://link.aps.org/doi/10.1103/PhysRevB.84.195452
37. S. Mao, Y. Kuramoto, K.-I. Imura, A. Yamakage, Analytic theory of edge modes in topological insulators. J. Phys. Soc. Jpn. **79**, 124709 (2010). https://journals.jps.jp/doi/pdf/10.1143/JPSJ.79.124709
38. S.S. Hegde, S. Vishveshwara, Majorana wave-function oscillations, fermion parity switches, and disorder in Kitaev chains. Phys. Rev. B **94**, 115166 (2016). https://link.aps.org/doi/10.1103/PhysRevB.94.115166
39. G. Biczó, O. Fromm, J. Kouteck, A. Lee, Inversion-free formulation of the direct recursion (transfer matrix) method. Chem. phys. **98**, 51–58 (1985). https://doi.org/10.1016/0301-0104(85)80093-8
40. T.B. Boykin, Generalized eigenproblem method for surface and interface states: the complex bands of GaAs and AlAs. Phys. Rev. B **54**, 8107 (1996). https://link.aps.org/doi/10.1103/PhysRevB.54.8107
41. L.G. Molinari, Identities and exponential bounds for transfer matrices. J. Phys. A: Math. Theor. **46**, 254004 (2013). https://doi.org/10.1088/1751-8113/46/25/254004
42. E. Cobanera, A. Alase, G. Ortiz, L. Viola, Generalization of Bloch's theorem for arbitrary boundary conditions: interfaces and topological surface band structure. Phys. Rev. B **98**, 245423 (2018). https://link.aps.org/doi/10.1103/PhysRevB.98.245423

Chapter 3
Investigation of Topological Boundary States via Generalized Bloch Theorem

In this chapter, we apply the generalized Bloch theorem developed in Chap. 2 to several lattice systems in one and two spatial dimensions (see Table 3.1). The chapter also includes discussion of some topologically trivial systems, either because they are still relevant in the discussion of topological phases or because they provide important insights in the application of generalized Bloch theorem. In selecting the models, our attempt has been to choose representatives from a wide variety of systems, such as topological insulators, superconductors, and, notably, some gapless systems. We have also included a 1D example to demonstrate the utility of generalized Bloch theorem for Hamiltonian engineering. For some superconducting models, we analyze the Josephson response of the system, with or without the help of generalized Bloch theorem. This analysis is important for experimental detection of Majorana modes. The contributions made by our analysis of each of the models are listed below. In general, our analysis allows us to make some statements about the boundary physics of the models under consideration with a higher confidence level than what numerical analyses of the same models would afford. Since the generalized Bloch theorem leverages symmetry properties to the extent possible, it also allows a better prediction of the response of the systems to various perturbations. The latter kind of analysis, however, is not carried out in this chapter.

The outline of this chapter is as follows: We start Sect. 3.1.1 by revisiting the impurity model on a single-band chain that we presented as the motivating example at the beginning of Chap. 2. We then show how the Hamiltonian of the BCS superconductor subject to open BCs can be diagonalized by sine transform. Our next application illustrates the utility of generalized Bloch theorem for Hamiltonian engineering; we systematically construct a Hamiltonian that hosts a perfectly localized edge state with certain properties. This Hamiltonian is known in the literature to describe a periodic Anderson model. In Sect. 3.1.4, we analyze a model of dimerized chain that encompasses several models of interest, such as Su–Schrieffer–Heeger (SSH) model, Aubrey–André–Harper model, and Rice–Mele

A. Alase, *Boundary Physics and Bulk-Boundary Correspondence in Topological Phases of Matter*, Springer Theses, https://doi.org/10.1007/978-3-030-31960-1_3

Table 3.1 Summary of representative models analyzed using the generalized Bloch theorem. Some emerging key results are highlighted in the fourth column. DOS: density of states, SC: superconductor (or superconducting, depending on context). Additional models that are amenable to solution by our approach include Majorana chains with twisted BCs [7] or longer-range (e.g., next-nearest-neighbor) couplings [8], dimerized Kitaev chains [9], period-three hopping models [10], as well as time-reversal-invariant TSC wires with spin–orbit coupling [11], to name a few Table adapted with permission from Ref. [12]. Copyrighted by the American Physical Society

	BCs	Some key results
1D systems		
Single-band chain	Edge impurities	Full diagonalization
Periodic Anderson model	Open	Full diagonalization
BCS chain	Open	Full diagonalization
Su–Schrieffer–Heeger model	Reconstructed	Full diagonalization
Rice–Mele model	Reconstructed	Full diagonalization
Aubry–André–Harper model (period-two)	Reconstructed	Full diagonalization
Majorana Kitaev chain	Open	Full diagonalization
		Power-law Majorana modes
Creutz ladder	Open	Power-law topological modes
Two-band s-wave TSC	Open/twisted	4π-periodic supercurrent without parity switch
SNS junction	Junction	Andreev bound states
2D systems		
Graphene (including modulated on-site potential)	Zigzag-bearded	Full diagonalization
	Armchair	Full diagonalization
Chiral p+ip TSC	Open (ribbon)	Closed-form edge bands and states
		Power-law surface modes
Two-band s-wave TSC	Open/twisted	$\mathbf{k}_{\parallel}$-resolved DOS
		Localization length at zero energy
		Enhanced 4π-periodic supercurrent

model. We then discuss several interesting features of Kitaev's Majorana chain that can be explored using generalized Bloch theorem, including Majorana wavefunction oscillations and Majorana wavefunction being exponentially localized with a linear prefactor in certain parameter regimes. We then study a two-band time-reversal invariant s-wave topological superconducting wire introduced in [1, 2], and employ indicator of bulk-boundary correspondence constructed by using the generalized Bloch theorem to analyze the Josephson response of this system. Remarkably, we find that the system shows a 4π-periodic Josephson current *without* a conventional fermionic parity switch. We explain this based on a suitable transformation of the Hamiltonian into two decoupled systems, each undergoing a parity switch. In Sect. 3.2.1, we include an exact calculation of the Andreev bound states in a simple model of a clean superconducting-normal-superconducting (SNS) junction, complementing the detailed numerical investigations reported in Ref. [3].

Moving to 2D systems, we first consider graphene ribbons with two types of edges, "zigzag-bearded" and "armchair" (in the terminology of Ref. [4]), in order to also provide an opportunity for direct comparison within our method and other analytical calculations in the literature. As a more advanced application, we compute in closed-form the surface band structure of the chiral $p + ip$ topological superconductor [5]. This problem is well under control within the continuum approximation [6], but not on the lattice. This distinction is important because the phase diagram of lattice models is richer than one would infer from the continuum approximation. Our analysis of $p + ip$ topological superconductor is the first of its kind to calculate chiral edge states in a topological superconductor completely analytically. As a technically harder example of a surface band-structure calculation, we investigate a two-band, gapless s-wave topological superconductor that can host symmetry-protected Majorana flat bands (MFBs) and is distinguished by a non-unique, anomalous bulk-boundary correspondence [1, 2]. We also show that the Josephson current in this case carries an extensive contribution of the 4π-periodic component.

3.1 Topological Boundary States in 1D Models

3.1.1 The Impurity Model Revisited

The single-particle Hamiltonian of the impurity model of Sect. 2.1 is the corner-modified, BBT matrix $H_{\text{tot}} = H_N + W$, with

$$H_N = -\mathsf{t}(T + T^\dagger), \quad \text{and} \quad W = \mathsf{w}P_\partial.$$

The boundary consists of two sites, so that $P_\partial = |1\rangle\langle 1| + |N\rangle\langle N|$, for any $N > 2$. Likewise, $R = 1 = d$. The first step in diagonalizing H_{tot} is solving the bulk equation. Since the reduced bulk Hamiltonian $H(z) = -\mathsf{t}(z + z^{-1})$,

$$p(\epsilon, z) = z\,(H(z) - \epsilon) = -\mathsf{t}\left(z^2 + \frac{\epsilon}{\mathsf{t}}z + 1\right). \tag{3.1}$$

Thus, every value of ϵ is regular and yields two (= the number of boundary degrees of freedom) solutions of the bulk equation. If $\epsilon \neq \pm 2\mathsf{t}$, the solutions are $|z_\ell, 1\rangle$, with

$$z_\ell = -\frac{\epsilon}{2\mathsf{t}} + (-1)^\ell \sqrt{\frac{\epsilon^2}{4\mathsf{t}^2} - 1}, \quad \ell = 1, 2,$$

with $z_1 z_2 = 1$ and $\epsilon = -\mathsf{t}(z_1 + z_2)$. The special values $\epsilon = \pm 2\mathsf{t}$ for which H_N yields only one of the two bulk solution have an interpretation as the edges of the energy band. If $\epsilon = 2\mathsf{t}$, then $H(z)$ yields only $|z_1 = -1, 1\rangle$, whereas if $\epsilon = -2\mathsf{t}$, it yields only $|z_1 = 1, 1\rangle$. In order to obtain the missing bulk solution in each case, one must consider the second order reduced bulk Hamiltonian (Eq. (2.20))

$$H_2(z) = -\mathsf{t}\begin{bmatrix} z+z^{-1} & 1-z^{-2} \\ 0 & z+z^{-1} \end{bmatrix}.$$

One may check that $H_2(z_1) - \epsilon \mathbb{1}_2 = 0$ if $\epsilon = \pm 2\mathsf{t}$, $z_1 = \mp 1$. Thus, the two linearly independent solutions of the bulk equation at these energies are $|z_1 = 1, v\rangle$, $v = 1, 2$, if $\epsilon = -2\mathsf{t}$, and $|z_1 = -1, v\rangle$, $v = 1, 2$, if $\epsilon = 2\mathsf{t}$.

For the purpose of solving the boundary equation, and hence the full diagonalization problem, it is convenient to organize the solutions of the bulk equation as

$$|\epsilon\rangle = \begin{cases} \alpha_1|z_1, 1\rangle + \alpha_2|z_2, 1\rangle & \text{if } \epsilon \neq \pm 2\mathsf{t} \\ \alpha_1|z_1 = -1, 1\rangle + \alpha_2|z_1 = -1, 2\rangle & \text{if } \epsilon = 2\mathsf{t} \\ \alpha_1|z_1 = 1, 1\rangle + \alpha_2|z_1 = 1, 2\rangle & \text{if } \epsilon = -2\mathsf{t} \end{cases}.$$

For comparison with Sect. 2.1, one should think of $z_1 = e^{ik}$ and $z_2 = e^{-ik}$. Because the ansatz is naturally broken into three pieces, so is the boundary matrix. For instance, when $\epsilon \neq \pm 2\mathsf{t}$, direct calculation yields

$$B(\epsilon) = \begin{bmatrix} -\mathsf{t}z_1^2 + (\mathsf{w}-\epsilon)z_1 & -\mathsf{t}z_2^2 + (\mathsf{w}-\epsilon)z_2 \\ -\mathsf{t}z_1^{N-1} + (\mathsf{w}-\epsilon)z_1^N & -\mathsf{t}z_2^{N-1} + (\mathsf{w}-\epsilon)z_2^N \end{bmatrix}.$$

However, from Eq. (3.1) it follows that

$$-\mathsf{t}(z_\ell + z_\ell^{-1}) - \epsilon = 0, \quad \ell = 1, 2. \tag{3.2}$$

This allows a simpler form to be obtained, by effectively changing the argument of the boundary matrix from ϵ to z_ℓ (or k). The complete final expression reads

$$B(\epsilon) = \begin{cases} \begin{bmatrix} \mathsf{t} + \mathsf{w}z_1 & \mathsf{t} + \mathsf{w}z_2 \\ (z_1\mathsf{t} + \mathsf{w})z_1^N & (z_2\mathsf{t} + \mathsf{w})z_2^N \end{bmatrix} & \text{if } \epsilon \neq \pm 2t \\ \begin{bmatrix} \mathsf{t} - \mathsf{w} & \mathsf{w} \\ (-1)^{N-1}(\mathsf{t} - \mathsf{w}) & (-1)^N(N(\mathsf{t} - \mathsf{w}) + \mathsf{t}) \end{bmatrix} & \text{if } \epsilon = 2t \\ \begin{bmatrix} \mathsf{w} + \mathsf{t} & \mathsf{w} \\ \mathsf{w} + \mathsf{t} & (\mathsf{w} + \mathsf{t})N + \mathsf{t} \end{bmatrix} & \text{if } \epsilon = -2\mathsf{t} \end{cases}. \tag{3.3}$$

Notice that if ϵ approaches $\pm 2\mathsf{t}$, the two distinct roots collide at $z_1 = z_2 = \mp 1$, and $B(\epsilon)$ becomes, trivially, a rank-one matrix, signaling the discontinuous behavior anticipated in Sect. 2.3.1. Furthermore, it follows from Eq. (2.23) that the power-law solution at $\epsilon = \pm 2\mathsf{t}$ may be written as $\partial_z(|z_1, 1\rangle) = |z_1, 2\rangle$. The entries of the second column of the corresponding boundary matrices satisfy $\langle b|H_\epsilon|z_1, 2\rangle = \partial_{z_2}\langle b|H_\epsilon|z_2, 1\rangle|_{z_2=z_1}$, where z_1 is the double root. Thus, the entries in the second column of $B(\epsilon)$ for $\epsilon = \pm 2\mathsf{t}$ may be obtained by differentiating with respect to z_2

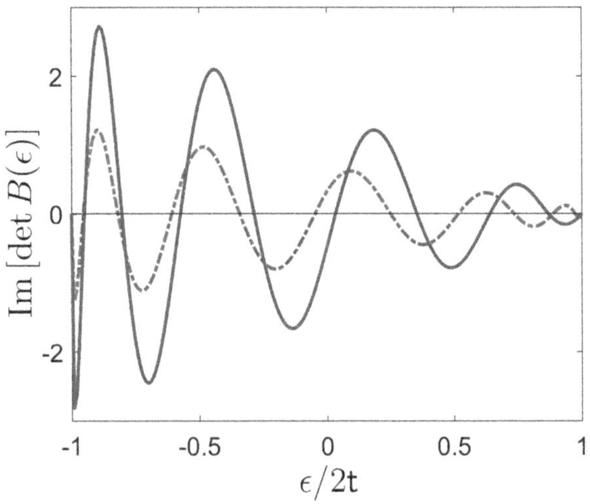

Fig. 3.1 Imaginary part of det $B(\epsilon)$ for $N = 10$ as a function of the dimensionless parameter $\epsilon/2\mathsf{t}$. Here, $B(\epsilon)$ is numerically evaluated from the top expression in Eq. (3.3), $\epsilon \neq \pm 2\mathsf{t}$. Its real part vanishes identically in this range of energies. The impurity potential is $\mathsf{w} = 0.7 > |\mathsf{w}_*|$ for the solid blue curve, and $\mathsf{w} = 0.3 < |\mathsf{w}_*|$ for the dashed red curve. In the regime $\mathsf{w} < \mathsf{w}_*$ ($\mathsf{w} > \mathsf{w}_*$), the system hosts zero (two) edge modes, which is reflected in the number of zeroes (N and $N-2$) of the respective curves, in the energy range $-1 < \epsilon < 1$. In both cases, the crossings through zero at $\epsilon = \pm 1$ do not have associated eigenstates of the Hamiltonian. The origin of such fictitious zeroes was discussed in Sect. 2.3.1. Figure adapted with permission from Ref. [13]. Copyrighted by the American Physical Society

the second column of $B(\epsilon)$ for other (generic) values of ϵ, an observation we will use in other examples as well (see, e.g., Sect. 3.1.3). We now analyze separately different regimes (see also Fig. 3.1 for illustration).

Vanishing Impurity Potential

If $\mathsf{w} = 0$, then $B(\epsilon = 2\mathsf{t})$ and $B(\epsilon = -2\mathsf{t})$ have a trivial kernel; the exotic states $|\epsilon = \pm 2\mathsf{t}\rangle$ cannot possibly arise as physical eigenvectors. For other energies, we find that the kernel of the boundary matrix

$$B(\epsilon) = \mathsf{t}\begin{bmatrix} 1 & 1 \\ z_1^{N+1} & z_2^{N+1} \end{bmatrix} \quad (\mathsf{w} = 0)$$

is non-trivial only if $z_1^{N+1} = z_2^{N+1}$, in which case we can take $\alpha_1 = 1$ and $\alpha_2 = -1$. From Eq. (3.1), it also follows that $z_1 z_2 = 1$. Hence, there are $2N + 2$ solutions,

$$z_1 = z_2^{-1} = e^{i\frac{\pi q}{N+1}}, \quad q = -N-1, -N, \ldots, N.$$

Of the associated $2N + 2$ (unnormalized) ansatz vectors

$$|\epsilon_q\rangle = |z_1, 1\rangle - |z_2, 1\rangle = 2i \sum_{j=1}^{N} \sin\left(\frac{\pi q}{N+1} j\right) |j\rangle,$$

two vanish identically ($q = -N - 1$ and $q = 0$). For $q = \pm 1, \ldots, \pm N$, it is immediate to check that $|\epsilon_{-q}\rangle = -|\epsilon_q\rangle$. This means that the ansatz yields exactly N linearly independent energy eigenvectors, of energy

$$\epsilon_q = -\mathsf{t}(z_1 + z_2) = -2t \cos\left(\frac{\pi q}{N+1}\right), \quad q = 1, \ldots, N.$$

This is precisely the result of Sect. 2.1, where the solutions were labeled in terms of allowed quantum numbers $k = \pi q/(N + 1),\ q = 1, \ldots, N$.

According to our general theory, the eigenspaces of H are in one-to-one correspondence with the zeroes of $\det B(\epsilon)$. For this system then, there should be at most N zeroes. The reason we find $2N + 2$ zeroes is due to the abovementioned (quadratic) change of argument in the boundary matrix from ϵ to k. Such a change of variables is advantageous for analytic work, and the associated redundancy is always rectified at the level of the ansatz.

Power-Law Solutions

What would it take for $|\epsilon = \pm 2t\rangle$ to become eigenvectors? The kernel of $B(\epsilon = 2\mathsf{t})$ is non-trivial only if

$$\mathsf{w} = \mathsf{t} \quad \text{or} \quad \mathsf{w} = \mathsf{t}\frac{N+1}{N-1}.$$

These two values coincide up to corrections of order $1/N$, but remember that our analysis is exact for any $N > 2$. Similarly, the kernel of $B(\epsilon = -2\mathsf{t})$ is non-trivial only if

$$\mathsf{w} = -\mathsf{t} \quad \text{or} \quad \mathsf{w} = -\mathsf{t}\frac{N+1}{N-1}.$$

Only one of these conditions can be met: for fixed w, either $|\epsilon = 2\mathsf{t}\rangle$ is an energy eigenstate or $|\epsilon = -2\mathsf{t}\rangle$ is, but not both. Let us look more closely at the state at the bottom of the energy band. As we just noticed, this state will be a valid eigenstate for either of the two values of w. Let us pick $\mathsf{w} = \mathsf{w}_* \equiv -\mathsf{t}(N + 1)/(N - 1)$, since it yields the most interesting ground state. Then,

$$B(\epsilon = -2t) = \begin{bmatrix} \mathsf{w}_* + \mathsf{t} & \mathsf{w}_* \\ \mathsf{w}_* + \mathsf{t} & \mathsf{w}_* \end{bmatrix},$$

so that one can set $\alpha_1 = 1/(\mathsf{w}_* + \mathsf{t})$, $\alpha_2 = -1/\mathsf{w}_*$, and

$$|\epsilon = -2t\rangle = \sum_{j=1}^{N} \Big(\frac{1}{\mathsf{w}_* + \mathsf{t}} - \frac{j}{\mathsf{w}_*}\Big)|j\rangle.$$

Notice that $\langle j|\epsilon = -2\mathsf{t}\rangle = -\langle N-j+1|\epsilon = -2\mathsf{t}\rangle$; that is, the power-law eigenvector of the impurity problem is an eigenstate of inversion symmetry.

Strong Impurity Potential

Lastly, consider the regime where $\mathsf{t} \ll |\mathsf{w}|$, for large N. Then, the values $\epsilon = \pm 2\mathsf{t}$ are excluded from the physical spectrum, and the eigenstates of the system can be determined from $\det B(\epsilon) = 0$. We expect bound states of energy w to leading order and well-localized at the edges, so that $0 < |z_1| < 1 < |z_2|$, say, with z_1 (z_2) associated with the left (right) edge. It is convenient to take advantage of this feature and modify the original ansatz to

$$|\epsilon\rangle = \alpha_1|z_1, 1\rangle + \alpha_2 z_2^{-N}|z_2, 1\rangle,$$

so that $|z_1, 1\rangle$ ($z_2^{-N}|z_2, 1\rangle$) peaks at the left (right) edge, respectively. The boundary matrix becomes

$$\tilde{B}(\epsilon) = \begin{bmatrix} \mathsf{t} + z_1\mathsf{w} & (\mathsf{t} + \mathsf{w}z_2)z_2^{-N} \\ (z_1\mathsf{t} + \mathsf{w})z_1^N & z_2\mathsf{t} + \mathsf{w} \end{bmatrix} \approx \begin{bmatrix} \mathsf{t} + \mathsf{w}z_1 & 0 \\ 0 & z_2\mathsf{t} + \mathsf{w} \end{bmatrix},$$

since $|z_1|^N \approx 0 \approx |z_2|^{-N}$. Keeping in mind that $z_1 z_2 = 1$, we see that the kernel of $\tilde{B}(\epsilon)$ is two-dimensional for

$$z_1 = -\frac{\mathsf{t}}{\mathsf{w}} = z_2^{-1}, \quad \epsilon_b = -\mathsf{t}(z_1 + z_2) = \mathsf{w} - \frac{\mathsf{t}^2}{\mathsf{w}^2},$$

and otherwise trivial. The corresponding energy eigenstates can be chosen to be

$$|\epsilon_b, 1\rangle = \sum_{j=1}^{N} \Big(-\frac{\mathsf{t}}{\mathsf{w}}\Big)^j |j\rangle, \quad |\epsilon_b, 2\rangle = \sum_{j=1}^{N} \Big(-\frac{\mathsf{w}}{\mathsf{t}}\Big)^{j-N} |j\rangle.$$

Notice that $|\epsilon_b, 2\rangle$ is the mirror image of $|\epsilon_b, 1\rangle$, up to normalization. This calculation can be made more rigorous with the help of arguments in Sect. 2.3.2.

The remaining $(N-2)$ eigenstates consist of standing waves. They can be computed from the original boundary matrix, approximated for $\mathsf{t} \ll |\mathsf{w}|$ as

$$B(\epsilon \neq \epsilon_b) \approx \mathrm{w}\begin{bmatrix} z_1 & z_2 \\ z_1^N & z_2^N \end{bmatrix}.$$

This boundary matrix has a non-trivial kernel only if

$$z_1 = z_2^{-1} = e^{i\frac{\pi s}{N-1}}, \quad s = 0, \ldots, 2(N-1)-1,$$

in which case one may choose $\alpha_1 = z_2,\ \alpha_2 = -z_1$. Then,

$$|\epsilon_s\rangle = \sum_{j=1}^{N}(z_1^{j-1} - z_2^{j-1})|j\rangle = 2i\sum_{j=2}^{N-1}\sin\Big(\frac{\pi s(j-1)}{N-1}\Big)|j\rangle.$$

Moreover, $|\epsilon_s\rangle = -|\epsilon_{N-1+s}\rangle$, $s = 1, \ldots, N-2$. Hence, as needed, we have obtained $(N-2)$ linearly independent eigenvectors of energy $\epsilon_s = -2\mathrm{t}\cos[\pi s/(N-1)]$.

The above discussion is further illustrated in Fig. 3.1, where the determinant of the exact boundary matrix is displayed as a function of energy.

3.1.2 The BCS Chain

A tight-binding BCS chain with N lattice sites can be modeled by a Hamiltonian [3]

$$\widehat{H}_{\text{tot}} = \widehat{H}_N = -\sum_{j,\sigma}\Big(\mathrm{t}\hat{\Phi}^\dagger_{j\sigma}\hat{\Phi}_{j+1\sigma} + \frac{\mu}{2}\hat{\Phi}^\dagger_{j\sigma}\hat{\Phi}_{j\sigma} + \text{H.c.}\Big) - \sum_j\Big(\Delta\hat{\Phi}^\dagger_{j\uparrow}\hat{\Phi}^\dagger_{j\downarrow} + \text{H.c.}\Big).$$

The single-particle Hamiltonian associated with $\widehat{H}_N$ is $H_N = h_0 + (T + T^\dagger)h_1$, where we assume open BCs, with $h_0 = -\mu\tau_z - \Delta\tau_y\sigma_y$, $h_1 = -\mathrm{t}\tau_z$. H_N commutes with $\mathcal{S} = \sigma_y$ because total spin is conserved. Thus, following the discussion in Sect. 2.4.1, we can block-diagonalize H_N as

$$H_N = \sum_{\mathrm{s}=\pm 1} H_{N,\mathrm{s}}|\mathrm{s}\rangle\langle\mathrm{s}|,$$

where $|\mathrm{s}\rangle$ denotes the eigenstate of σ_y for the eigenvalue $\mathrm{s} = \pm 1$. The internal matrices for $H_{N,\mathrm{s}}$ are $h_{\mathrm{s},0} = -\mu\tau_z - \mathrm{s}\,\Delta\tau_y$, $h_{\mathrm{s},1} = -\mathrm{t}\tau_z$, and the action of the particle-hole symmetry on the blocks is

$$\mathcal{P}H_{N,\mathrm{s}}|\mathrm{s}\rangle\langle\mathrm{s}|\mathcal{P}^{-1} = \tau_x H^*_{N,s}\tau_x(|\mathrm{s}\rangle\langle\mathrm{s}|)^* = -H_{N,-\mathrm{s}}|-\mathrm{s}\rangle\langle-\mathrm{s}|. \tag{3.4}$$

Hence, the two blocks are exchanged by particle-hole symmetry, whereas the full Hamiltonian only changes sign. Note that, taken individually, these blocks do *not* respect the particle-hole symmetry because of Eq. (3.4). Therefore, the many-body Hamiltonian does *not* decouple into two blocks.

The non-trivial spatial structure of each of the two blocks is encoded in the matrix $T+T^\dagger$. According to the discussion in Sect. 2.2.3, this fact suffices to guarantee the absence of edge modes and goes a long way towards analytic solvability. For open BCs the eigenstates are

$$|\epsilon_{\mathrm{n},q},\mathrm{s}\rangle = \sum_{j=1}^{N} \sin\left(\frac{\pi q j}{N+1}\right)|j\rangle \begin{bmatrix} \dfrac{i\mathrm{s}\Delta}{\epsilon_{\mathrm{n},q}+\mu+2\mathrm{t}\cos\left(\frac{\pi q}{N+1}\right)} \end{bmatrix},$$

with $q = 1, 2, \ldots, N$ and $\mathrm{n} = 1, 2$ the band index for spin s along the y direction. The energy $\epsilon_{\mathrm{n},q}$ satisfies the relation

$$\epsilon_{\mathrm{n},q} = (-1)^{\mathrm{n}}\sqrt{\left(\mu+2\mathrm{t}\cos\left(\frac{\pi q}{N+1}\right)\right)^2+|\Delta|^2}. \tag{3.5}$$

3.1.3 A Periodic Anderson Model

The goal of this section is to bring to the fore the utility of generalized Bloch theorem from the point of view of *Hamiltonian engineering*. We will design from basic principles a "comb" model, see Fig. 3.2, with the peculiar property of exhibiting a perfectly localized mode at zero energy while all other modes are dispersive. The zero mode is distributed over two sites on the same end of the comb, with weights determined by a ratio of hopping amplitudes. The model we construct is known in the literature as periodic Anderson model.

The starting point is the single-particle Hamiltonian $H = H_N = Th_1 + T^\dagger h_1^\dagger$. In order to have perfectly localized eigenvectors at zero energy, the bulk equation must bear emergent solutions. Therefore, we assume that h_1 is non-invertible. Let $|u^+\rangle$ be in the kernel of $h_1^\dagger$. Since T annihilates $|j=1\rangle$,

$$\begin{aligned} H(|j=1\rangle|u^+\rangle) &= (Th_1 + T^\dagger h_1^\dagger)(|j=1\rangle|u^+\rangle) \\ &= T|j=1\rangle h_1|u^+\rangle + T^\dagger|j=1\rangle h_1^\dagger|u^+\rangle \\ &= 0. \end{aligned}$$

Similarly, if $|u^-\rangle$ is in the kernel of h_1, then $|j=N\rangle|u^-\rangle$ is also in the kernel of H. Therefore, $|j=1\rangle|u^+\rangle$ and $|j=N\rangle|u^-\rangle$ are perfectly localized zero-energy modes.

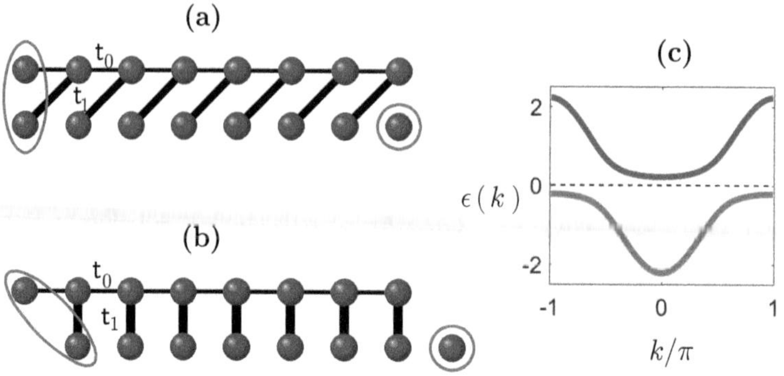

Fig. 3.2 Two variants of the periodic Anderson model. In (**a**), thin (thick) black lines indicate intra-ladder (diagonal) hopping with strength t_0 (t_1). Red ovals or circles show the support of the zero-energy edge modes. In (**b**), upon shifting the lower chain by one site to the right, t_1 can be interpreted as direct inter-ladder hopping strength. (**c**) Band structure for the parameter regime $t_1/t_0 = 0.7$. The (black) dashed line represents zero energy, which lies in the band gap. Figure adapted with permission from Ref. [13]. Copyrighted by the American Physical Society

A concrete example may be obtained by choosing

$$h_1 = -\begin{bmatrix} t_0 & 0 \\ t_1 & 0 \end{bmatrix} \quad \text{and} \quad h_1^\dagger = -\begin{bmatrix} t_0 & t_1 \\ 0 & 0 \end{bmatrix},$$

whose kernel is, respectively, spanned by

$$|u^-\rangle = \begin{bmatrix} 0 \\ 1 \end{bmatrix} \quad \text{and} \quad |u^+\rangle = \begin{bmatrix} -t_1 \\ t_0 \end{bmatrix}.$$

This example corresponds to a many-body Hamiltonian of two coupled fermionic chains, as illustrated in Fig. 3.2:

$$\widehat{H}_{\text{tot}} = -\sum_{j=1}^{N-1} (t_0 \hat{\Phi}_{j,1}^\dagger \hat{\Phi}_{j+1,1} + t_1 \hat{\Phi}_{j+1,1}^\dagger \hat{\Phi}_{j,2} + \text{H.c.}), \tag{3.6}$$

where $\hat{\Phi}_{j,1}$ and $\hat{\Phi}_{j,2}$ denote the jth fermions in the upper and lower chain, t_0 is the intra-ladder hopping in one of the chains, and t_1 is the diagonal hopping strength between the two chains of the ladder, respectively. Physically, this "topological comb model" is closely related to the 1D periodic Anderson model in its non-interacting (spinless) limit, see Ref. [14].

Zero-Energy Modes

The perfectly localized zero-energy modes in this case are $|j = 1\rangle|u^+\rangle$ and $|j = N\rangle|u^-\rangle$ that translate, after normalization, into the fermionic operators

$$\hat{\eta}_1^\dagger = \frac{1}{\sqrt{\mathrm{t}_0^2 + \mathrm{t}_1^2}}(\mathrm{t}_1 \hat{\Phi}_{1,1}^\dagger - \mathrm{t}_0 \hat{\Phi}_{1,2}^\dagger), \quad \eta_2^\dagger = \hat{\Phi}_{N,2}^\dagger. \tag{3.7}$$

The operator $\hat{\eta}_2^\dagger$ trivially describes a zero-energy mode, since it corresponds to the last fermion on the lower chain that is decoupled from the rest. However, $\hat{\eta}_1^\dagger$ corresponds to a non-trivial zero-energy mode, localized over the first sites of the two chains. For large values of $|\mathrm{t}_0/\mathrm{t}_1|$, $\hat{\eta}_1^\dagger$ is localized mostly on the f-chain, whereas for small values it is localized mostly on the c-chain.

Remarkably, such a non-trivial zero-energy mode is *robust* against arbitrary fluctuations in hopping strengths, despite the absence of a protecting chiral symmetry. Imagine that in Eq. (3.6) the hopping strengths $\mathrm{t}_{0,j}$ and $\mathrm{t}_{1,j}$ are position-dependent. Then, $\widehat{H}_{\text{tot}}$ may be written as $\widehat{H}_{\text{tot}} = -(\mathrm{t}_{0,1}\hat{\Phi}_{1,1}^\dagger \hat{\Phi}_{2,1} + \mathrm{t}_{1,1}\hat{\Phi}_{2,1}^\dagger \hat{\Phi}_{1,2} + \text{H.c.}) + \widehat{G}$, where $\widehat{G}$ does not contain terms involving $\hat{\Phi}_{1,1}$ and $\hat{\Phi}_{1,2}$, so that $[\widehat{G}, \hat{\Phi}_{1,1}] = 0 = [\widehat{G}, \hat{\Phi}_{1,2}]$. Then it is easy to verify that the expression for the zero-energy mode is obtained from $\hat{\eta}_1^\dagger$ in Eq. (3.7) after substituting $\mathrm{t}_0 \mapsto \mathrm{t}_{0,1}$ and $\mathrm{t}_1 \mapsto \mathrm{t}_{1,1}$. We conclude that the zero-energy edge mode is protected by an "emergent symmetry" that has a non-trivial action only on the sites corresponding to $j = 1$. Likewise, assume for concreteness that $\mathrm{t}_0 = \pm\mathrm{t}_1$, and consider the inter-chain perturbation described by

$$\widehat{H}_1 \equiv \mu \sum_{j=1}^{N} (\hat{\Phi}_{j,1}^\dagger \pm \hat{\Phi}_{j,2}^\dagger)(\hat{\Phi}_{j,1} \pm \hat{\Phi}_{j,2}), \quad \mu \in \mathbb{R}.$$

In this case, the corresponding single-particle Hamiltonian is $H_N = h_0 + Th_1 + T^\dagger h_1^\dagger$ with

$$h_0 = \mu \begin{bmatrix} 1 & \pm 1 \\ \pm 1 & 1 \end{bmatrix}.$$

Nevertheless, the zero-energy mode corresponding to $|1\rangle|u^+\rangle$ is still an emergent solution for $\epsilon = 0$, and can be verified to satisfy the boundary equation as well.

Complete Closed-Form Solution

We now obtain a complete closed-form solution of the eigenvalue problem corresponding to Eq. (3.6) (open BCs). The reduced bulk Hamiltonian is

$$H(z) = -\begin{bmatrix} \mathrm{t}_0(z+z^{-1}) & \mathrm{t}_1 z^{-1} \\ \mathrm{t}_1 z & 0 \end{bmatrix},$$

with the associated polynomial ($R = 1, d = 2$)

$$p(\epsilon, z) = z^2[\epsilon^2 + \epsilon \mathrm{t}_0(z+z^{-1}) - \mathrm{t}_1^2]. \tag{3.8}$$

The model has two energy bands with a gap containing $\epsilon = 0$, and no chiral symmetry. Because H is real, this enforces the symmetry $z \leftrightarrow z^{-1}$ of the non-zero roots of $p(\epsilon, z)$ that satisfy $z_1 z_2 = 1$. For generic $\epsilon \neq 0$, there are two distinct non-zero roots and, therefore, two extended bulk solutions. The eigenvector of $H(z)$ may be generically expressed as

$$|u(\epsilon, z)\rangle = \begin{bmatrix} \epsilon \\ -\mathrm{t}_1 z \end{bmatrix}.$$

Using Eq. (2.24), the number of emergent bulk solutions is $2Rd - 2 = 2 = 2s_0$, one localized on each edge. As $K_+(\epsilon) = h_1^\dagger$ and $K_-(\epsilon) = h_1$, such solutions are found from their kernels, spanned by $|u^+\rangle$ and $|u^-\rangle$, independently of ϵ. The boundary matrix

$$B(\epsilon) = \begin{bmatrix} \mathrm{t}_0\epsilon - \mathrm{t}_1^2 z_1 & \mathrm{t}_0\epsilon - \mathrm{t}_1^2 z_2 & 0 & \epsilon \mathrm{t}_1 \\ 0 & 0 & 0 & -\epsilon \mathrm{t}_0 \\ z_1^{N+1}\mathrm{t}_0\epsilon & z_2^{N+1}\mathrm{t}_0\epsilon & 0 & 0 \\ z_1^{N+1}\mathrm{t}_1\epsilon & z_2^{N+1}\mathrm{t}_1\epsilon & -\epsilon & 0 \end{bmatrix},$$

whose kernel is non-trivial only if

$$\epsilon \mathrm{t}_0(z_1^{N+1} - z_2^{N+1}) - \mathrm{t}_1^2 z_1 z_2 (z_1^N - z_2^N) = 0.$$

In this case, since $z_1 z_2 = 1$, we may reduce this system to one variable by substituting $z_2 = z_1^{-1}$, which then yields the polynomial equation

$$\epsilon \mathrm{t}_0 z_1^{2N+2} - \mathrm{t}_1^2 z_1^{2N+1} + \mathrm{t}_1^2 z_1 - \epsilon \mathrm{t}_0 = 0. \tag{3.9}$$

The algebraic system of Eqs. (3.8) and (3.9) determine the "dispersing" extended-support bulk modes of the system. When these equations are both satisfied, the kernel of the boundary matrix is spanned by

$$\boldsymbol{\alpha} = \frac{i}{2}\begin{bmatrix} z_1^{-(N+1)} & -z_1^{N+1} & 0 & 0 \end{bmatrix}^{\mathsf{T}},$$

and the corresponding eigenvectors of $H_{\rm tot}$ are given by

$$|\epsilon\rangle = \frac{iz_1^{-(N+1)}}{2}|z_1, 1\rangle\begin{bmatrix}\epsilon \\ -\mathrm{t}_1 z_1\end{bmatrix} - \frac{iz_1^{N+1}}{2}|z_1^{-1}, 1\rangle\begin{bmatrix}\epsilon \\ -\mathrm{t}_1 z_1^{-1}\end{bmatrix},$$

which, upon substituting $z_1 = e^{ik}$, can be recast as[1]

$$|\epsilon\rangle = \sum_{j=1}^{N}|j\rangle\begin{bmatrix}\epsilon \sin k(N+1-j) \\ -\mathrm{t}_1 \sin k(N-j)\end{bmatrix}. \tag{3.10}$$

To check whether $|\epsilon\rangle$ in Eq. (3.10) indeed satisfies the eigenvalue equation, notice that

$$\langle j|H_{\rm tot} - \epsilon I_N|\epsilon\rangle = \begin{cases} -\epsilon\langle 1|\epsilon\rangle + h_1\langle 2|\epsilon\rangle & \text{if } j=1 \\ h_1^\dagger\langle j-1|\epsilon\rangle + h_1\langle j+1|\epsilon\rangle - \epsilon\langle j|\epsilon\rangle & \text{if } 2 \le j \le N-1 \\ h_1^\dagger\langle N-1|\epsilon\rangle - \epsilon\langle N|\epsilon\rangle & \text{if } j = N \end{cases}.$$

Using the expression for $|\epsilon\rangle$, $\langle N|H_{\rm tot} - \epsilon I_N|\epsilon\rangle$ vanishes trivially, while, for $j = 1$,

$$\langle 1|H_{\rm tot} - \epsilon I_N|\epsilon\rangle = -\begin{bmatrix}\epsilon\mathrm{t}_0 \sin k(N-1) + \epsilon^2 \sin kN \\ 0\end{bmatrix},$$

which is seen to vanish from the relation

$$\epsilon\mathrm{t}_0 \sin k(N-1) + \epsilon^2 \sin kN = \sin kN[\epsilon^2 - \mathrm{t}_1^2 + 2\epsilon\mathrm{t}_0 \cos k] + [-\epsilon\mathrm{t}_0 \sin k(N+1) + \mathrm{t}_1^2 \sin kN].$$

The first term on the right-hand side is equal to $p(\epsilon, e^{ik}) = 0$, whereas the second term vanishes due to Eq. (3.9). Finally, for $2 \le j \le N-1$, we get

$$\langle j|H_{\rm tot} - \epsilon I_N|\epsilon\rangle = -\begin{bmatrix}\sin k(N+1-j)[\epsilon^2 - \mathrm{t}_1^2 + 2\epsilon\mathrm{t}_0 \cos k] \\ 0\end{bmatrix},$$

which equals zero, completing the argument.

Next, we find the values of ϵ for which Eq. (3.8) has a double root. The discriminant of $p(\epsilon, z)$ is $\mathscr{D}(p(\epsilon, z)) = (\epsilon^2 - \mathrm{t}_1^2)^2 - 4\epsilon^2\mathrm{t}_0^2$, and vanishes for $\epsilon = -\mathrm{t}_0 \pm \sqrt{\mathrm{t}_0^2 + \mathrm{t}_1^2}$ and $\epsilon = \mathrm{t}_0 \pm \sqrt{\mathrm{t}_0^2 + \mathrm{t}_1^2}$, for which the corresponding double

[1] Alternatively, we could have substituted the analytic expression for either of the roots $z_1(\epsilon)$ or $z_2(\epsilon)$ of Eq. (3.8) in Eq. (3.9), to obtain a single equation in ϵ, whose roots coincide with the eigenvalues of H.

roots are $z_1 = +1$ and $z_1 = -1$, respectively. In these cases, the bulk equation may have power-law solutions. While one could construct the reduced bulk Hamiltonian $H_2(z)$ to identify these solutions, another quick way to proceed is suggested by Eq. (2.23), as already remarked in Sect. 3.1.1. A power-law solution may now be written as

$$\partial_{z_1}(|z_1, 1\rangle|u(\epsilon, z_1)\rangle) = |z_1, 2\rangle|u(\epsilon, z_1)\rangle + |z_1, 1\rangle\partial_{z_1}|u(\epsilon, z_1)\rangle,$$

where z_1 is the double root corresponding to ϵ. The first column of the new boundary matrix remains the same as the original one, while its second column is determined from the derivative of the second column of the original boundary matrix with respect to z_2, computed at $z_2 = z_1$. For $\epsilon = -\mathrm{t}_0 \pm \sqrt{\mathrm{t}_0^2 + \mathrm{t}_1^2}$, we have $z_1 = 1$ and

$$B(\epsilon) = \begin{bmatrix} \mathrm{t}_0\epsilon - \mathrm{t}_1^2 & -\mathrm{t}_1^2 & 0 & \epsilon\mathrm{t}_1 \\ 0 & 0 & 0 & -\epsilon\mathrm{t}_0 \\ \mathrm{t}_0\epsilon & (N+1)\mathrm{t}_0\epsilon & 0 & 0 \\ \mathrm{t}_1\epsilon & (N+1)\mathrm{t}_1\epsilon & -\epsilon & 0 \end{bmatrix}.$$

Some algebra reveals that $\det B(\epsilon) \neq 0$, so that these values of ϵ do not appear in the spectrum of H for any values of parameters $\mathrm{t}_0, \mathrm{t}_1$. Similar analysis for $\epsilon = \mathrm{t}_0 \pm \sqrt{\mathrm{t}_0^2 + \mathrm{t}_1^2}$ yields the same conclusion. Therefore, there are no power-law solutions compatible with open BCs.

We now derive the perfectly localized zero-energy modes described in Sect. 3.1.3. Notice that for $\epsilon = 0$, the only possible roots of $p(\epsilon, z)$ are $z_0 = 0$, and from its degree it follows that there are $s_0 = 2$ emergent solutions on each edge. In this case,

$$K_+(0) = \begin{bmatrix} h_1^\dagger & 0 \\ 0 & h_1^\dagger \end{bmatrix},$$

with its kernel spanned by

$$|u_1^+\rangle = \begin{bmatrix}|u^+\rangle & 0\end{bmatrix}^\mathsf{T} \quad \text{and} \quad |u_2^+\rangle = \begin{bmatrix}0 & |u^+\rangle\end{bmatrix}^\mathsf{T}.$$

Similarly, the kernel of $K^+(0)$ is spanned by

$$|u_1^-\rangle = \begin{bmatrix}|u^-\rangle & 0\end{bmatrix}^\mathsf{T} \quad \text{and} \quad |u_2^-\rangle = \begin{bmatrix}0 & |u^-\rangle\end{bmatrix}^\mathsf{T}.$$

Thus, the ansatz for $\epsilon = 0$ consists of all four perfectly localized solutions (see Eqs. (2.29) and (2.30)). The boundary matrix in this case is

$$B(\epsilon = 0) = \begin{bmatrix} 0 & 0 & 0 & \mathsf{t}_1\mathsf{t}_0 \\ 0 & 0 & 0 & \mathsf{t}_1^2 \\ -\mathsf{t}_1 & 0 & 0 & 0 \\ 0 & 0 & 0 & 0 \end{bmatrix},$$

which has a two-dimensional kernel, spanned by

$$\boldsymbol{\alpha}_1 = \begin{bmatrix} 0 & 0 & 1 & 0 \end{bmatrix}^\mathsf{T}, \quad \boldsymbol{\alpha}_2 = \begin{bmatrix} 0 & 1 & 0 & 0 \end{bmatrix}^\mathsf{T}.$$

The corresponding two zero-energy edge modes are then

$$|\epsilon = 0, \boldsymbol{\alpha}_1\rangle = |1\rangle|u^+\rangle, \quad |\epsilon = 0, \boldsymbol{\alpha}_2\rangle = |N\rangle|u^-\rangle,$$

consistent with Eq. (3.7). The eigenvector $|\epsilon = 0, \boldsymbol{\alpha}_1\rangle$ has support only on the first site of the two band chain. Since $|N\rangle|u^-\rangle = |N\rangle[0 \quad 1]^\mathsf{T}$, the eigenvector $|\epsilon = 0, \boldsymbol{\alpha}_2\rangle$ represents the decoupled degree of freedom at the right end of the chain, as shown in Fig. 3.2a, b.

3.1.4 Dimerized Chains

The 1D model Hamiltonian of the form

$$\widehat{H} = \sum_{j=1}^{2N} [\mathsf{v} - (-1)^j \delta\mathsf{v}] \hat{\Phi}_j^\dagger \hat{\Phi}_j - \sum_{j=1}^{2N-1} [(\mathsf{t} - (-1)^j \delta\mathsf{t}) \hat{\Phi}_j^\dagger \hat{\Phi}_{j+1} + \text{H.c.}],$$

where the parameters

$$\mathsf{t} = \frac{\mathsf{t}_1 + \mathsf{t}_2}{2}, \quad \delta\mathsf{t} = \frac{\mathsf{t}_1 - \mathsf{t}_2}{2}, \quad \mathsf{v} = \frac{\mathsf{v}_1 + \mathsf{v}_2}{2}, \quad \delta\mathsf{v} = \frac{\mathsf{v}_1 - \mathsf{v}_2}{2},$$

subsumes several interesting spin-insensitive phenomena of 1D electronic matter. At half-filling, the model is mostly insulating (the gap only closes only if $\delta\mathsf{t} = 0 = \delta\mathsf{v}$), and has been used for investigating solitons in polyenes (the Rice–Mele model at $\mathsf{v} = 0$), ferroelectricity, and charge fractionalization (the SSH model [15], or even Peierls chain sometimes, at $\mathsf{v} = 0 = \delta\mathsf{v}$); see Ref. [16] and the references therein. If $\delta\mathsf{t} = 0$, our dimerized chain can also be regarded as a special instance of the Aubrey–Harper family of Hamiltonians.

At present, the SSH model is regarded as the simplest particle-conserving topological state of independent electrons (see again Ref. [16] for a discussion of the Berry phase if $\mathsf{v} = 0$). In this sense, it is the natural counterpart of the Kitaev's Majorana chain, and more is in fact true: *as a many-body Hamiltonian*, the SSH model is dual to the Majorana chain at vanishing chemical potential

[17, 18]. In contrast, if $\delta \mathsf{t} = 0 \neq \delta \mathsf{v}$ (we will informally call this regime the Aubrey–Harper chain), the model is topologically trivial. The Aubrey–Harper chain is exactly solvable for open BCs. For generic parameters, the full model is *not* analytically solvable for open BCs, but we will introduce distorted open BCs that yield analytic rather than just exact solvability. Fortunately, these unconventional open BCs map by duality to the standard ones for the Majorana chain. In all cases, a very precise picture of intra-gap states can be obtained in a well-controlled large-size approximation that does not remove the geometric inversion operation $j \leftrightarrow N+1-j$, as passing to a half-infinite system geometry does.

The single-particle Hamiltonian for our dimerized chain subject to open BCs is

$$H_N = h_0 + (Th_1 + \text{H.c.}), \qquad h_0 = \begin{bmatrix} \mathsf{v}_1 & -\mathsf{t}_1 \\ -\mathsf{t}_1^* & \mathsf{v}_2 \end{bmatrix}, \quad h_1 = \begin{bmatrix} 0 & 0 \\ -\mathsf{t}_2 & 0 \end{bmatrix}, \quad \mathsf{v}_1, \mathsf{v}_2, \mathsf{t}_2 \in \mathbb{R}.$$

For $\mathsf{v}_1 = \mathsf{v}_2 = 0$, the Hamiltonian has a chiral symmetry

$$\mathcal{C}_1 = \mathbb{1}_N \begin{bmatrix} 1 & 0 \\ 0 & -1 \end{bmatrix}.$$

For real values of t_1 and $\mathsf{v}_2 = -\mathsf{v}_1$, the system has another non-local chiral symmetry,

$$\mathcal{C}_2 = \sum_{j=1}^{N} |N+1-j\rangle\langle j| \begin{bmatrix} 0 & -i \\ i & 0 \end{bmatrix},$$

which is, however, absent in the limit $N \to \infty$. The analytic continuation of the Bloch Hamiltonian is then

$$H(z) = \begin{bmatrix} \mathsf{v}_1 & -\mathsf{t}_1 - \mathsf{t}_2 z^{-1} \\ -\mathsf{t}_1^* - \mathsf{t}_2 z & \mathsf{v}_2 \end{bmatrix}, \tag{3.11}$$

so that the condition $\det(H(z) - \epsilon \mathbb{1}_2) = 0$ is equivalent to the "dispersion relation" $p(\epsilon, z) = 0$, where

$$p(\epsilon, z) = z^2[(\epsilon - \mathsf{v}_1)(\epsilon - \mathsf{v}_2) - (\mathsf{t}_1 + \mathsf{t}_2 z^{-1})(\mathsf{t}_1^* + \mathsf{t}_2 z)]. \tag{3.12}$$

Before we continue investigating this model with the aid of the generalized Bloch theorem, it is convenient to isolate the occurrence of flat bands. For the dimerized chain, flat bands are only possible if $\mathsf{t}_1 = 0$ or $\mathsf{t}_2 = 0$. The diagonalization of the system is then trivial. We will not pursue it further, assuming from now on that $\mathsf{t}_1, \mathsf{t}_2 \neq 0$. In order to be able to diagonalize our dimerized chain in closed-form, we will impose BCs

$$W = |N\rangle\langle N| \begin{bmatrix} 0 & \mathsf{t}_1 \\ \mathsf{t}_1^* & 0 \end{bmatrix}.$$

Since the range of hopping is $R = 1$ and the number of internal states is $d = 2$ (two atoms per unit cell), the number of boundary degrees of freedom is $2Rd = 4$. This number coincides with the number of solutions of the bulk equation for each value of the ansatz parameter ϵ. There are two emergent bulk solutions, and two extended ones labeled by the roots z_ℓ, $\ell = 1, 2$ of Eq. (3.12). The two roots coincide, that is, $z_1 = z_2$, only if ϵ takes one of the four values

$$\left\{\mathsf{v} \pm \sqrt{\delta\mathsf{v}^2 + (|\mathsf{t}_1| + \mathsf{t}_2)^2},\ \mathsf{v} \pm \sqrt{\delta\mathsf{v}^2 + (|\mathsf{t}_1| - \mathsf{t}_2)^2}\right\}.$$

For these special values of the energy, one of the extended solutions shows power-law behavior.

Ignoring power-law solutions for the moment, the propagating solutions are $|z_\ell, 1\rangle|u_\ell\rangle$, $\ell = 1, 2$, with

$$|u_\ell\rangle = |u(\epsilon, z_\ell)\rangle \equiv \begin{bmatrix} \mathsf{t}_1 + \mathsf{t}_2 z_\ell^{-1} \\ \mathsf{v}_1 - \epsilon \end{bmatrix}, \tag{3.13}$$

such that $(H(z_\ell) - \epsilon \mathbb{1}_2)|u_\ell\rangle = 0$. Notice that for $\epsilon = \mathsf{v}_1$ and $z_2 = -\mathsf{t}_2/\mathsf{t}_1$, the vector $|u(\mathsf{v}_1, -\mathsf{t}_2/\mathsf{t}_1)\rangle$ vanishes. Thus, we deal with the case $\epsilon = \mathsf{v}_1$ separately. For our dimerized chain, the matrices $K_\pm$ that determine the emergent solutions are simply $K_+ = h_1^\dagger$ and $K_- = h_1$, so that the solutions themselves are $|\psi_0\rangle = |1\rangle|u^+\rangle$ and $|\psi_3\rangle = |N\rangle|u^+\rangle$, with

$$|u^+\rangle = \begin{bmatrix} 1 \\ 0 \end{bmatrix}, \quad |u^-\rangle = \begin{bmatrix} 0 \\ 1 \end{bmatrix},$$

independently of ϵ. We emphasize that emergent solutions are *not* always independent of ϵ. Then our ansatz for energy eigenstates of the dimerized chain is

$$|\epsilon\rangle = \alpha_0|1\rangle|u^+\rangle + \sum_{\ell=1}^{2} \alpha_\ell |z_\ell, 1\rangle|u(\epsilon, z_\ell)\rangle + \alpha_3|N\rangle|u^-\rangle.$$

Our generalized Bloch theorem guarantees that the eigenstates of the model are necessarily contained in the ansatz, with amplitudes $\boldsymbol{\alpha} = [\alpha_0\ \alpha_1\ \alpha_2\ \alpha_3]^\mathsf{T}$ determined by the boundary matrix

$$B(\epsilon) = \begin{bmatrix} \mathsf{v}_1 - \epsilon & \mathsf{t}_2(\mathsf{v}_1 - \epsilon) & \mathsf{t}_2(\mathsf{v}_1 - \epsilon) & 0 \\ -\mathsf{t}_1^* & 0 & 0 & 0 \\ 0 & z_1^N \mathsf{t}_1(\mathsf{v}_1 - \epsilon) & z_2^N \mathsf{t}_1(\mathsf{v}_1 - \epsilon) & 0 \\ 0 & z_1^N(\mathsf{v}_1 - \epsilon)(\mathsf{v}_2 - \epsilon) & z_2^N(\mathsf{v}_1 - \epsilon)(\mathsf{v}_2 - \epsilon) & \mathsf{v}_2 - \epsilon \end{bmatrix} \tag{3.14}$$

as $B(\epsilon)\boldsymbol{\alpha} = 0$. The first and the last columns of the boundary matrix are contributed by the emergent modes, and the remaining two by the propagating modes. The kernel of the boundary matrix is non-trivial only if

$$\epsilon = \mathsf{v}_2 \quad \text{or} \quad z_1^N = z_2^N \quad \text{or} \quad \epsilon = \mathsf{v}_1. \tag{3.15}$$

Out of these three possibilities, $\epsilon = \mathsf{v}_2$ yields the kernel vector $\boldsymbol{\alpha} = [0\ 0\ 0\ 1]^\mathsf{T}$ of the boundary matrix, which represents the decoupled fermion at site $j = N$,

$$|\epsilon = \mathsf{v}_2, \boldsymbol{\alpha}\rangle = |N\rangle \begin{bmatrix} 0 \\ 1 \end{bmatrix}.$$

In order to solve the second equation ($z_1^N = z_2^N$) fully, it is necessary to notice a "symmetry" of Eq. (3.12): The roots of the dispersion relation must satisfy the constraint

$$z_1 z_2 = \mathsf{t}_1^*/\mathsf{t}_1 \equiv e^{-2i\phi}. \tag{3.16}$$

This leads to the allowed values

$$z_1 = e^{2i\phi} z_2^{-1} = e^{i\frac{\pi}{N}q - i\phi}, \quad q = -N-1, \ldots, N. \tag{3.17}$$

Combining this equation with Eq. (3.12), we find that the corresponding energy values are

$$\epsilon_\mathrm{n}(q) = \mathsf{v} + (-1)^\mathrm{n}\sqrt{\delta_\mathsf{v}^2 + |\mathsf{t}(q)|^2}, \quad \mathrm{n} = 1, 2,$$
$$|\mathsf{t}(q)|^2 \equiv |\mathsf{t}_1|^2 + \mathsf{t}_2^2 + 2|\mathsf{t}_1|\mathsf{t}_2 \cos(\pi q/N),$$

independent of ϕ. The last step is putting together the stationary-wave states associated with Eq. (3.17). The kernel of the boundary matrix is spanned by $\boldsymbol{\alpha} = \begin{bmatrix} 0\ 1\ -1\ 0 \end{bmatrix}^\mathsf{T}$. The actual eigenstates are

$$|\epsilon_\mathrm{n}(q), \boldsymbol{\alpha}\rangle = |\chi_1(q)\rangle \begin{bmatrix} \mathsf{t}_1 \\ \mathsf{v}_1 - \epsilon_\mathrm{n}(q) \end{bmatrix} + |\chi_2(q)\rangle \begin{bmatrix} \mathsf{t}_2 \\ 0 \end{bmatrix},$$

with

$$|\chi_1(q)\rangle \equiv 2i \sum_{j=1}^{N} \sin(\pi q j/N) e^{-i\phi j} |j\rangle, \tag{3.18}$$

$$|\chi_2(q)\rangle \equiv 2i \sum_{j=1}^{N} \sin(\pi q (j-1)/N) e^{-i\phi (j-1)} |j\rangle. \tag{3.19}$$

Notice that the eigenvectors $|\epsilon_n(q), \boldsymbol{\alpha}\rangle$ and $|\epsilon_n(-q), \boldsymbol{\alpha}\rangle$ obtained in this way are identical. Further, the energies corresponding to $q = \{0, N\}$ are precisely the ones that have associated power-law bulk solutions, and the boundary matrix in Eq. (3.14) is not valid for these energies. Therefore, the above analysis has revealed only $2(N-1)$ bulk eigenstates (along with the one localized eigenstate at $\epsilon = \mathsf{v}_2$ found earlier). We are still missing one eigenstate, because we have not yet analyzed the case $\epsilon = \mathsf{v}_1$, and also not considered the situation corresponding to power-law modes.

Let us now focus on the case $\epsilon = \mathsf{v}_1$. $B(\epsilon)$ in Eq. (3.14) is not the correct boundary matrix for $\epsilon = \mathsf{v}_1$, since $|u(\epsilon = \mathsf{v}_1, z_2 = -\mathsf{t}_2/\mathsf{t}_1)\rangle$ vanishes as mentioned before. For this energy and z_2 (which satisfy $p(\epsilon, z_2) = 0$), we have

$$H(-\mathsf{t}_2/\mathsf{t}_1) - \mathsf{v}_1\mathbb{1}_2 = \begin{bmatrix} 0 & 0 \\ -\mathsf{t}_1^* + \mathsf{t}_2^2/\mathsf{t}_1 & \mathsf{v}_2 - \mathsf{v}_1, \end{bmatrix},$$

whose kernel is spanned by

$$|u_2\rangle = \begin{bmatrix} \mathsf{t}_1(\mathsf{v}_2 - \mathsf{v}_1) \\ |\mathsf{t}_1|^2 - \mathsf{t}_2^2 \end{bmatrix}.$$

We can still use $|u_1\rangle = |u(\epsilon = \mathsf{v}_1, z_1 = -\mathsf{t}_1^*/\mathsf{t}_2)\rangle$, and the two emergent solutions ($|\psi^+\rangle$ and $|\psi^-\rangle$) are as before. Then the boundary matrix for $\epsilon = \mathsf{v}_1$ is

$$B(\epsilon = \mathsf{v}_1) = \begin{bmatrix} 0 & 0 & \mathsf{t}_2(|\mathsf{t}_1|^2 - \mathsf{t}_2^2) & 0 \\ -\mathsf{t}_1^* & 0 & 0 & 0 \\ 0 & 0 & (-\mathsf{t}_2/\mathsf{t}_1)^N\mathsf{t}_1(|\mathsf{t}_1|^2 - \mathsf{t}_2^2) & 0 \\ 0 & 0 & (-\mathsf{t}_2/\mathsf{t}_1)^N(\mathsf{v}_2 - \mathsf{v}_1)(|\mathsf{t}_1|^2 - \mathsf{t}_2^2) & \mathsf{v}_2 - \mathsf{v}_1 \end{bmatrix}.$$

The kernel of $B(\epsilon = \mathsf{v}_1)$ is one-dimensional, and is spanned by $\boldsymbol{\alpha} = [0\ 1\ 0\ 0]^\mathsf{T}$. Thus, there is only one eigenstate at $\epsilon = \mathsf{v}_1$,

$$|\epsilon = \mathsf{v}_1, \boldsymbol{\alpha}\rangle = |z_1 = -\mathsf{t}_1^*/\mathsf{t}_2, 1\rangle \begin{bmatrix} (|\mathsf{t}_1|^2 - \mathsf{t}_2^2)/\mathsf{t}_1 \\ 0 \end{bmatrix}.$$

For $\mathsf{t}_2 < \mathsf{t}_1$, this energy eigenstate is exponentially localized on the right edge, whereas for $\mathsf{t}_2 > \mathsf{t}_1$, it is localized on the left edge. This behavior is characteristic of the topological phase transition that occurs at $\mathsf{t}_2 = \mathsf{t}_1$. It is not possible to continue this eigenvector into the parameter regime $\mathsf{t}_2 = \mathsf{t}_1$. With this localized eigenstate, we have found all $2N$ eigenstates of $H_N + W$. According to these results, the emergent solution on the left does not enter the physical spectrum for open BCs. The one on the right does, at energy $\epsilon = \mathsf{v}_2$.

Since we have already found the eigenbasis of $H_N + W$ in terms of the ansatz pertaining to those ϵ for which $z_1 \neq z_2$, the boundary matrix calculated at those values of ϵ which bear coinciding roots should produce no more eigenvectors. It is instructive to check explicitly that this is the case. By looking at the discriminant of Eq. (3.12), we find that such double roots appear for the energy values in the set $\{\mathsf{v}\pm\sqrt{\delta\mathsf{v}^2+|\mathsf{t}(0)|^2},\ \mathsf{v}\pm\sqrt{\delta\mathsf{v}^2+|\mathsf{t}(N)|^2}\}$. Let us consider $\epsilon_\pm=\mathsf{v}\pm\sqrt{\delta\mathsf{v}^2+|\mathsf{t}(0)|^2}$, for which $z_1=z_2=e^{-i\phi}$ is a double root. In addition to the generic solution $|\psi_1\rangle=|z_1\rangle|u(\epsilon,z_1)\rangle$, the bulk equation also has a power-law solution in this case, which is

$$|\psi_2\rangle=\partial_{z_1}|\psi_1\rangle=\big(|z_1\rangle\partial_{z_1}+|z_1,2\rangle\big)|u(\epsilon,z_1)\rangle.$$

This effectively replaces each entry in the second column of the boundary matrix $B(\epsilon)$ in Eq. (3.14) by its derivative with respect to z_2. Therefore, the resulting boundary matrix is

$$B(\epsilon_\pm)=\begin{bmatrix}\mathsf{v}_1-\epsilon_\pm & \mathsf{t}_2(\mathsf{v}_1-\epsilon_\pm) & 0 & 0\\ \mathsf{t}_1e^{i\phi} & 0 & 0 & 0\\ 0 & \mathsf{t}_1e^{-iN\phi}(\mathsf{v}_1-\epsilon_\pm) & N(\mathsf{v}_1-\epsilon_\pm)\mathsf{t}_1e^{i(N-1)\phi} & 0\\ 0 & e^{-iN\phi}(\mathsf{v}_1-\epsilon_\pm)(\mathsf{v}_2-\epsilon_\pm) & Ne^{-i(N-1)\phi}(\mathsf{v}_1-\epsilon_\pm)(\mathsf{v}_2-\epsilon_\pm) & \mathsf{v}_2-\epsilon_\pm\end{bmatrix}.$$

This boundary matrix is found to have a non-trivial kernel if and only if $\epsilon_\pm=\mathsf{v}_1$. But since this analysis pertains to the points away from the phase transition ($|\mathsf{t}_1|\neq|\mathsf{t}_2|$), neither of the conditions $\epsilon_\pm=\mathsf{v}_1$ can be satisfied. A similar analysis for the energy values $\mathsf{v}\pm\sqrt{\delta\mathsf{v}^2+|\mathsf{t}(N)|^2}$, for which $z_1=z_2=-e^{-i\phi}$ is the double root, leads to the conclusion that, away from the critical points, no eigenvector of H takes contributions from power-law solutions. This is a particular feature of this Hamiltonian.

3.1.5 The p-Wave Majorana Chain

Kitaev's Majorana chain [19] is a prototypical model of p-wave topological superconductivity [20, 21]. In terms of spinless fermions, the relevant many-body Hamiltonian in the absence of disorder and under open BCs reads

$$\widehat{H}_N=-\sum_{j=1}^{N}\mu\,\hat{\Phi}_j^\dagger\hat{\Phi}_j-\sum_{j=1}^{N-1}\left(\mathsf{t}\,\hat{\Phi}_j^\dagger\hat{\Phi}_{j+1}-\Delta\,\hat{\Phi}_j^\dagger\hat{\Phi}_{j+1}^\dagger+\text{H.c.}\right),$$

where $\mu,\mathsf{t},\Delta\in\mathbb{R}$ denote the chemical potential, hopping amplitude, and pairing strengths, respectively. This Hamiltonian, expressed in spin language via a Jordan–Wigner transformation, describes the well-known anisotropic XY spin chain, which has a long history in quantum magnetism, including analysis of boundary effects for both open and periodic BCs [22, 23].[2,3] The corresponding single-particle Hamiltonian is

$$H_N = h_0 + (Th_1 + T^\dagger h_1^\dagger), \quad h_0 = \begin{bmatrix} -\mu & 0 \\ 0 & \mu \end{bmatrix}, \quad h_1 = \begin{bmatrix} -\mathsf{t} & \Delta \\ -\Delta & \mathsf{t} \end{bmatrix}. \tag{3.20}$$

We will now investigate the eigenstates and resulting features of this Hamiltonian in various parameter regimes.

Generic Parameter Values

The reduced bulk Hamiltonian for this model is

$$H(z) = \begin{pmatrix} \mathsf{a}(z) & \mathsf{b}(z) \\ -\mathsf{b}(z) & -\mathsf{a}(z) \end{pmatrix}, \quad \mathsf{a}(z) = \mu + \mathsf{t}(z + z^{-1}), \quad \mathsf{b}(z) = \Delta(z - z^{-1}).$$

The equation

$$p(\epsilon, z) = z^2[(z+z^{-1})^2(\mathsf{t}^2 - \Delta^2) + (z+z^{-1})(2\mu\mathsf{t}) + (\mu^2 + 4\Delta^2 - \epsilon^2)] = 0 \tag{3.21}$$

is an analytic continuation of the standard dispersion relation $\epsilon^2 = (\mu + 2\mathsf{t}\cos k)^2 + (2\Delta \sin k)^2$, where real values of k force z to be on the unit circle. The roots of the characteristic equation for a given value of ϵ satisfy

$$z_\ell + 1/z_\ell = \frac{-\mu\mathsf{t} \pm \sqrt{\mu^2\mathsf{t}^2 - (\mathsf{t}^2 - \Delta^2)(\mu^2 + 4\Delta^2 - \epsilon^2)}}{(\mathsf{t}^2 - \Delta^2)}.$$

Note that the roots z_ℓ always appear in reciprocal pairs. Therefore, we can rewrite the roots z_3, z_4 in terms of z_1, z_2 and denote the full set of roots by $\{z_1, z_1^{-1}, z_2, z_2^{-1}\}$, where we have chosen $|z_1|, |z_2| \le 1$. The corresponding eigenvectors of $H(z_\ell)$ may be written as

$$|u(\epsilon, z_\ell)\rangle = \begin{bmatrix} \mathsf{b}(z_\ell) \\ \epsilon - \mathsf{a}(z_\ell) \end{bmatrix}.$$

[2] Explicitly, the Jordan–Wigner mapping yields the Hamiltonian $\widehat{H}_{XY} = -\sum_{j=1}^{N} B_z(\sigma_z^j + 1) - \sum_{j=1}^{N-1}(J_x\sigma_x^j\sigma_x^{j+1} + J_y\sigma_x^j\sigma_x^{j+1})$, where $\sigma_x^j, \sigma_y^j, \sigma_z^j$ are Pauli matrices for spin j, $B_z = \mu/2$ is the strength of the magnetic field along the z-direction, and $J_x = \mathsf{t} - \Delta$, $J_y = \mathsf{t} + \Delta$ are coupling strengths along x and y, respectively.

[3] If $|\epsilon\rangle$ is an eigenstate of H with energy ϵ, satisfying $\mathcal{S}_1|\epsilon\rangle = |\epsilon\rangle$, then $\mathcal{S}_2|\epsilon\rangle$ is also an eigenstate with the same energy. Further, $\mathcal{S}_2|\epsilon\rangle$ is orthogonal to $|\epsilon\rangle$, as the relation $\{\mathcal{S}_1, \mathcal{S}_2\} = 0$ leads to $\mathcal{S}_1(\mathcal{S}_2|\epsilon\rangle) = -(\mathcal{S}_2|\epsilon\rangle)$.

We identify

$$\mathcal{S} = -\sum_{j=1}^{N} |N+1-j\rangle\langle j| \begin{bmatrix} 1 & 0 \\ 0 & -1 \end{bmatrix}$$

to be a symmetry of the Hamiltonian H_N. It commutes with the bulk projector P_B, so that

$$[\mathcal{S}, P_B(H_N - \epsilon)] = 0.$$

Following Sect. 2.4.1, this allows us to partition the bulk solution space into $\mathsf{s} = +1$ and $\mathsf{s} = -1$ eigenspaces of $\mathcal{S}$. Notice that under this transformation,

$$\mathcal{S}|z_\ell, 1\rangle|u(\epsilon, z_\ell)\rangle = \sum_{j=1}^{N} |N+1-j\rangle\langle j|z_\ell, 1\rangle \begin{bmatrix} -\mathsf{b}(z_\ell) \\ \epsilon - \mathsf{a}(z_\ell) \end{bmatrix}$$

$$= z_\ell^{N+1}|z_\ell^{-1}, 1\rangle \begin{bmatrix} \mathsf{b}(z_\ell^{-1}) \\ \epsilon - \mathsf{a}(z_\ell^{-1}) \end{bmatrix} = z_\ell^{N+1}|z_\ell^{-1}, 1\rangle|u(\epsilon, z_\ell^{-1})\rangle.$$

This equation is a consequence of the symmetry

$$\sigma_z \widetilde{H}(z)\sigma_z = \widetilde{H}(z^{-1}), \quad \sigma_z = \begin{bmatrix} 1 & 0 \\ 0 & -1 \end{bmatrix}$$

of the reduced bulk Hamiltonian. Therefore, for each energy, we obtain two ansätze,

$$|\epsilon, \mathsf{s}, \boldsymbol{\alpha}\rangle = \sum_{\ell=1,2} \alpha_\ell \{|z_\ell, 1\rangle|u(\epsilon, z_\ell)\rangle + \mathsf{s} z_\ell^{N+1}|z_\ell^{-1}, 1\rangle|u(\epsilon, z_\ell^{-1})\rangle\},$$

corresponding to the eigenvalues $\mathsf{s} = \pm 1$. Each of these ansätze, with only two free parameters, is representative of the two-dimensional bulk solution space compatible with the corresponding eigenvalue of the symmetry.

The next step is to construct the boundary matrices corresponding to $\mathsf{s} = \pm 1$. We need to find a basis of the boundary subspace in which s is block-diagonal. One such basis is $\{|\mathsf{s}, m\rangle,\ \mathsf{s} = 1, -1, m = 1, 2\}$, where

$$|\mathsf{s}, 1\rangle \equiv \frac{1}{\sqrt{2}}(|1\rangle - \mathsf{s}|N\rangle) \begin{bmatrix} 1 \\ 0 \end{bmatrix}, \quad |\mathsf{s}, 2\rangle \equiv \frac{1}{\sqrt{2}}(|1\rangle + \mathsf{s}|N\rangle) \begin{bmatrix} 0 \\ 1 \end{bmatrix}.$$

The two boundary matrices are then

$$B(\epsilon, \mathsf{s}) = -\sqrt{2}h_1^\dagger \begin{bmatrix} \mathsf{b}(z_1)(1 - \mathsf{s}z_1^{N+1}) & \mathsf{b}(z_2)(1 - \mathsf{s}z_2^{N+1}) \\ (\epsilon - \mathsf{a}(z_1))(1 + \mathsf{s}z_1^{N+1}) & (\epsilon - \mathsf{a}(z_2))(1 + \mathsf{s}z_2^{N+1}) \end{bmatrix}, \quad \mathsf{s} = \pm 1.$$

This leads to the conditions

$$\mathsf{f}(z_1, \mathsf{s}) = \mathsf{f}(z_2, \mathsf{s}), \qquad \mathsf{f}(z, \mathsf{s}) = \frac{\mathsf{b}(z)}{\epsilon - \mathsf{a}(z)}\left[\frac{1 - z^{N+1}}{1 + z^{N+1}}\right]^{\mathsf{s}}.$$

In simplifying the boundary matrix, we have used

$$\langle N|\langle m|(H_N - \epsilon)|\epsilon, s, \boldsymbol{\alpha}\rangle = (-1)^m s\langle 1|\langle m|(H_N - \epsilon)|\epsilon, s, \boldsymbol{\alpha}\rangle,$$

which follows from the symmetry $\mathcal{S}$ of H_N, and $(h_0 - \epsilon\mathbb{1}_2 + z_\ell h_1)|u(\epsilon, z_\ell)\rangle = -z_\ell^{-1}h_1^\dagger|u(\epsilon, z_\ell)\rangle$, which follows from the bulk equation.

Majorana Wavefunction Oscillations

Recently, it was shown [24] that, inside the so-called circle of oscillations, namely the parameter regime

$$\left(\frac{\mu}{2\mathsf{t}}\right)^2 + \left(\frac{\Delta}{\mathsf{t}}\right)^2 = 1, \tag{3.22}$$

the Majorana wavefunction *oscillates while decaying* in space. Such oscillations in Majorana wavefunction are not observed outside this circle. This observation has consequences on the fermionic parity of the ground state [24]. Because of duality, spin excitations in the XY chain show a similar behavior in the corresponding parameter regime $B_z^2 = \mathsf{t}^2 - \Delta^2 = J_x J_y$. We now analyze this phenomenon by leveraging the analysis of Sect. 2.2. For simplicity, we address directly the large-N limit.

Clearly, whether a wavefunction oscillates in space depends on the nature of the extended bulk solutions that contribute to the wavefunction. In particular, let $|\psi\rangle = |z, 1\rangle|u\rangle$ be one such bulk solution. For a wavefunction to be decaying asymptotically, we must have $|z| < 1$. Further, if $z \in \mathbb{R}$, then $|\psi_j\rangle = z|\psi_{j-1}\rangle$ implies that the part of the wavefunction associated with this bulk solution simply decays exponentially without any oscillations. On the other hand, if $z \equiv |z|e^{i\phi}$ with non-zero phase, then a linear combination of vectors

$$|z, 1\rangle + |z^*, 1\rangle = \sum_{j=1}^{N} 2|z|^j \cos(\phi j)|j\rangle$$

can show oscillatory behavior while decaying. This is precisely the phenomenon observed in this case. For $\epsilon = 0$, Eq. (3.21) admits four distinct roots in general, out of which two lie inside the unit circle and contribute to the Majorana mode on the left edge. Whether any of these two roots is complex decides if the Majorana wavefunction oscillates for those parameter values. Notice that the characteristic equation is quadratic in the variable $\omega = z + z^{-1}$. We get the two values of ω to be

$$\omega_{\pm} = \frac{-\mu \mathrm{t} \pm \Delta\sqrt{\mu^2 - 4(\mathrm{t}^2 - \Delta^2)}}{(\mathrm{t}^2 - \Delta^2)}.$$

Likewise, notice that for $\mu^2 < 4(\mathrm{t}^2 - \Delta^2)$, we get both ω_+ and ω_- to be complex, which necessarily means that both z_1, z_2 inside the unit circle are also necessarily complex. Further, the symmetry of Eq. (3.21) forces that $z_2 = z_1^*$. This leads to the oscillatory behavior of the Majorana wavefunction in the regime $\mu^2 < 4(\mathrm{t}^2 - \Delta^2)$, that is, *inside* the circle defined by Eq. (3.22). Thus, the spatial behavior of Majorana excitations in this regime is formally similar to the solution of an underdamped classical harmonic oscillator (see Fig. 3.3). *Outside* the circle, the roots $\omega_\pm$ are real. With some algebra, it can be shown that $|\omega_\pm| > 2$ in this regime, which also means that both z_1, z_2 are real roots. This is why oscillations are not observed in this parameter regime, in agreement with the results of Ref. [24]. The Majorana wavefunction in this case resembles qualitatively the solution of an overdamped harmonic oscillator.

The situation when the parameters lie precisely *on* the circle is particularly interesting. In this case, we find that $\omega_+ = \omega_- \equiv \omega_0 = -4\mathrm{t}/\mu$. Let us assume $\mathrm{t}/\Delta > 0$ for simplicity. It then follows that $z_1 = z_2 = -2(\mathrm{t} - \Delta)/\mu$, which rightly indicates appearance of a power-law solution. Let us specifically analyze the case of open BCs on one end (for $N \gg 1$ as stated). One of the two decaying bulk solutions is $|\psi_{1,1}\rangle = |z_1, 1\rangle|u(z_1)\rangle$, where

$$|u(z)\rangle = \begin{bmatrix} \Delta(z - z^{-1}) \\ \mu + \mathrm{t}(z + z^{-1}) \end{bmatrix}.$$

The other bulk solution is obtained from

$$|\psi_{1,2}\rangle = \partial_{z_1}|\psi_{11}\rangle = z_1^{-1}|z_1, 1\rangle \begin{bmatrix} \Delta(z_1 + z_1^{-1}) \\ \mathrm{t}(z_1 - z_1^{-1}) \end{bmatrix} + |z_1, 2\rangle \begin{bmatrix} \Delta(z_1 - z_1^{-1}) \\ \mu + \mathrm{t}(z_1 + z_1^{-1}) \end{bmatrix}.$$

The relevant boundary matrix,

$$B(\epsilon = 0) \equiv \begin{bmatrix} (2\mathrm{t}z_1 + \mu)\Delta & 2\mathrm{t}\Delta \\ -\mu\mathrm{t} - z_1(\mathrm{t}^2 + \Delta^2) - z_1^{-1}(\mathrm{t}^2 - \Delta^2) & -(\mathrm{t}^2 + \Delta^2) + z_1^{-2}(\mathrm{t}^2 - \Delta^2) \end{bmatrix},$$

may be computed by relating its second column to the partial derivative of the first column at $z = z_1$ as also done previously. We also used Eq. (3.21) for simplification. Some algebra reveals that $B(0)$ has a one-dimensional kernel, spanned by the vector

$$\boldsymbol{\alpha} = \left[-\mu\mathrm{t} \; 2\Delta(\mathrm{t} - \Delta)\right]^{\mathrm{T}}.$$

This leads to the power-law Majorana wavefunction

$$|\epsilon = 0\rangle = -\mu\mathrm{t}\,|\psi_{1,1}\rangle + 2\Delta(\mathrm{t} - \Delta)\,|\psi_{1,2}\rangle$$

$$= \frac{8\Delta^2(\mathrm{t}-\Delta)}{\mu} \sum_{j=1}^{\infty} j z_1^{j-1} |j\rangle \begin{bmatrix} 1 \\ -1 \end{bmatrix}, \tag{3.23}$$

which decays exponentially with a linear prefactor (see Fig. 3.3). In principle, the existence of such exotic Majorana modes could be probed in the proposed Kitaev chain realizations based on linear quantum dot arrays [25], which are expected to afford tunable control on all parameters.

The points on the circle of oscillations, according to Sect. 2.3.3, should also correspond to the parameter values at which the transfer matrix possess generalized eigenvectors of rank two, failing to be diagonalizable. Let us verify this explicitly. Except for the points $\mu = 0$, $\Delta/\mathrm{t} = \pm 1$ in this regime, the matrix h_1 in Eq. (3.20) is invertible. The transfer matrix is then

$$t(\epsilon = 0) = \frac{1}{\mu^2} \begin{bmatrix} 0 & 0 & \mu^2 & 0 \\ 0 & 0 & 0 & \mu^2 \\ -4(\mathrm{t}^2+\Delta^2) & -8\mathrm{t}\Delta & -4\mathrm{t}\mu & -4\Delta\mu \\ -8\mathrm{t}\Delta & -4(\mathrm{t}^2+\Delta^2) & -4\Delta\mu & -4\mathrm{t}\mu \end{bmatrix},$$

where μ, t, and Δ satisfy Eq. (3.22). It can be checked that $t(\epsilon = 0)$ has only two eigenvalues, namely $z_\ell = -2(\mathrm{t} + (-1)^\ell \Delta)/\mu$, $\ell = 1, 2$, each of algebraic multiplicity two, and that both of these eigenvalues have only one eigenvector, given by

$$P_{1,2}|z_\ell, 1\rangle|u_\ell\rangle = \begin{bmatrix} z_\ell \\ (-1)^\ell z_\ell \\ z_\ell^2 \\ (-1)^\ell z_\ell^2 \end{bmatrix},$$

hence geometric multiplicity equal to one. Both z_1, z_2 are then defective, making $t(\epsilon = 0)$ not diagonalizable. In fact, $t(\epsilon = 0)$ has one generalized eigenvector of rank two corresponding to each eigenvalue, given by

$$P_{1,2}|z_\ell, 2\rangle|u_\ell\rangle = \begin{bmatrix} 1 \\ (-1)^\ell \\ (2z_\ell) \\ (-1)^\ell (2z_\ell) \end{bmatrix}.$$

The Parameter Regime $|\mathrm{t}| = |\Delta|$, $\mu \neq 0$

For concreteness, we assume $\mathrm{t} = \Delta$, but a similar analysis may be repeated for the case $\mathrm{t} = -\Delta$. The reduced bulk Hamiltonian in this case is

$$H(z) = \begin{bmatrix} -\mu - \mathrm{t}(z+z^{-1}) & \mathrm{t}(z-z^{-1}) \\ -\mathrm{t}(z-z^{-1}) & \mu + \mathrm{t}(z+z^{-1}) \end{bmatrix},$$

with associated polynomial

$$p(\epsilon, z) = -z^2[2\mu\mathrm{t}(z+z^{-1}) + (\mu^2 + 4\mathrm{t}^2 - \epsilon^2)]. \tag{3.24}$$

As in the topological comb example, for generic values of ϵ the above has two distinct non-zero roots z_1 and z_2, which implies a two-dimensional space of extended bulk solutions and one emergent solution on each edge. Let the two extended solutions be labeled by z_1 and $z_2 = z_1^{-1}$, with $|z_1| \leq 1$. Then, we get

$$|u(\epsilon, z_\ell)\rangle = \begin{bmatrix} \mathrm{t}(z_\ell - z_\ell^{-1}) \\ \epsilon + \mu + \mathrm{t}(z_\ell + z_\ell^{-1}) \end{bmatrix}, \quad \ell = 1, 2.$$

The two emergent solutions are obtained from the one-dimensional kernels of the matrices $K_+(\epsilon) = h_1^\dagger$ and $K_-(\epsilon) = h_1$, which are spanned by

$$|u_1^+\rangle = \begin{bmatrix} 1 \\ -1 \end{bmatrix} \quad \text{and} \quad |u_1^-\rangle = \begin{bmatrix} 1 \\ 1 \end{bmatrix},$$

respectively. Following Eq. (2.35), the boundary matrix is

$$B(\epsilon) = \begin{bmatrix} 2\mathrm{t}^2 z_1 + \mathrm{t}(\epsilon+\mu) & 2\mathrm{t}^2 z_1^{-1} + \mathrm{t}(\epsilon+\mu) & 0 & -\mu-\epsilon \\ -2\mathrm{t}^2 z_1 + \mathrm{t}(\epsilon+\mu) & -2\mathrm{t}^2 z_1^{-1} + \mathrm{t}(\epsilon+\mu) & 0 & -\mu+\epsilon \\ z_1^{N+1}[-2\mathrm{t}^2 z_1^{-1} - \mathrm{t}(\epsilon-\mu)] & z_1^{-(N+1)}[-2\mathrm{t}^2 z_1 - \mathrm{t}(\epsilon-\mu)] & -\mu-\epsilon & 0 \\ z_1^{N+1}[-2\mathrm{t}^2 z_1^{-1} - \mathrm{t}(\epsilon-\mu)] & z_1^{-(N+1)}[-2\mathrm{t}^2 z_1 - \mathrm{t}(\epsilon-\mu)] & \mu-\epsilon & 0 \end{bmatrix}.$$

A closer look at this boundary matrix reveals that open BCs do not allow any contributions from the emergent solutions in the energy eigenstates, which turn out to be linear combinations of the two extended solutions. The condition for ϵ to be an energy eigenvalue is $\det B(\epsilon) = 0$, which simplifies to

$$2\mathrm{t}z_1 + \epsilon + \mu = \pm z_1^{(N+1)}(2\mathrm{t}z_1^{-1} + \epsilon + \mu). \tag{3.25}$$

Explicitly, as long as $\epsilon \notin \mathcal{R} \equiv \{\mu \pm 2\mathrm{t}, -\mu \pm 2\mathrm{t}\}$, the corresponding eigenstate is

$$|\epsilon\rangle = |z_1, 1\rangle|u(\epsilon, z_1)\rangle \mp z_1^{N+1}|z_1^{-1}, 1\rangle|u(\epsilon, z_1^{-1})\rangle.$$

The above equation is particularly interesting for zero energy, since it dictates the *necessary and sufficient* conditions for the existence of Majorana modes. For $\epsilon = 0$, the root z_1 takes values

$$z_1 = \begin{cases} -\mu/2\mathsf{t} & \text{if } |\mu| < 2|\mathsf{t}| \\ -2\mathsf{t}/\mu & \text{if } |\mu| > 2|\mathsf{t}| \end{cases}.$$

In the large-N limit, the factor z_1^{N+1} in the right-hand side of Eq. (3.25) vanishes, thanks to our choice of $|z_1| < 1$. However, the left-hand side vanishes *only* in the topologically non-trivial regime characterized by $|\mu| < 2|\mathsf{t}|$, giving rise to a localized Majorana excitation. The unnormalized Majorana wavefunction in this limit is characterized by an exact exponential decay (see also Fig. 3.3), namely

$$|\epsilon = 0\rangle = \left(\frac{4\mathsf{t}^2 - \mu^2}{2\mu}\right) \sum_{j=1}^{\infty} z_1^j |j\rangle \begin{bmatrix} 1 \\ -1 \end{bmatrix}.$$

For the analysis of the non-generic energy values in $\mathcal{R}$, we return to the finite system size N. For such ϵ, $p(\epsilon, z)$ has double roots at $z_1 = 1$ and $z_1 = -1$, so that the bulk equation has one power-law solution in each case. We illustrate the case

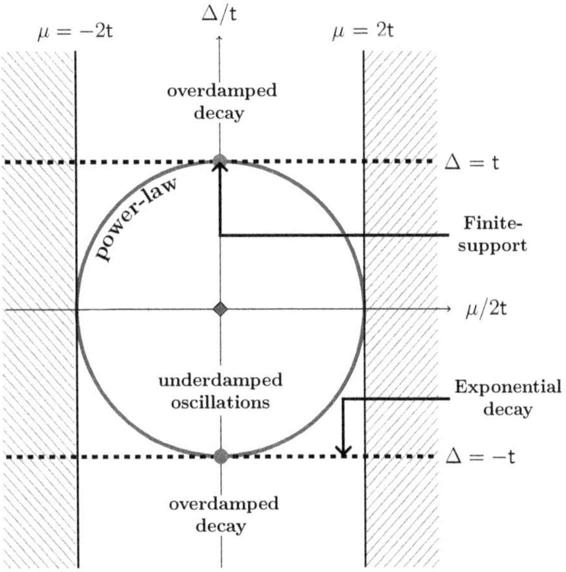

Fig. 3.3 Spatial behavior of Majorana wavefunctions for various parameter regimes of the Kitaev chain under open BCs in the large-N limit. The origin (blue diamond), $\mu = 0$, $\Delta = 0$, corresponds to a metal at half-filling. The region shaded in black pattern is the trivial regime, which does not host Majoranas, and is separated from the non-trivial phase by solid black lines indicating the critical points. The interior of the circle of oscillations (Eq. (3.22)) (shaded in light blue) hosts Majoranas whose wavefunction decays with oscillations, whereas the region outside shows a behavior similar to overdamped decay of a classical harmonic oscillator. On the circle, the wavefunction decays exponentially with a power-law prefactor. The "sweet spots" (red dots) host perfectly localized Majorana modes on the edge. Figure adapted with permission from Ref. [13]. Copyrighted by the American Physical Society

$\epsilon = \mu + 2t$, since the analysis is nearly identical for the other values in $\mathcal{R}$. Since $z_1 = 1$ is now a root of multiplicity two, we construct the rank two reduced bulk Hamiltonian,

$$H_2(z) - \epsilon \mathbb{1}_4 = \begin{bmatrix} H(z=1) - \epsilon \mathbb{1}_2 & H^{(1)}(1) \\ 0 & H(z=1) - \epsilon \mathbb{1}_2 \end{bmatrix}$$
$$= 2 \begin{bmatrix} -2t - \mu & 0 & -2t & \mu & t \\ 0 & 0 & -t & 0 \\ 0 & 0 & -2t - \mu & 0 \\ 0 & 0 & 0 & 0 \end{bmatrix}.$$

The kernel of this matrix is spanned by $\{|u_{1,1}\rangle, |u_{1,2}\rangle\}$, where

$$|u_{1,1}\rangle \equiv \begin{bmatrix} 0 \\ 1 \\ 0 \\ 0 \end{bmatrix}, \quad |u_{1,2}\rangle \equiv \begin{bmatrix} 1 \\ 0 \\ 0 \\ (2+\mu/t) \end{bmatrix}, \quad |1\rangle, |2\rangle \in \mathbb{C}^2,$$

which correspond to the extended solutions

$$|\psi_{1,1}\rangle = |1,1\rangle \begin{bmatrix} 0 \\ 1 \end{bmatrix}, \quad |\psi_{1,2}\rangle = |1,2\rangle \begin{bmatrix} 0 \\ 1 \end{bmatrix} + (2+\mu/t)|1,1\rangle \begin{bmatrix} 1 \\ 0 \end{bmatrix}$$

of the bulk equation. The boundary matrix is found to be

$$B = 2 \begin{bmatrix} -t-\mu & 0 & t & t \\ t & 0 & -t & -t \\ 0 & -t-\mu & -t & -(2N+1)t - (N+1)\mu \\ 0 & -t & -t & -(2N+1)t - (N-1)\mu \end{bmatrix},$$

which has a non-trivial kernel only if the parameter μ, t satisfy

$$(2N+1)t + (N+1)\mu - 1 = 0.$$

Explicitly, the eigenstates corresponding to eigenvalues $\epsilon = \pm(\mu + 2\mathrm{t})$ are then

$$|\epsilon = \mu + 2\mathrm{t}\rangle = \sum_{j=1}^{N} |j\rangle \begin{bmatrix} 1 \\ -1 + \frac{2j}{N+1} \end{bmatrix}, \quad |\epsilon = -\mu - 2\mathrm{t}\rangle = \sum_{j=1}^{N} |j\rangle \begin{bmatrix} -1 + \frac{2j}{N+1} \\ 1 \end{bmatrix}.$$

Note that the contribution to the eigenvector in this case comes solely from extended solutions. Remarkably, the eigenvector has contribution of the power-law solution $|\psi_{1,2}\rangle$. Similar conclusions hold for other values of $\epsilon \in \mathcal{R}$.

The Parameter Regime $|\mathrm{t}| = |\Delta|$, $\mu = 0$

This regime, sometimes affectionately called the "sweet spot," is remarkable. Since the analytic continuation of the Bloch Hamiltonian is

$$H(z) = \mathrm{t}\begin{bmatrix} -(z+z^{-1}) & z-z^{-1} \\ -(z-z^{-1}) & z+z^{-1} \end{bmatrix},$$

one finds that $\det(H(z) - \epsilon \mathbb{1}_2) = \epsilon^2 - 4\mathrm{t}^2$. Thus, the energies $\epsilon = \pm 2\mathrm{t}$ realize a flat band and its charge conjugate. From the point of view of the generalized Bloch theorem, these two energies are singular. According to Sect. 2.2.2, they necessarily belong to the physical spectrum of the Kitaev chain *regardless of BCs*, each yielding $\mathcal{O}(N)$ corresponding bulk-localized eigenvectors.

In order to construct such eigenvectors, note that for $\epsilon = \pm 2\mathrm{t}$, the adjugate of $H(z) - \epsilon \mathbb{1}_d$ is the matrix

$$\mathrm{adj}(H(z) \mp 2\mathrm{t}\mathbb{1}_d) = \mathrm{t}\begin{bmatrix} z+z^{-1} \mp 2 & -z+z^{-1} \\ z-z^{-1} & -z-z^{-1} \mp 2 \end{bmatrix},$$

which immediately provides two kernel vectors

$$|v_{1,\pm}(z)\rangle = \begin{bmatrix} 1+z^{-2} \pm 2z^{-1} \\ 1-z^{-2} \end{bmatrix}, \quad |v_{2,\pm}(z)\rangle = \begin{bmatrix} -1+z^{-2} \\ -1-z^{-2} \pm 2z^{-1} \end{bmatrix}.$$

In this case, we see that the kernel vectors contain polynomials in z^{-1} of degree $2 < \delta_0 = (d-1)2Rd = 4$ (recall Eq. (2.31)). For a suitable range of lattice coordinates js, the compactly supported sequences

$$\Psi_{j1,\pm} = |j\rangle \begin{bmatrix} 1 \\ 1 \end{bmatrix} \pm 2|j+1\rangle \begin{bmatrix} 1 \\ 0 \end{bmatrix} + |j+2\rangle \begin{bmatrix} 1 \\ -1 \end{bmatrix},$$

$$\Psi_{j2,\pm} = -|j\rangle \begin{bmatrix} 1 \\ 1 \end{bmatrix} \pm 2|j+1\rangle \begin{bmatrix} 0 \\ 1 \end{bmatrix} + |j+2\rangle \begin{bmatrix} 1 \\ -1 \end{bmatrix}$$

yield non-zero solutions $|\Psi_{j\mu,\pm}\rangle = \boldsymbol{P}_{1,N}\Psi_{j\mu,\pm}$, $\mu = 1, 2$, of the bulk equation. However, it is not a priori clear how many of these are linearly independent. For example, it is immediate to check that $\Psi_{j1,\pm} + \Psi_{j2,\pm} = \mp(\Psi_{j+1,2,\pm} - \Psi_{j+1,1,\pm})$. In this case, a basis of compactly supported solutions can be chosen from the states

$$|\tilde{\Psi}_0\rangle = |1\rangle \begin{bmatrix} -1 \\ 1 \end{bmatrix} \qquad \text{if } j = 0,$$

$$|\tilde{\Psi}_{j,\pm}\rangle = |j\rangle \begin{bmatrix}1\\1\end{bmatrix} \pm |j+1\rangle \begin{bmatrix}1\\-1\end{bmatrix} \quad \text{if } 1 \le j \le N-1,$$

$$|\tilde{\Psi}_N\rangle = |N\rangle \begin{bmatrix}1\\1\end{bmatrix} \quad \text{if } j = N.$$

Out of these $N+1$ states, the ones corresponding to $j = 1, \ldots, N-1$ can be immediately checked to be eigenstates of energy $\epsilon \pm 2t$.[4] In contrast, $|\tilde{\Psi}_0\rangle$ and $|\tilde{\Psi}_N\rangle$ are *not* eigenstates: they do not satisfy the boundary equation trivially like other states localized in the bulk. We have thus found $2N-2$ eigenstates of the Hamiltonian, $N-1$ for each band $\epsilon = \pm 2t$.

The two missing eigenstates appear at $\epsilon = 0$, which is a regular value of energy and so it is controlled by the generalized Bloch theorem. To verify rigorously that $\epsilon = 0$ are the only eigenvalues other than the flat bands, we will start our analysis with a generic value of ϵ other than $\pm 2t$. In this case, $p(\epsilon, z)$ has no non-zero roots, and hence there are no extended bulk solutions. Therefore, we only need to find the kernels of the matrices K_+ and K_-, where

$$K_+ = \begin{bmatrix} h_1^\dagger & h_0 - \epsilon \\ 0 & h_1^\dagger \end{bmatrix}, \quad K_- = \begin{bmatrix} h_1 & 0 \\ h_0 - \epsilon & h_1 \end{bmatrix}.$$

They are found to be spanned by $\{|u_1^+\rangle, |u_2^+\rangle\}$ and $\{|u_1^-\rangle, |u_2^-\rangle\}$, respectively, where

$$|u_1^-\rangle \equiv \begin{bmatrix}1\\-1\\0\\0\end{bmatrix}, \quad |u_2^-\rangle \equiv \begin{bmatrix}-\epsilon\\-\epsilon\\2t\\-2t\end{bmatrix}, \quad |u_1^+\rangle \equiv \begin{bmatrix}0\\0\\1\\1\end{bmatrix}, \quad |u_2^+\rangle \equiv \begin{bmatrix}2t\\2t\\-\epsilon\\\epsilon\end{bmatrix},$$

corresponding to the emergent solutions

$$|\psi_{01}\rangle = |1\rangle \begin{bmatrix}1\\-1\end{bmatrix}, \quad |\psi_{02}\rangle = -\epsilon|1\rangle \begin{bmatrix}1\\1\end{bmatrix} + 2t|2\rangle \begin{bmatrix}1\\-1\end{bmatrix},$$

$$|\psi_{11}\rangle = |N\rangle \begin{bmatrix}1\\1\end{bmatrix}, \quad |\psi_{12}\rangle = -\epsilon|N\rangle \begin{bmatrix}1\\-1\end{bmatrix} + 2t|N-1\rangle \begin{bmatrix}1\\1\end{bmatrix}.$$

[4] Observe that the states corresponding to $j = 1, \ldots, N-2$ are related to the basis states $\Psi_{j1,\pm}$ and $\Psi_{j2,\pm}$ as follows: $\Psi_{j1,\pm} = \tilde{\Psi}_{j,\pm} \pm \tilde{\Psi}_{j+1,\pm}$ and $\Psi_{j2,\pm} = -\tilde{\Psi}_{j,\pm} \pm \tilde{\Psi}_{j+1,\pm}$.

The boundary matrix reads

$$B(\epsilon) = \begin{bmatrix} -\epsilon & (\epsilon^2 - 4t^2) & 0 & 0 \\ \epsilon & (\epsilon^2 - 4t^2) & 0 & 0 \\ 0 & 0 & -\epsilon & (\epsilon^2 - 4t^2) \\ 0 & 0 & -\epsilon & -(\epsilon^2 - 4t^2) \end{bmatrix}.$$

Since we have assumed $\epsilon \neq \pm 2t$, the boundary matrix has a non-trivial kernel only if $\epsilon = 0$, in which case it is spanned by $\{\boldsymbol{\alpha}_1 = [1\ 0\ 0\ 0]^{\mathsf{T}}, \boldsymbol{\alpha}_2 = [0\ 0\ 1\ 0]^{\mathsf{T}}\}$. Therefore, the zero-energy eigenspace is spanned by

$$|\epsilon = 0, 1\rangle = \frac{1}{\sqrt{2}}|1\rangle \begin{bmatrix} 1 \\ -1 \end{bmatrix}, \quad |\epsilon = 0, 2\rangle = \frac{1}{\sqrt{2}}|N\rangle \begin{bmatrix} 1 \\ 1 \end{bmatrix}.$$

Since these solutions are perfectly localized on the two edges, they exist for any $N > 2$ (see also Fig. 3.3). Interestingly, the above states also appeared as solutions of the bulk equation at the singular energies $\epsilon = \pm 2$t, and failed to satisfy the BCs at those values of energy. We do not know whether this fact is just a coincidence or has some deeper significance.

3.1.6 An *s*-Wave Topological Superconductor

The s-wave, spin-singlet, two-band superconductor model introduced and analyzed in Ref. [1, 2] derives its topological nature from the interplay between a Dimmock-type intra-band spin–orbit coupling and interband hybridization terms. Due to the spin degree of freedom in each of the two relevant orbitals, the Nambu basis corresponding to an atom at position j consists of eight fermionic operators that we write as the vector

$$\hat{\Psi}_j^\dagger = \left[\hat{\Phi}_{j1\uparrow}^\dagger\ \hat{\Phi}_{j1\downarrow}^\dagger\ \hat{\Phi}_{j2\uparrow}^\dagger\ \hat{\Phi}_{j2\downarrow}^\dagger\ \hat{\Phi}_{j1\uparrow}\ \hat{\Phi}_{j1\downarrow}\ \hat{\Phi}_{j2\uparrow}\ \hat{\Phi}_{j2\downarrow}\right].$$

The indices 1 and 2 indicate the orbital, whereas $\uparrow$, $\downarrow$ denote the eigenstates of spin σ_z operator. In this basis, the single-particle Hamiltonian under open BCs is given by

$$\widehat{H}_N = \sum_{j=1}^{N} \hat{\Psi}_j^\dagger h_0 \hat{\Psi}_j + \sum_{j=1}^{N-1} (\hat{\Psi}_j^\dagger h_1 \hat{\Psi}_{j+1} + \text{H.c.}),$$

with

$$h_0 = \begin{bmatrix} -\mu & \mathsf{u} & -i\Delta\sigma_y & 0 \\ \mathsf{u} & -\mu & 0 & i\Delta\sigma_y \\ i\Delta\sigma_y & 0 & \mu & -\mathsf{u} \\ 0 & -i\Delta\sigma_y & -\mathsf{u} & \mu \end{bmatrix} = -\mu\tau_z + \mathsf{u}\tau_z\nu_x + \Delta\tau_y\nu_z\sigma_y,$$

$$h_1 = \begin{bmatrix} i\lambda\sigma_x & -\mathsf{t} & 0 & 0 \\ -\mathsf{t} & -i\lambda\sigma_x & 0 & 0 \\ 0 & 0 & i\lambda\sigma_x & \mathsf{t} \\ 0 & 0 & \mathsf{t} & -i\lambda\sigma_x \end{bmatrix} = -t\tau_z\nu_x + i\lambda\nu_z\sigma_x,$$

where the real parameters μ, u, t, λ, Δ denote the chemical potential, the interband hybridization, hopping, spin–orbit coupling, and pairing potential strengths, respectively, and $\tau_\alpha, \nu_\alpha, \sigma_\alpha, \alpha = \{x, y, z\}$, are Pauli matrices in Nambu, orbital, and spin spaces. The BdG Hamiltonian can be expressed as $H_N = h_0 + (Th_1 + T^\dagger h_1^\dagger)$ in terms of the matrices h_0, h_1.

The topological properties of the above Hamiltonian were analyzed in Ref. [2]. The system is time-reversal invariant, which places it in the symmetry class DIII. The topological phases may thus be distinguished by a $\mathbb{Z}_2$-invariant, given by the parity of the sum of the Berry phases for the two occupied negative bands in one of the Kramers' sectors *only* [2]. For open BCs and for non-vanishing pairing, the system in its trivial phases was found to host zero or two pairs of Majoranas on each edge, in contrast to the topologically non-trivial phase supporting one pair of Majoranas per edge. Similar to the 2D version of the model, one may see that the existence of such Majorana modes is protected by a non-trivial chiral symmetry, of the form $\tau_y\sigma_z$. The single-particle Hamiltonian H_N for open BCs can be exactly diagonalized as described in Sect. 2.3.1. In the large N-limit, the boundary matrix $B_\infty(\epsilon = 0)$ calculated by using the ansatz in Eq. (2.39) yields degeneracy $\mathscr{K}_0 = 0, 4, 8$ in the no-pair, one-pair, and two-pair phases, respectively, verifying the bulk-boundary correspondence previously established through numerical diagonalization.

Josephson Response

We now consider the system in Josephson ring configuration, in which the first and the last sites of the open chain are coupled by the same hopping and spin–orbit terms as in the rest of the chain, only weaker by a factor of $1/\mathsf{w}$. A flux ϕ is introduced between the two ends via this weak link. In the large-N limit, this link acts as a junction, with the corresponding tunneling term in the many-body Hamiltonian being given by

$$\widehat{H}_\partial(\phi) = \hat{\Psi}_N^\dagger(\mathsf{w}h_1 U_\phi)\hat{\Psi}_1 + \text{H.c.}, \qquad U_\phi = \begin{bmatrix} e^{i\phi/2}\mathbb{1}_4 & 0 \\ 0 & e^{-i\phi/2}\mathbb{1}_4 \end{bmatrix}.$$

The total Hamiltonian is then $\widehat{H}_{\text{tot}}(\phi) = \widehat{H}_N + \widehat{H}_\partial(\phi)$. The Hamiltonian displays fractional Josephson effect in the topologically non-trivial phase, as inferred from its 4π-periodic many-body ground state energy (Fig. 3.4a), with the phenomenon being observed *only* if the open-chain Hamiltonian correspondingly hosts an odd number of Majorana pairs per edge. The physics behind the 4π-periodicity can be explained in terms of the crossing of a positive and a negative single-particle energy level happening at precisely zero energy as a function of flux ϕ.

The singular behavior, at $\phi = \pi, 3\pi$, of the indicator $\mathscr{I}_{\epsilon=0}(\phi)$ is shown in Fig. 3.4c. Some physical insight may be gained by looking at the dependence of quasiparticle energy upon flux ϕ: as seen in the top panel, the 4π-periodicity is associated with a crossing of a positive and a negative quasiparticle energy level; this crossing occurs precisely at zero energy, indicating the presence of a pair of Majorana modes for the value of flux at the crossing (solid black lines). In contrast, in the trivial phase with two pairs of Majorana modes (middle panel), the level crossing does *not* occur at zero energy, leading to the standard 2π-periodicity of $E(\phi)$, also found when no Majorana mode is present (bottom panel). Since these quasiparticle energy levels lie in the gap, they are localized, and we can carry out the analysis in the thermodynamic limit. This reveals that, in the non-trivial phase, the boundary equation is satisfied at $\phi = \pi, 3\pi$, confirming the presence of exact zero-energy modes at those values. Even more interestingly, the proposed indicator $\mathcal{D}$ has been numerically evaluated and plotted (dotted red lines): singularities clearly emerge *only* in the topologically non-trivial phase, as claimed.

Parity Switch and Decoupling Transformation

Despite the 4π-periodic Josephson response witnessed in the topologically non-trivial phase, it turns out that the ground state fermionic parity *remains unchanged for all flux values*. In the non-trivial regime of interest, we may focus on the three low-lying energy levels. Specifically, for values of $\phi < \pi$, let $|\Omega(\phi)\rangle$ denote the many-body ground state, with energy $E_0(\phi)$, as in Fig. 3.4a. As we will show, there are two degenerate quasiparticle excitations, say, $\hat{\eta}_1(\phi)$, $\hat{\eta}_2(\phi)$, with small positive energy $\epsilon_0(\phi)$. This results in a *twofold degenerate first excited many-body state*, with energy $E_1(\phi) = E_0(\phi) + \epsilon_0(\phi)$, and a corresponding eigenspace is spanned by $\{\hat{\eta}_1^\dagger(\phi)|\Omega(\phi)\rangle, \hat{\eta}_2^\dagger(\phi)|\Omega(\phi)\rangle\}$. The second excited state, $\hat{\eta}_1^\dagger(\phi)\hat{\eta}_2^\dagger(\phi)|\Omega(\phi)\rangle$, is not degenerate and has energy $E_2(\phi) = E_0(\phi) + 2\epsilon_0(\phi)$. Note that this state has the same (even) fermionic parity as the ground state. At $\phi = \pi$, the quasiparticle excitation has exactly zero energy, $\epsilon_0(\pi) = 0$, causing all three energy levels to become degenerate. As ϕ crosses π, $\epsilon_0(\phi)$ becomes negative. Therefore, for $\pi < \phi < 3\pi$, we find that $E_2(\phi) < E_1(\phi) < E_0(\phi)$. The continuation of the state $\hat{\eta}_1^\dagger(\phi)\hat{\eta}_2^\dagger(\phi)|\Omega(\phi)\rangle$ with energy $E_2(\phi)$ thus becomes the new ground state, whereas

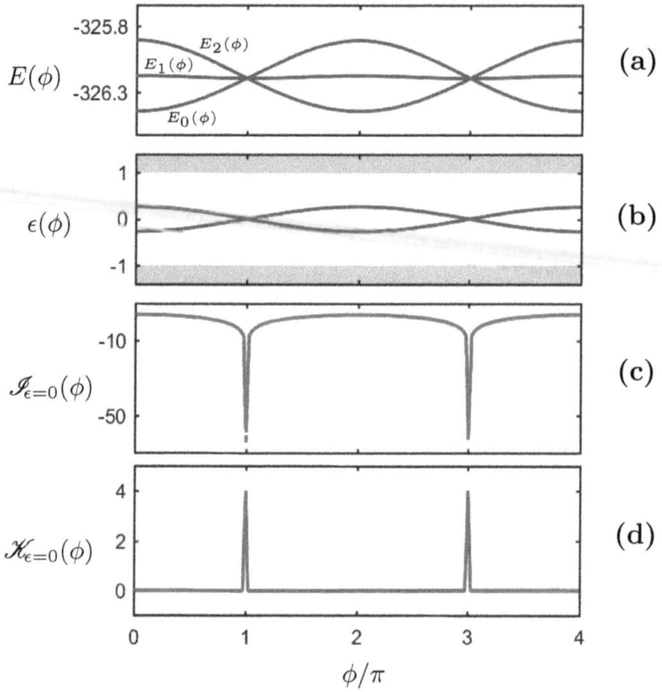

Fig. 3.4 (**a**) Low-lying many-body energy eigenvalues in the Josephson ring configuration, as a function of flux ϕ. The energy level $E_1(\phi)$ is doubly degenerate. (**b**) Energy of the bound mode and its anti-particle excitation. The shaded (blue) area denotes the continuum of energy states in the bulk. (**c**) The indicator of Eq. (2.40) (solid red line) in the topologically non-trivial phase. (**d**) Degeneracy of the zero-energy level inferred from the dimension of the kernel of $B_\infty(\epsilon = 0, \phi)$. The parameters are $\mathsf{w} = 0.2$, $\mu = 0$, $\mathsf{u} = \mathsf{t} = \lambda = 1$, $\Delta = 2$, $N = 60$ in (**a**) and (**b**). Figure adapted with permission from Ref. [13]. Copyrighted by the American Physical Society

the continuation of the original ground state $|\Omega(\phi)\rangle$ now attains the maximum energy among these three levels. Since the new ground state has the same parity as the original one, the system shows no parity switch, with a similar analysis holding for the crossover at $\phi = 3\pi$. We conclude that the absence of a fermionic parity switch originates from the twofold degeneracy of the single-particle energy levels.

While the system under open BCs is time-reversal invariant, away from $\phi = 0, 2\pi$ this symmetry is broken by the tunneling term $\widehat{H}_\partial(\phi)$. Therefore, Kramer's theorem is not responsible in general for the degeneracy in the single-particle levels. Instead, we now explain the physical origin of this degeneracy in terms of a "decoupling transformation" in real space, thanks to which the system in the Josephson bridge configuration is mapped into *two decoupled systems* in the same configuration, each with half the number of internal degrees of freedom as the original one. Although each of these smaller systems does undergo a parity switch, the total parity being the sum of individual parities remains unchanged.

Observe that the Hamiltonian $\widehat{H}_{\rm tot}(\phi)$ is invariant under the unitary symmetries $\hat{\mathcal{S}}_1$ and $\hat{\mathcal{S}}_2$, defined by the action

$$\hat{\mathcal{S}}_1 : \hat{\Phi}_{j1\uparrow}(\hat{\Phi}_{j2\uparrow}) \mapsto \hat{\Phi}_{j2\uparrow}(\hat{\Phi}_{j1\uparrow}), \quad \hat{\Phi}_{j1\downarrow}(\hat{\Phi}_{j2\downarrow}) \mapsto -\hat{\Phi}_{j2\downarrow}(-\hat{\Phi}_{j1\downarrow}),$$
$$\hat{\mathcal{S}}_2 : \hat{\Phi}_{j1\uparrow}(\hat{\Phi}_{j2\uparrow}) \mapsto i\hat{\Phi}_{j1\downarrow}(i\hat{\Phi}_{j2\downarrow}), \quad \hat{\Phi}_{j1\downarrow}(\hat{\Phi}_{j2\downarrow}) \mapsto i\hat{\Phi}_{j1\uparrow}(i\hat{\Phi}_{j2\uparrow}).$$

We can use the eigenbasis of $\hat{S}_1$ to decouple $\widehat{H}_{\rm tot}(\phi)$ into two independent Hamiltonians. Consider, for each site $j = 1, \ldots, N$, the canonical transformation

$$\hat{\Upsilon}_{j1\sigma} \equiv \frac{\hat{\Phi}_{j\sigma} + \hat{\Phi}_{2j\sigma}}{\sqrt{2}}, \quad \hat{\Upsilon}_{j2\sigma} \equiv \frac{\hat{\Phi}_{j\sigma} - \hat{\Phi}_{2j\sigma}}{\sqrt{2}}, \quad \sigma = \uparrow, \downarrow, \tag{3.26}$$

and let $\hat{U}_1$ be the unitary change of basis defined by $\hat{U}_1 : \hat{\Psi}_j^\dagger \mapsto \left[\hat{\Gamma}_{1j}^\dagger \; \hat{\Gamma}_{2j}^\dagger\right]$, where

$$\hat{\Gamma}_{1j}^\dagger \equiv \left[\hat{\Upsilon}_{j1\uparrow}^\dagger \quad \hat{\Upsilon}_{j2\downarrow}^\dagger \quad \hat{\Upsilon}_{j1\uparrow} \quad \hat{\Upsilon}_{j2\downarrow}\right],$$
$$\hat{\Gamma}_{2j}^\dagger \equiv \left[\hat{\Upsilon}_{j1\downarrow}^\dagger \; -\hat{\Upsilon}_{j2\uparrow}^\dagger \quad \hat{\Upsilon}_{j1\downarrow} \; -\hat{\Upsilon}_{j2\uparrow}\right].$$

By letting $\hat{\Gamma}_{\rm i}^\dagger \equiv \left[\hat{\Gamma}_{\rm i1}^\dagger \; \ldots \; \hat{\Gamma}_{{\rm i}N}^\dagger\right]$ for $\rm i = 1, 2$, the action of $\hat{U}_1$ decouples $\widehat{H}_{\rm tot}(\phi)$ according to

$$\widehat{H}_{\rm tot}(\phi) \equiv \widehat{H}_{\rm tot,1}(\phi) + \widehat{H}_{\rm tot,2}(\phi) = \hat{\Gamma}_1^\dagger H_{\rm tot,1}(\phi)\hat{\Gamma}_1 + \hat{\Gamma}_2^\dagger H_{\rm tot,2}(\phi)\hat{\Gamma}_2,$$

where $\widehat{H}_\pm(\phi)$ describes two smaller systems, each in a Josephson ring configuration, with hopping and pairing amplitudes given by

$$h_{\rm i0} = \begin{bmatrix} -\mu + \mathsf{u}\tilde{\sigma}_z & -i\Delta\tilde{\sigma}_y \\ i\Delta\tilde{\sigma}_y & \mu - \mathsf{u}\tilde{\sigma}_z \end{bmatrix} = -\mu\tau_z + \mathsf{u}\tau_z\tilde{\sigma}_z + \Delta\tau_y\tilde{\sigma}_y,$$

$$h_{\rm i1} = \begin{bmatrix} -(-1)^{\rm i} i\lambda\tilde{\sigma}_x - t\tilde{\sigma}_z & 0 \\ 0 & -(-1)^{\rm i} i\lambda\tilde{\sigma}_x + \mathsf{t}\tilde{\sigma}_z \end{bmatrix} = -(-1)^{\rm i} i\lambda\tilde{\sigma}_x - \mathsf{t}\tau_z\tilde{\sigma}_z,$$

with $\rm i = 1, 2$ and $\tilde{\sigma}_\alpha$ denoting Pauli matrices in the modified spin basis. The decoupling transformation in Eq. (3.26) is close in spirit to the one already employed under periodic BCs [1, 2]. Indeed, it is worth remarking that $\hat{\Gamma}_{1j}$ and $\hat{\Gamma}_{2j}$ are still time-reversals of each other, in the sense that $\mathcal{T}\hat{\Gamma}_{1j}^\dagger\mathcal{T}^{-1} = \hat{\Gamma}_{2j}^\dagger$, with $\mathcal{T}$ being the anti-unitary time-reversal operator for the system. Because of the tunneling term, however, the two decoupled (commuting) Hamiltonians $\widehat{H}_{\rm tot,i}(\phi)$ for $\rm i = 1, 2$ are related by the relation $\mathcal{T}\widehat{H}_{\rm tot,1}(\phi)\mathcal{T}^{-1} = \widehat{H}_{\rm tot,2}(4\pi - \phi)$.

It now remains to show that $\widehat{H}_{\rm tot,1}(\phi)$ and $\widehat{H}_{\rm tot,2}(\phi)$ have identical single-particle energy spectrum, and therefore lead to the desired degeneracy in the energy levels of $\widehat{H}_{\rm tot}(\phi)$. This follows by examining the symmetries of the single-particle BdG Hamiltonian $H_{\rm tot}(\phi)$. Corresponding to $\hat{\mathcal{S}}_1$, $H_{\rm tot}(\phi)$ has a unitary symmetry $\mathcal{S}_1 = \nu_x\sigma_z$, and thus gets block-diagonalized into two blocks, $H_{\text{tot,i}}(\phi)$. Similarly, corresponding to $\hat{\mathcal{S}}_2$, $H(\phi)$ has another unitary symmetry $\mathcal{S}_2 = i\tau_z\sigma_x$. Further, $\mathcal{S}_1$ and $\mathcal{S}_2$ satisfy the anticommutation relation $\{\mathcal{S}_1, \mathcal{S}_2\} = 0$, which is responsible for the doubly degenerate eigenvalue spectrum (see second footnote in Sect. 3.1.5). In fact, one can also verify directly that $\widehat{H}_{\text{tot},1}(\phi)$ and $\widehat{H}_{\text{tot},2}(\phi)$ satisfy $\hat{\mathcal{S}}_2\widehat{H}_{\text{tot},1}(\phi)\hat{\mathcal{S}}_2^\dagger = \widehat{H}_{\text{tot},2}(\phi)$. This explains the origin of the double degeneracy of each single-particle energy level, and hence of the absence of fermionic parity switch.

3.2 Interfaces and 2D Systems

3.2.1 Andreev Bound States of an SNS Junction

We illustrate the generalized Bloch theorem for interfaces by outlining an analytical calculation of the Andreev bound states for an idealized SNS junction. The equilibrium Josephson effect, namely the phenomenon of supercurrent flowing through a junction of two superconducting leads connected via a normal link, is of great importance for theoretical understanding of superconductivity, as well as for its applications in superconducting circuits. A question this phenomenon poses is to understand how exactly a weak link with induced band gap due to superconducting proximity effect can carry a supercurrent. An answer to this question invokes the formation of bound states in the band gap of the weak link, known as the "Andreev bound states" that allow transport of Cooper pairs [26].

We model a basic $D = 1$ SNS junction as a system formed by attaching a finite metallic chain (a "normal dot," denoted by N) to two semi-infinite superconducting chains ("superconducting leads," denoted by $\mathsf{S1}$ and $\mathsf{S2}$). Following Ref. [3], we describe the superconducting leads in terms of a $D = 1$ BCS pairing Hamiltonian,

$$\widehat{H}_{\mathsf{S}} = -\sum_{j,\sigma} \mathsf{t}\hat{\Phi}^\dagger_{j\sigma}\hat{\Phi}_{j+1\sigma} - \sum_{j} \Delta\hat{\Phi}^\dagger_{j\uparrow}\hat{\Phi}^\dagger_{j\downarrow} + \text{H.c.}, \tag{3.27}$$

where we have assumed zero chemical potential. This Hamiltonian can be diagonalized analytically for open BCs, see Sect. 3.1.2 (see also Refs. [27–29] for a critical discussion of $D = 1$ models of superconductivity). The normal dot is modeled by NN hopping of strength t. The links connecting the superconducting regions to the metallic one have a weaker hopping strength, $\mathsf{t}' < \mathsf{t}$. The Hamiltonian of the full system is thus $\widehat{H}_{\mathsf{SNS}} = \widehat{H}_{\mathsf{S1}} + \widehat{H}_{\mathsf{S2}} + \widehat{H}_{\mathsf{T}} + \widehat{H}_{\mathsf{N}}$, where $\widehat{H}_{\mathsf{S1}}$ and $\widehat{H}_{\mathsf{S2}}$ denote the superconducting Hamiltonians for the leads, $\widehat{H}_{\mathsf{N}}$ describes the normal metal, and $\widehat{H}_{\mathsf{T}}$ is the tunneling Hamiltonian, of the form

$$\widehat{H}_{\mathsf{T}} = -\sum_{\sigma=\pm1} \big[\mathsf{t}'(\hat{\Phi}^\dagger_{-2L\sigma}\hat{\Phi}_{-2L+1\sigma} + \hat{\Phi}^\dagger_{2L-1\sigma}\hat{\Phi}_{2L\sigma}) + \text{H.c.}\big]. \tag{3.28}$$

The region S1 extends from $j = -\infty$ on the left to $j = -2L$, whereas S2 extends from $j = 2L$ to $j = \infty$, so that the length of the metallic chain is $N \equiv 4L - 1$.

With reference to Sect. 2.4.2, our aim is to find the exact Andreev bound states that form on the normal region. The block-diagonalization in spin space reduces to solving the boundary value problem for the blocks with reduced internal space. Because of Eq. (3.4), note that each spin block does *not* individually describe an SNS junction Hamiltonian. The SNS junction is modeled as the system formed by attaching a finite metallic N chain to two semi-infinite superconducting chains, S1 and S2, with the length of the metallic chain being $N = 4L - 1$ for some positive integer L. The projectors corresponding to the left and right semi-infinite S1 and S2 regions are

$$P_1 = \sum_{j=-\infty}^{-2L} |j\rangle\langle j|, \quad P_2 = \sum_{j=2L}^{\infty} |j\rangle\langle j|,$$

whereas the region N is finite with an associated projector

$$P_3 = \sum_{j=-2L+1}^{2L-1} |j\rangle\langle j|.$$

The single-particle Hamiltonian H_{SNS} of the junction is block-diagonalized in the basis of the spin operator σ_y, and the two blocks are related to each other by the particle-hole symmetry in the same way as described by Eq. (3.4). Let us focus on the s $= +1$ block, and denote it by H_+. This system has three translation-invariant regions (bulks) connected by two internal boundaries. The energy eigenvector ansatz in this case is obtained by extending the ansatz in Eq. (2.61) to a system of three bulks.

Consider first the case with no phase difference between the two superconducting leads S1 and S2, that is, $\Delta_1 = \Delta_2 = \Delta$ for a real value of Δ. Note that H_+ obeys a mirror symmetry about $j = 0$,

$$\mathcal{S}_1 = \sum_{j\in\mathbb{Z}} |-j\rangle\langle j| \otimes \mathbb{1}_2,$$

and another local symmetry,

$$\mathcal{S}_2 = \sum_{j\in\mathbb{Z}} (-1)^j |j\rangle\langle j| \otimes \tau_y.$$

Since we are only interested in the states bound on the metal N region, we restrict the value of energy to be in the band gap of the superconductors, which is $(-\Delta, \Delta)$. For these bound states to carry a superconducting current, they must be of extended nature on the metallic region, which is allowed by energies such that $|\epsilon| < |\mathsf{t}|$. The eigenstate ansatz for any such energy in each of the three bulks will be in terms of the roots of Eq. (3.5), with $\mu = 0$. Noting appropriate symmetries of the polynomial, we denote the four roots in the bulks of $\mathsf{S1}$ and $\mathsf{S2}$ by $\{z_1, z_1^{-1}, -z_1, -z_1^{-1}\}$. Without loss of generality, we can choose $|z_1| > 1$ and $\mathsf{t}(z_1 + z_1^{-1}) = i\mathsf{D}$, with $\mathsf{D} \equiv \sqrt{\Delta^2 - \epsilon^2}$. From Eq. (3.5) and the above constraints, we find that

$$z_1 = -\frac{\mathsf{D} + \sqrt{\mathsf{D}^2 + 4\mathsf{t}^2}}{2i\mathsf{t}}. \tag{3.29}$$

For an exponentially decaying mode in the $\mathsf{S1}$ and $\mathsf{S2}$ region, the ansatz is given by

$$P_1|\epsilon, \mathrm{s}_1, \mathrm{s}_2, \boldsymbol{\alpha}\rangle = \alpha_1 \left(P_1|z_1, 1\rangle \begin{bmatrix} i\Delta \\ \epsilon + i\mathsf{D} \end{bmatrix} + \mathrm{s}_2 P_1| - z_1, 1\rangle \begin{bmatrix} \mathsf{D} - i\epsilon \\ -\Delta \end{bmatrix} \right),$$

$$P_2|\epsilon, \mathrm{s}_1, \mathrm{s}_2, \boldsymbol{\alpha}\rangle = \alpha_1 \left(\mathrm{s}_1 P_2|1/z_1, 1\rangle \begin{bmatrix} i\Delta \\ \epsilon + i\mathsf{D} \end{bmatrix} + \mathrm{s}_1\mathrm{s}_2 P_2| - 1/z_1, 1\rangle \begin{bmatrix} \mathsf{D} - i\epsilon \\ -\Delta \end{bmatrix} \right),$$

respectively, where $\mathrm{s}_1, \mathrm{s}_2$ denote eigenvalues of symmetries $\mathcal{S}_1$ and $\mathcal{S}_2$, respectively. For the metallic region, since $\Delta = 0$, all four roots lie on the unit circle. We denote them by $\{w_1, w_1^{-1}, -w_1, -w_1^{-1}\}$, with the convention $\mathsf{t}(w_1 + w_1^{-1}) = -\epsilon$. Then the ansatz for the N region can be written as

$$P_3|\epsilon, \mathrm{s}_1, \mathrm{s}_2, \boldsymbol{\alpha}\rangle = \alpha_2 \Bigg(P_3|w_1, 1\rangle + \mathrm{s}_1 P_3|1/w_1, 1\rangle \begin{bmatrix} 1 \\ 0 \end{bmatrix} + \mathrm{s}_2 P_3| - w_1, 1\rangle$$
$$+ \mathrm{s}_1\mathrm{s}_2 P_3| - 1/w_1, 1\rangle \Bigg) \begin{bmatrix} 0 \\ \mathrm{i} \end{bmatrix}.$$

Therefore, we have obtained four eigenstate ansätze corresponding to the four cases $\{\mathrm{s}_1 = \pm 1,\ \mathrm{s}_2 = \pm 1\}$, which we denote by $|\epsilon, \mathrm{s}_1, \mathrm{s}_2, \boldsymbol{\alpha}\rangle$, where $\boldsymbol{\alpha} = [\alpha_1\ \alpha_2]^\mathsf{T}$ are the free parameters. The BCs are provided by the weak links, that is, $j = \pm 2L, \pm(2L-1)$. We choose the basis $\{|2L, \mathrm{s}_1, \mathrm{s}_2\rangle,\ |2L-1, \mathrm{s}_1, \mathrm{s}_2\rangle,\ \mathrm{s}_1, \mathrm{s}_2 = 1, -1\}$ of the boundary subspace, where

$$|2L, \mathrm{s}_1, \mathrm{s}_2\rangle \equiv \frac{1}{2}(| - 2L\rangle + \mathrm{s}_1|2L\rangle) \begin{bmatrix} 1 \\ i\mathrm{s}_2 \end{bmatrix},$$

$$|2L-1, \mathrm{s}_1, \mathrm{s}_2\rangle \equiv \frac{1}{2}(| - 2L + 1\rangle + \mathrm{s}_1|2L - 1\rangle) \begin{bmatrix} 1 \\ -i\mathrm{s}_2 \end{bmatrix}.$$

There will be four boundary matrices $B(\epsilon, \mathrm{s}_1, \mathrm{s}_2)$, for $\mathrm{s}_1, \mathrm{s}_2 = 1, -1$, arising from the equations

$$\langle 2L, \mathrm{s}_1, \mathrm{s}_2|(H - \epsilon \mathbb{1}_2))|\epsilon, \mathrm{s}_1, \mathrm{s}_2, \boldsymbol{\alpha}\rangle = 0,$$

$$\langle 2L-1, \mathsf{s}_1, \mathsf{s}_2 | (H - \epsilon \mathbb{1}_2)) | \epsilon, \mathsf{s}_1, \mathsf{s}_2, \boldsymbol{\alpha} \rangle = 0.$$

The boundary matrix corresponding to $(\mathsf{s}_1, \mathsf{s}_2)$ is

$$B(\epsilon, \mathsf{s}_1, \mathsf{s}_2) = \begin{bmatrix} \mathsf{t} z_1^{-2L+1}(i\Delta - \mathsf{s}_2(\mathsf{D} - i\epsilon)) & -\mathsf{t}'(w_1^{-2L+1} + s_1 w_1^{2L-1}) \\ -\mathsf{t}' z_1^{-2L}(i\Delta + \mathsf{s}_2(\mathsf{D} - i\epsilon)) & \mathsf{t}(w_1^{-2L} + s_1 w_1^{2L}) \end{bmatrix}.$$

In writing the above, we made use of the identity $(h_0 - \epsilon \mathbb{1} + z_\ell h_1)|u_\ell\rangle = -z_\ell^{-1} h_1^\dagger |u_\ell\rangle$, which follows from the bulk equation. The condition for non-trivial kernel of the boundary matrix in the four cases leads, after simplification using Eq. (3.29), to the four boundary equations

$$F_{\mathsf{s}_2}(\epsilon) = g_{\mathsf{s}_1}(\epsilon), \quad \mathsf{s}_1, \mathsf{s}_2 = \pm 1, \tag{3.30}$$

where

$$F_{\mathsf{s}_2}(\epsilon) = -\left(\frac{\mathsf{t}'}{\mathsf{t}}\right)^2 \left(\frac{2\mathsf{t}}{\epsilon + \mathsf{s}_2 \Delta}\right) \left(1 + \sqrt{1 + \frac{4\mathsf{t}^2}{\Delta^2 - \epsilon^2}}\right)^{-1} =$$

and

$$g_{\mathsf{s}_1}(\epsilon) = \begin{cases} \cos k(2L-1)/\cos k(2L) & \text{if } \mathsf{s}_1 = +1 \\ \sin k(2L-1)/\sin k(2L) & \text{if } \mathsf{s}_1 = -1 \end{cases}.$$

where $e^{ik} \equiv w_1$. Whenever any one of these conditions is satisfied, ϵ is an eigenvalue. The coefficients α_1, α_2, that completely determine the eigenstates in the four cases, in turn satisfy

$$\frac{\alpha_2}{\alpha_1} = \left(\frac{\mathsf{t}'}{\mathsf{t}}\right) \frac{z_1^{-2L}(i\Delta + \mathsf{s}_2(\mathsf{D} - i\epsilon))}{2\cos k(2L)}, \quad \text{if } \mathsf{s}_1 = +1, \mathsf{s}_2 = \pm 1,$$

$$\frac{\alpha_2}{\alpha_1} = \left(\frac{\mathsf{t}'}{\mathsf{t}}\right) \frac{z_1^{-2L}(i\Delta + \mathsf{s}_2(\mathsf{D} - i\epsilon))}{2\sin k(2L)}, \quad \text{if } \mathsf{s}_1 = -1, \mathsf{s}_2 = \pm 1.$$

1. The number of Andreev bound states *increases with dot size*. For fixed parameters, increasing the size of the normal dot N generically increases the number of solutions of the boundary equations and so the number of Andreev bound states. This feature is illustrated by the last three columns of Fig. 3.5, read from top to bottom.
2. For $\Delta > 2|\mathsf{t}|$, *each* state of the isolated normal dot becomes an Andreev bound state. When the metal strip is completely disconnected from the superconductors, that is, $\mathsf{t}' = 0$, the bound states corresponding to $\mathsf{s}_1 = -1$ in the metal are given by

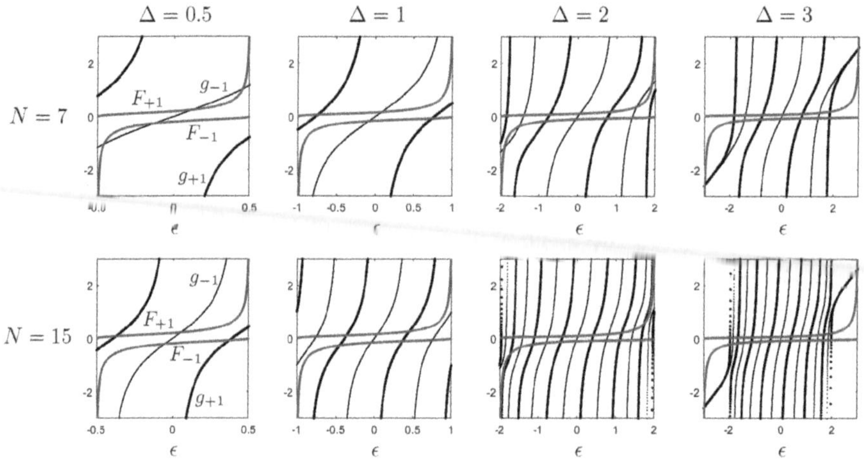

Fig. 3.5 Andreev bound states of an SNS junction. The energies of the states are those values of ϵ (in units of t) for which two independent constraints become compatible. These constraints, derived from the boundary matrix, are equations of the form $F_{\pm 1}(\epsilon) = g_{\pm 1}(\epsilon)$ and $F_{\pm 1}(\epsilon) = g_{\mp 1}(\epsilon)$, see Eq. (3.30). The solid (blue) lines are the curves traced by $F_{\pm 1}(\epsilon)$, the thick black dotted lines are the curves traced by $g_{+1}(\epsilon)$, and the thin black dotted lines are the curves traced by $g_{-1}(\epsilon)$. Each intersection corresponds to an Andreev bound state. The top and bottom panels correspond to $N = 4\mathrm{L} - 1$ sites for the normal dot, with $L = 2$ and $L = 4$, respectively. Other parameters are $\mathrm{t} = 1, \mathrm{t}' = 0.5$ (a.u.) throughout. Figure adapted with permission from Ref. [12]. Copyrighted by the American Physical Society

$$k = \frac{\pi q}{2L - 1}, \quad q = 0, 1, \ldots, 2L - 2.$$

Notice that each of these lies singularly between the poles of the function $g_{+1}(\epsilon)$, as can be seen from the relation

$$\frac{\pi(q + 1)}{2L} > \frac{\pi q}{2L - 1} > \frac{\pi q}{2L}.$$

Further, $g_{-1}(\epsilon)$ assumes all real values between any two adjacent poles. Suppose that the energy of original metallic state at $k = \pi q/2L - 1$ was in the energy gap of the superconductors that is in the interval $(-\Delta, \Delta)$. Now when we turn on the weak tunneling ($\mathrm{t}' > 0$), the new value of k must lie between the two adjacent poles $\pi(q + 1)/2L$ and $\pi q/2L$. This analysis proves that each metallic state at energy $|\epsilon| < \Delta$ gets converted into a bound state with slightly different value of energy in the presence of weak tunneling. This analytical result is consistent with the numerical observations of Ref. [3] (notice that in our setup the gate voltage is set to zero). In Fig. 3.5, the last column, for $\Delta = 3$, corresponds to this regime for two different dot sizes and, as one can see, there are exactly twice as many Andreev bound states as normal sites very close to the positions of the states of the isolated normal dot. These states correspond to the points where the black lines cross from negative to positive values in the figure. The role of the connection to the leads is to change the horizontal line at zero into the two blue lines.

3. For $\Delta > 2|\mathsf{t}|$, there are bound states with very large penetration length into the leads. In addition to the $2N$ Andreev bound states just discussed, there are four other bound states for $|\Delta| > 2|\mathsf{t}|$ *and only then*. These states are pinned in energy near $\epsilon = \pm\Delta$. They show a characteristically large penetration depth into the superconducting leads and a decaying profile inside the normal dot. Unlike the Andreev bound states that arise from the states of the normal dot, these pinned bound states resemble the bulk scattering states of the superconductor, but they happen to be normalizable.

3.2.2 Graphene Ribbons

In this section we investigate NN tight-binding models on the honeycomb (hexagonal) lattice, with graphene as the prime motivation [30]. The surface band structure of graphene sheets or ribbons is well understood, even analytically in limiting cases [4, 31, 32].[5] As emphasized in Ref. [34], a perturbation that breaks inversion symmetry can have interesting effects on these surface bands. With this in mind, in our analysis below we include a sublattice potential and show that the Hamiltonian for a ribbon subject to zigzag-bearded BCs can be fully diagonalized in closed-form.

Zigzag-Bearded Boundary Conditions

The honeycomb lattice is bipartite, with triangular sublattices A and B displaced by d relative to each other, see Fig. 3.6 (left). We parameterize the lattice sites $\mathbf{R}$ as

$$\mathbf{R}(j_1, j, m) = \begin{cases} j_1\mathbf{m}_1 + j\mathsf{s} + \mathbf{d} & \text{if } m = 1 \\ j_1\mathbf{m}_1 + j\mathsf{s} & \text{if } m = 2 \end{cases},$$

$$\mathbf{m}_1 \equiv a\begin{bmatrix}1\\0\end{bmatrix}, \quad \mathsf{s} = \frac{a}{2}\begin{bmatrix}1\\\sqrt{3}\end{bmatrix}, \quad \mathbf{d} = -\frac{a}{2\sqrt{3}}\begin{bmatrix}\sqrt{3}\\1\end{bmatrix},$$

with $j_1, j \in \mathbb{Z}$, $a = 1$ being the lattice parameter and $m = 1$ ($m = 2$) denoting the A (B) sublattice. The localized (basis) states are $|j\rangle|m = 1\rangle$ and $|j\rangle|m = 2\rangle$, and so the sublattice label plays the role of a pseudospin-1/2 degree of freedom. The ribbon we consider is translation-invariant in the $\mathbf{m}_1$ direction and terminated along s, with single-particle Hamiltonian $H_N = \sum_{\mathbf{k}_\parallel \in \text{SBZ}} |\mathbf{k}_\parallel\rangle\langle\mathbf{k}_\parallel| H_{\mathbf{k}_\parallel, N}$, where

[5] It is worth noting that eigenfunctions may also be obtained by means of Lie-algebraic methods that are in principle applicable beyond quadratic Hamiltonians, see Ref. [33].

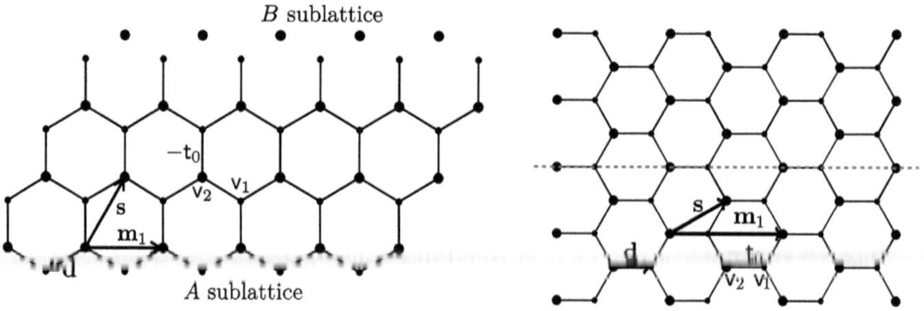

Fig. 3.6 Graphene ribbon, periodic or infinite in the horizontal m_1 direction. Left: The ribbon is terminated in the vertical direction by a zigzag edge on the bottom and a "bearded" edge on top. The decoupled B sites at the top are auxiliary degrees of freedom. Right: The ribbon is terminated by armchair edges. The system has mirror symmetry about the dashed (red) line. In both cases, on-site potentials v_1 and v_2 are associated with the A and B sublattice, respectively. Figure adapted with permission from Ref. [12]. Copyrighted by the American Physical Society

$$H_{\mathbf{k}_\parallel,N} = \mathbb{1}_N \begin{bmatrix} \mathsf{v}_1 & -\mathsf{t}_0(1+e^{-ik_1}) \\ -\mathsf{t}_0(1+e^{ik_1}) & \mathsf{v}_2 \end{bmatrix} + \left(T \begin{bmatrix} 0 & 0 \\ -\mathsf{t}_0 & 0 \end{bmatrix} + \text{H.c.} \right),$$

$\mathbf{k}_\parallel = k_1 \mathbf{n}_1$, and the 2×2 matrices act on the sublattice degree of freedom. Notice that H_N is chirally symmetric if the on-site potentials $\mathsf{v}_1 = 0 = \mathsf{v}_2$, and the edges of the ribbon are of the zigzag type, see Fig. 3.6 (left). While in the following we shall set $\mathsf{v}_1 = 0$ for simplicity, it is easy to restore v_1 anywhere along the way if desired. In particular, $\mathsf{v}_1 = -\mathsf{v}_2$ is an important special case [34].

The analytic continuation of the Bloch Hamiltonian is

$$H_{\mathbf{k}_\parallel}(z) = \begin{bmatrix} 0 & -\mathsf{t}_1(\mathbf{k}_\parallel)e^{-i\phi_{\mathbf{k}_\parallel}} - \mathsf{t}_0 z^{-1} \\ -\mathsf{t}_1(\mathbf{k}_\parallel)e^{i\phi_{\mathbf{k}_\parallel}} - \mathsf{t}_0 z & \mathsf{v}_2 \end{bmatrix},$$

$$\mathsf{t}_1(\mathbf{k}_\parallel) \equiv \mathsf{t}_0\sqrt{2(1+\cos(\mathbf{k}_\parallel))}, \quad e^{i\phi_{\mathbf{k}_\parallel}} \equiv \mathsf{t}_0(1+e^{i\mathbf{k}_\parallel})/\mathsf{t}_1(\mathbf{k}_\parallel).$$

This analysis reveals the formal connection between graphene and the SSH model: just compare the above $H_{\mathbf{k}_\parallel}(z)$ with $H(z)$ in Eq. (3.11).

We impose BCs in terms of an operator W such that

$$\langle \mathbf{k}_\parallel | W | \mathbf{k}'_\parallel \rangle = \delta_{\mathbf{k}_\parallel,\mathbf{k}'_\parallel} |N\rangle\langle N| \begin{bmatrix} 0 & \mathsf{t}_1(\mathbf{k}_\parallel)e^{-i\phi_{\mathbf{k}_\parallel}} \\ \mathsf{t}_1(\mathbf{k}_\parallel)e^{i\phi_{\mathbf{k}_\parallel}} & 0 \end{bmatrix}.$$

In real space, this corresponds to

$$W = |N\rangle\langle N| \begin{bmatrix} 0 & -\mathsf{t}_0 \\ -\mathsf{t}_0 & 0 \end{bmatrix} + \left(\mathbf{T}|N\rangle\langle N| \begin{bmatrix} 0 & 0 \\ -\mathsf{t}_0 & 0 \end{bmatrix} + \text{H.c.} \right),$$

where $\mathbf{T}$ implements shift of $-\mathbf{m}_1$. The meaning of these BCs is as follows: for the modified ribbon Hamiltonian described by $H = H_N + W$, the sites $|j_1\rangle|j = N\rangle|m = 2\rangle$ are decoupled from the rest of the system and each other, see Fig. 3.6 (left). The termination of the actual ribbon, consisting of the sites connected to each other, is of the zigzag type on the lower edge, and "bearded" on the upper edge. From a geometric perspective, this ribbon is special because every B site is connected to exactly three A sites, but not the other way around.

At this point we may borrow results from dimerized chains that we solved in Sect. 3.1.4, to which we refer for full detail. The energy eigenstates that are perfectly localized on the upper edge (consisting of decoupled sites) constitute a flat surface band at energy v_2. For $|\mathbf{k}_\parallel| > 2\pi/3$, the energy eigenstates localized on the lower edge constitute a flat surface band at $\mathrm{v}_1 = 0$ energy. Explicitly, these zero modes are

$$|\epsilon = 0, \mathbf{k}_\parallel\rangle = |\mathbf{k}_\parallel\rangle|z_1(\mathbf{k}_\parallel)\rangle \begin{bmatrix} (\mathrm{t}_1(\mathbf{k}_\parallel)^2 - \mathrm{t}_0^2)e^{-i\phi_{\mathbf{k}_\parallel}}/\mathrm{t}_1(\mathbf{k}_\parallel) \\ 0 \end{bmatrix},$$

$$z_1(\mathbf{k}_\parallel) \equiv -e^{i\phi_{\mathbf{k}_\parallel}} \frac{\mathrm{t}_1(\mathbf{k}_\parallel)}{\mathrm{t}_0} = -(1 + e^{i\mathbf{k}_\parallel}).$$

While their energy is insensitive to $\mathbf{k}_\parallel$, their characteristic localization length is not:

$$\mathscr{L}_{\mathrm{loc}}(\mathbf{k}_\parallel) = -\frac{1}{\ln(|z_1(\mathbf{k}_\parallel)|)} = -\frac{2}{\ln(2 + 2\cos(\mathbf{k}_\parallel))}. \tag{3.31}$$

For $\mathbf{k}_\parallel \neq \pm\frac{2\pi}{3}$, the bulk states are

$$|\epsilon_{\mathrm{n}}(\mathbf{k}_\parallel, q)\rangle = |\mathbf{k}_\parallel\rangle|\chi_1(\mathbf{k}_\parallel, q)\rangle \begin{bmatrix} \mathrm{t}_1(\mathbf{k}_\parallel)e^{-i\phi_{\mathbf{k}_\parallel}} \\ -\epsilon_{\mathrm{n}}(\mathbf{k}_\parallel, q) \end{bmatrix} + |\mathbf{k}_\parallel\rangle|\chi_2(\mathbf{k}_\parallel, q)\rangle \begin{bmatrix} \mathrm{t}_0 \\ 0 \end{bmatrix},$$

with

$$|\chi_1(\mathbf{k}_\parallel, q)\rangle \equiv 2i \sum_{j=1}^{N} \sin\big(\pi q j/N\big) e^{-i\phi_{\mathbf{k}_\parallel} j} |j\rangle, \tag{3.32}$$

$$|\chi_2(\mathbf{k}_\parallel, q)\rangle \equiv 2i \sum_{j=1}^{N} \sin\big(\pi q (j-1)/N\big) e^{-i\phi_{\mathbf{k}_\parallel}(j-1)} |j\rangle, \tag{3.33}$$

$$\epsilon_{\mathrm{n}}(\mathbf{k}_\parallel, q) = \frac{\mathrm{v}_2}{2} + (-1)^{\mathrm{n}} \sqrt{\frac{\mathrm{v}_2^2}{4} + \mathrm{t}_1(\mathbf{k}_\parallel)^2 + \mathrm{t}_0^2 + 2\mathrm{t}_1(\mathbf{k}_\parallel) t_2 \cos\big(\frac{\pi}{N} q\big)}, \tag{3.34}$$

for n $= 1, 2$. Since $\mathsf{t}_1(\mathbf{k}_\parallel = \pm\frac{2\pi}{3}) = \mathsf{t}_0$, the virtual chains $H_{\mathbf{k}_\parallel,N}$ are gapless if $\mathsf{v}_2 = 0$, reflecting the fact that graphene is a semimetal. The energy eigenstates are similar but simpler than the ones just described.

Armchair Terminations

The graphene ribbon with zigzag terminations can be described in terms of smooth terminations of the triangular Bravais lattice with two atoms per unit cell. In contrast, armchair terminations require a fairly different description of the underlying atomic array. Figure 3.6 (right) shows how to describe this system in terms of a "centered rectangular" Bravais lattice [35] with two atoms per unit cell and smooth parallel terminations. In this case, we parameterize the lattice sites $\mathbf{R}$ as

$$\mathbf{R}(j_1, j, m) = \begin{cases} j_1\mathbf{m}_1 + j\mathsf{s} + \mathbf{d} & \text{if } m = 1 \\ j_1\mathbf{m}_1 + j\mathsf{s} & \text{if } m = 2 \end{cases},$$

$$\mathbf{m}_1 \equiv a\begin{bmatrix}\sqrt{3}\\0\end{bmatrix}, \quad \mathsf{s} = \frac{a}{2}\begin{bmatrix}\sqrt{3}\\1\end{bmatrix}, \quad \mathbf{d} = \frac{a}{\sqrt{3}}\begin{bmatrix}1\\0\end{bmatrix},$$

where as before $j_1, j \in \mathbb{Z}$, $a = 1$, and $m \in \{1, 2\}$ labels the sublattice. The total single-particle Hamiltonian can now be taken to read $H = H_N + W$, with $W = 0$ and $H_N = \sum_{\mathbf{k}_\parallel \in \text{SBZ}} |\mathbf{k}_\parallel\rangle\langle\mathbf{k}_\parallel| H_{\mathbf{k}_\parallel,N}$, where

$$H_{\mathbf{k}_\parallel,N} = \mathbb{1}_N \begin{bmatrix}\mathsf{v}_1 & -\mathsf{t}_0\\ -\mathsf{t}_0 & \mathsf{v}_2\end{bmatrix} + \left(T\begin{bmatrix}0 & -\mathsf{t}_0 e^{-ik_1}\\ -\mathsf{t}_0 & 0\end{bmatrix} + \text{H.c.}\right), \quad \mathbf{k}_\parallel = k_1\mathbf{n}_1,$$

and the analytic continuation of the Bloch Hamiltonian for each $\mathbf{k}_\parallel$ is

$$H_{\mathbf{k}_\parallel}(z) = \begin{bmatrix}\mathsf{v}_1 & -\mathsf{t}_0(1 + ze^{-ik_1 l} + z^{-1})\\ -\mathsf{t}_0(1 + z^{-1}e^{ik_1} + z) & \mathsf{v}_2\end{bmatrix}.$$

The diagonalization of the Hamiltonian proceeds from here on as before. There is, however, a shortcut based on Sect. 2.2.3, which explains in addition the absence of edge modes in this system. Let $T_{\mathbf{k}_\parallel} \equiv e^{-ik_1/2}T$. In terms of this $\mathbf{k}_\parallel$-dependent matrix,

$$H_{\mathbf{k}_\parallel,N} = \mathbb{1}_N \begin{bmatrix}\mathsf{v}_1 & -\mathsf{t}_0\\ -\mathsf{t}_0 & \mathsf{v}_2\end{bmatrix} - \mathsf{t}_0\left(T_{\mathbf{k}_\parallel} + T^\dagger_{\mathbf{k}_\parallel}\right)\begin{bmatrix}0 & e^{-ik_1/2}\\ e^{ik_1/2} & 0\end{bmatrix}.$$

It follows that the (unnormalized) energy eigenstates of the graphene ribbon with armchair terminations are

$$|\epsilon_{q,\mathrm{n}}\rangle = |\mathbf{k}_{\parallel}\rangle \sum_{j=1}^{N} |j\rangle e^{ik_1 j/2} \sin[\pi q j/(N+1)] \times \left[\frac{-2\mathrm{t}_0(1+e^{-ik_1/2}\cos[\pi q/(N+1)])}{\epsilon_{q,\pm}}\right],$$

where $q = 1, \ldots, N$, and $\epsilon_{q,\mathrm{n}}$ for $\mathrm{n} = 1, 2$ are the two roots (in ϵ) of the quadratic equation

$$\epsilon^2 - \mathrm{v}_2\epsilon - \mathrm{t}_0^2 - 4\mathrm{t}_0^2\cos(k_1/2)\cos[\pi q/(N+1)] - 4\mathrm{t}_0^2\cos^2[\pi q/(N+1)] = 0.$$

These are the $2N$ energy eigenstates of the system for each value of $\mathbf{k}_{\parallel}$.

3.2.3 A Chiral $p + ip$ Superconductor

The spinless $p+ip$ superconductor of Ref. [5] is the prototype of spinless superconductivity in $D = 2$. The model may be regarded as the mean-field approximation to an exactly solvable (by the algebraic Bethe ansatz) pairing Hamiltonian [36]. It belongs to class D in the Altland–Zirnbauer classification, and thus, according to the tenfold way, it admits an integer ($\mathbb{Z}$) topological invariant. There has been hope for some time that the related phenomenon of triplet superconductivity is realized in layered perovskite strontium ruthenate (Sr_2RuO_4), but the matter remains controversial [37]. The many-body model Hamiltonian can be taken to be

$$\hat{H} = -\mathrm{t}\sum_{\mathbf{j}}(\hat{\Phi}^{\dagger}_{\mathbf{j}+\mathbf{s}}\hat{\Phi}_{\mathbf{j}} + \hat{\Phi}^{\dagger}_{\mathbf{j}+\mathbf{m}}\hat{\Phi}_{\mathbf{j}} + \text{H.c.}) - \Delta\sum_{\mathbf{j}}(\hat{\Phi}_{\mathbf{j}}\hat{\Phi}_{\mathbf{j}+\mathbf{s}} - i\hat{\Phi}_{\mathbf{j}}\hat{\Phi}_{\mathbf{j}+\mathbf{m}} + \text{H.c.}) - (\mu - 4\mathrm{t})\sum_{\mathbf{j}}\hat{\Phi}^{\dagger}_{\mathbf{j}}\hat{\Phi}_{\mathbf{j}},$$

on the square lattice of unit lattice spacing and with standard unit vectors $\mathbf{s}, \mathbf{m}$ pointing in the x and y directions, respectively. The parameters t, Δ are real numbers. The corresponding single-particle Hamiltonian is

$$H = -[(\mu - 4\mathrm{t})\mathbb{1} + \mathrm{t}(V_{\mathbf{s}} + V_{\mathbf{s}}^{\dagger}) + \mathrm{t}(V_{\mathbf{m}} + V_{\mathbf{m}}^{\dagger})]\tau_z + i\Delta(V_{\mathbf{s}} - V_{\mathbf{s}}^{\dagger})\tau_y + i\Delta(V_{\mathbf{m}} - V_{\mathbf{m}}^{\dagger})\tau_x,$$

in terms of infinite shift operators $V_{\mathbf{s}} \equiv \sum_{\mathbf{j}} |\mathbf{j}\rangle\langle\mathbf{j}+\mathbf{s}|$, $V_{\mathbf{m}} \equiv \sum_{\mathbf{j}} |\mathbf{j}\rangle\langle\mathbf{j}+\mathbf{m}|$.

Closed-Form Chiral Edge States

If energy is measured in units of t, then the parameter space of the model can be taken to be 2D after a gauge transformation that renders $\Delta > 0$. We shall focus

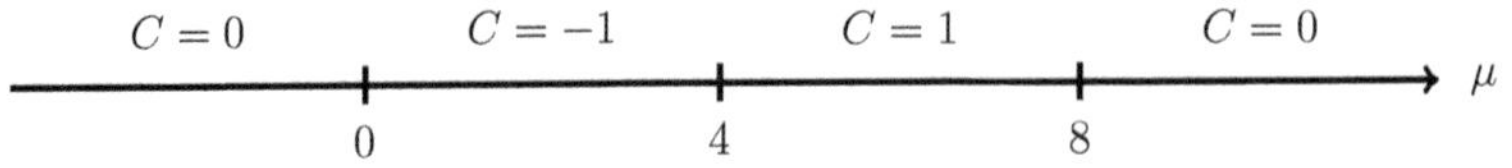

Fig. 3.7 The value of the Chern invariant as a function of μ in the parameter regime $\Delta = 1 = t$. Figure adapted with permission from Ref. [12]. Copyrighted by the American Physical Society

on the line $\Delta = 1 = t$, in which μ is the only variable parameter. The Bloch Hamiltonian is

$$H(\mathbf{k}) = \begin{bmatrix} \mathsf{e}(\mathbf{k}) & \Delta(\mathbf{k}) \\ \Delta(\mathbf{k})^* & -\mathsf{e}(\mathbf{k}) \end{bmatrix}, \quad \begin{array}{l} \Delta(\mathbf{k}) \equiv 2i \sin k_1 - 2 \sin k_2 \\ \mathsf{e}(\mathbf{k}) \equiv -2\cos k_1 - 2\cos k_2 - \mu + 4 \end{array},$$

for $\mathbf{k} = (k_1, k_2) \in [-\pi, \pi) \times [-\pi, \pi)$. The single-particle bulk dispersion then reads

$$\epsilon(k_1, k_2)^2 = \mu^2 - 8\mu + 24 + 4(\mu - 4)(\cos k_1 + \cos k_2) + 8 \cos k_1 \cos k_2,$$

and it is fully gapped unless $\mu = 0, 4, 8$. The gap closes at $k = 0$ if $\mu = 0$, $k = (-\pi, 0)$ and $k = (0, -\pi)$ if $\mu = 4$, and at $k = (-\pi, -\pi)$ if $\mu = 8$. For $0 < \mu < 4$ and $4 < \mu < 8$ the system is in the weak-pairing topologically non-trivial phase, with the value of the Chern invariant $C = -1$ and $C = 1$, respectively, see Fig. 3.7. The Chern number is given by the formula

$$C = \frac{1}{2\pi i} \int \mathcal{F}(\mathbf{k})\, d^2\mathbf{k},$$

in terms of the curvature $\mathcal{F}(\mathbf{k}) = \partial_{k_x} \mathcal{A}_y - \partial_{k_y} \mathcal{A}_x$ of the Berry connection $\mathcal{A}_\nu = \langle u_\mathbf{k} | \partial_{k_\nu} | u_\mathbf{k} \rangle$ associated with the negative band with wavefunctions $|u_\mathbf{k}\rangle$. We evaluated C by way of the numerically gauge-invariant formula given in Appendix D of Ref. [2].

We now impose open BCs in the x direction while keeping the y direction translation-invariant, that is, $(0, k_2) = \mathbf{k}_\parallel$. Accordingly, we need the analytic continuation of the Bloch Hamiltonian in k_1. Let us introduce the compact notation

$$\omega \equiv -2\cos k_2 - \mu + 4, \quad \xi \equiv -2 \sin k_2,$$

so that $H_{\mathbf{k}_\parallel}(z) = h_{\mathbf{k}_\parallel,0} + z h_1 + z^{-1} h_1^\dagger$, with

$$h_{\mathbf{k}_\parallel,0} = \begin{bmatrix} \omega & \xi \\ \xi & -\omega \end{bmatrix}, \quad h_1 = \begin{bmatrix} -1 & 1 \\ -1 & 1 \end{bmatrix}. \tag{3.35}$$

The condition $\det(H_{\mathbf{k}_\parallel}(z) - \epsilon \mathbb{1}_2) = 0$ is then equivalent to the equation

$$\epsilon^2 = \omega^2 + \xi^2 + 4 - 2\omega\,(z + z^{-1}). \tag{3.36}$$

Note that the replacement $z + z^{-1} \mapsto 2\cos k_1$ recovers the bulk dispersion relation. Moreover, if $2 < \mu < 6$, there are values of $\mathbf{k}_\parallel$ for which $\omega = 0$ and the dispersion relation becomes flat. From $H_{\mathbf{k}_\parallel}(z)$ it is immediate to reconstruct the family of virtual chain Hamiltonians

$$H_{\mathbf{k}_\parallel,N} = \mathbb{1}_N h_{\mathbf{k}_\parallel,0} + T h_1 + T^\dagger h_1^\dagger.$$

From the point of view of any one of these chains, mirror symmetry is broken by the NN pairing terms. This fact is important, because then the boundary matrix is *not* mirror-symmetric either, which will ultimately lead to surface states of opposite chirality on the left and right edges.

The number of edge degrees of freedom is $2Rd = 4$ for each value of $\mathbf{k}_\parallel$. Since h_1 (Eq. (3.35)) is not invertible, and Eq. (3.36) is a polynomial of degree 2 in z, the complete eigenstate ansatz is formed out of four independent states (one ansatz state for each $\mathbf{k}_\parallel$): two extended states associated with the roots $z_\ell = z_\ell(\epsilon, \mathbf{k}_\parallel)$, $\ell = 1, 2$, of Eq. (3.36) and two emergent states of finite support localized on the edges of the virtual chains $H_{\mathbf{k}_\parallel,N}$. With hindsight, we will ignore the emergent states and focus on the reduced ansatz, namely

$$|\epsilon\rangle = \alpha_1 |z_1, 1\rangle |u_1\rangle + \alpha_2 z_2^{-N+1} |z_2, 1\rangle |u_2\rangle.$$

The state $|z_1, 1\rangle|u_1\rangle$ should represent a surface state for the left edge, $z_2^{-N+1}|z_2, 1\rangle|u_2\rangle$ one for the right edge, with

$$|u_\ell\rangle = \begin{bmatrix} \xi + z_\ell - z_\ell^{-1} \\ -\omega + \epsilon + z_\ell + z_\ell^{-1} \end{bmatrix} \tag{3.37}$$

satisfying the equation $H_{\mathbf{k}_\parallel}(z_\ell)|u_\ell\rangle = \epsilon|u_\ell\rangle$. The boundary equations $P_\partial(H_{\mathbf{k}_\parallel,N} - \epsilon\mathbb{1}_{2N})|\epsilon\rangle = 0$ are encoded in the boundary matrices

$$B_{\mathbf{k}_\parallel}(\epsilon) = -\begin{bmatrix} h_1^\dagger |u_1\rangle & z_2^{-N-1} h_1^\dagger |u_2\rangle \\ z_1^{N+1} h_1 |u_1\rangle & h_1 |u_2\rangle \end{bmatrix},$$

which are, however, non-square 4×2 matrices as we have ignored the two emergent states that in principle appear in the ansatz. Nonetheless, since h_1 is a matrix of rank one, we can extract a square boundary matrix, namely

$$\tilde{B}_{\mathbf{k}_\parallel}(\epsilon) = \begin{bmatrix} z_1(\xi - \omega + \epsilon + 2z_1) & z_2^{-N}(\xi - \omega + \epsilon + 2z_2) \\ z_1^N(\xi + \omega - \epsilon - 2z_1^{-1}) & z_2(\xi + \omega - \epsilon - 2z_2^{-1}) \end{bmatrix},$$

that properly captures the BCs for our reduced trial states. Surface states are characterized by the condition $|z_1| = |z_2^{-1}| < 1$. Hence, in the large-N limit, one may set $z_1^N = z_2^{-N} = 0$. Within this approximation, the left and right edges are effectively decoupled by virtue of their large spatial separation.

In summary, the left surface band is determined by the polynomial system

$$\begin{cases} 0 = \xi - \omega + \epsilon + 2z_1 \\ 0 = \epsilon^2 + 2\omega\,(z_1 + z_1^{-1}) - (\omega^2 + \xi^2 + 4) \end{cases}. \tag{3.38}$$

In the following, we will focus on the cases $0 < \mu < 2$ or $6 < \mu < 8$ for simplicity (these parameter regimes are in the weak pairing phase and satisfy $\omega \neq 0$ for all values of $\mathbf{k}_\parallel$). Notice that

$$|u_1\rangle = (\xi + z_1 - z_1^{-1}) \begin{bmatrix} 1 \\ -1 \end{bmatrix} \tag{3.39}$$

due to the (top) boundary equation in Eq. (3.38) (recall also Eq. (3.37)). The physical solutions are surprisingly simple.[6] They are

$$\epsilon \equiv \epsilon_{\text{left}}(\mathbf{k}_\parallel) = -\xi - 2\sin k_2,$$
$$z_1 = z_1(\mathbf{k}_\parallel) = \frac{\omega}{2} = 2 - \frac{\mu}{2} - \cos k_2.$$

These functions of $\mathbf{k}_\parallel$ represent the dispersion relation and "complex momentum" of surface excitations on the left edge for those values of $\mathbf{k}_\parallel$ (and *only* those values) such that $|z_1(\mathbf{k}_\parallel)| < 1$ (see Fig. 3.8). Notice that *the edge band is chiral*. The surface band touches the bulk band at the two values of $\mathbf{k}_\parallel$ such that $|z_1(\mathbf{k}_\parallel)| = 1$. The (unnormalized) surface states are, for large N,

$$|\epsilon_{\text{left}}(\mathbf{k}_\parallel)\rangle = \sum_{j=1}^{N} \left(2 - \frac{\mu}{2} - \cos k_2\right)^j |k_\parallel\rangle |j\rangle \begin{bmatrix} 1 \\ -1 \end{bmatrix}.$$

Similarly, the right surface band is determined by the polynomial system

$$\begin{cases} 0 = \xi + \omega - \epsilon - 2z_2^{-1} \\ 0 = \epsilon^2 + 2\omega\,(z_2 + z_2^{-1}) - (\omega^2 + \xi^2 + 4) \end{cases}. \tag{3.40}$$

Due to the boundary equation,

$$|u_2\rangle = (\xi + z_2 - z_2^{-1}) \begin{bmatrix} 1 \\ 1 \end{bmatrix}, \tag{3.41}$$

---

[6] There are two other solutions of the system in Eq. (3.38), $z_{1,\pm} = -\frac{1}{2}(\xi \pm \sqrt{4+\xi^2})$. These solutions are excluded because the internal state $|u_\ell\rangle$ vanishes identically if evaluated at $z_\ell = z_{1,\pm}$, see Eq. (3.39). Similar remarks apply to the system in Eq. (3.40).

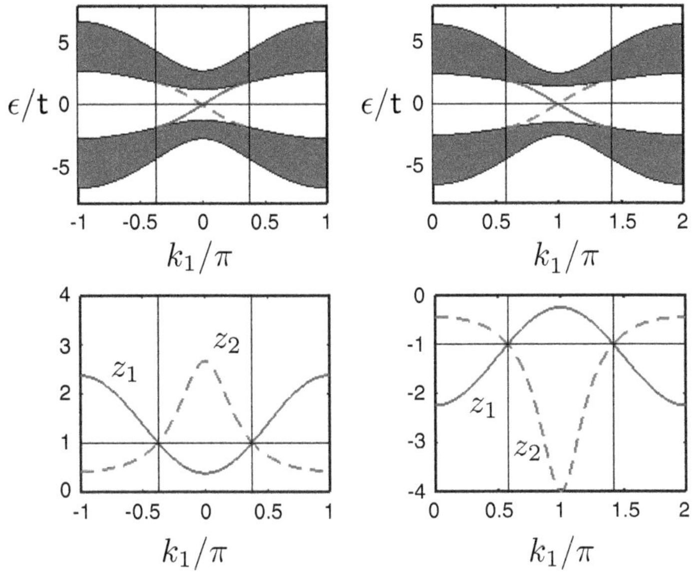

Fig. 3.8 Surface bands for $\mu = 1.5$, centered at $k_1 = 0$ (top left panel), and $\mu = 6.5$, centered at $k_1 = -\pi$ (top right panel). The shaded (gray) region shows the bulk bands. The electrons on the right edge (dashed red curve) propagate to the right only, and those on the left edge (solid blue curve) to the left only, that is, the surface bands are chiral. The lower panels show the behavior of z_1 (solid blue curve) and z_2 (dashed red curve) with k_1. Notice how z_1 (z_2) enters (exits) the unit circle precisely when the surface bands touch the bulk bands, as marked by vertical solid black lines. Figure adapted with permission from Ref. [12]. Copyrighted by the American Physical Society

the physical solutions are

$$\epsilon \equiv \epsilon_{\text{right}}(k_\parallel) = \xi = -2\sin k_2,$$

$$z_2 = z_2(k_\parallel) = \frac{2}{\omega} = \left(2 - \frac{\mu}{2} - \cos k_2\right)^{-1}.$$

This surface band is also chiral, but with the *opposite* chirality to that of the left edge. The right surface band touches the bulk band at the pair of values of $\mathbf{k}_\parallel$ such that $|z_2(\mathbf{k}_\parallel)| = 1$. These values of $\mathbf{k}_\parallel$ are the same as those computed for the surface band on the left edge, due to the fact that $z_1(\mathbf{k}_\parallel) = z_2(\mathbf{k}_\parallel)^{-1}$. It is not obvious from comparing Eqs. (3.38) and (3.40) that this basic relationship should hold, but the actual solutions do satisfy it. The (unnormalized) surface states are, for large N,

$$|\epsilon_{\text{right}}(\mathbf{k}_\parallel)\rangle = \sum_{j=1}^{N} \left(2 - \frac{\mu}{2} - \cos k_2\right)^{-(j-N+1)} |\mathbf{k}_\parallel\rangle |j\rangle \begin{bmatrix} 1 \\ 1 \end{bmatrix}.$$

The root $z_1(\mathbf{k}_\parallel)$ ($z_2(\mathbf{k}_\parallel)$) is entirely outside (inside) the unit circle if $\mu < 0$ or $\mu > 8$. This is a direct indication that the system does not host surface bands in these parameter regimes. In Fig. 3.8, we show the surface bands for two values of the chemical potential, one for each topologically non-trivial phase. The location of the surface bands in the Brillouin zone is not determined by the dispersion relation, which is itself independent of μ, but by the behavior of the wavefunctions as witnessed by $z_1(\mathbf{k}_\parallel) = z_2(\mathbf{k}_\parallel)^{-1}$.

Power-Law Zero Modes

Here we return to the basic model Hamiltonian with three parameters t, Δ, μ. We consider a sheet of material rolled into a cylinder along the y-direction and half-infinite in the x-direction. The virtual wires are

$$H_{\mathbf{k}_\parallel} = h_{\mathbf{k}_\parallel,0} + T h_{\mathbf{k}_\parallel,1} + T^\dagger h^\dagger_{\mathbf{k}_\parallel,1},$$

$$h_{\mathbf{k}_\parallel,0} = \begin{bmatrix} -(\mu - 4\mathrm{t}) - 2t\cos k_2 & -2\Delta \sin k_2 \\ -2\Delta \sin k_2 & (\mu - 4\mathrm{t}) + 2\mathrm{t}\cos k_2 \end{bmatrix},$$

$$h_{\mathbf{k}_\parallel,1} = \begin{bmatrix} -\mathrm{t} & \Delta \\ -\Delta & \mathrm{t} \end{bmatrix}.$$

The values $k_2 = -\pi, 0$ have special significance. Since the off-diagonal entries of h_0 vanish at these momenta, the virtual $D = 1$ systems can be interpreted as one-dimensional superconductors. In particular,

$$h_{0,0} = \begin{bmatrix} -(\mu - 2\mathrm{t}) & 0 \\ 0 & \mu - 2\mathrm{t} \end{bmatrix}, \quad h_{-\pi,0} = \begin{bmatrix} -(\mu - 6\mathrm{t}) & 0 \\ 0 & \mu - 6\mathrm{t} \end{bmatrix}$$

and so the virtual chains $H_{-\pi}$ and H_0 are precisely the Majorana chain of Kitaev, at two distinct values of an effective chemical potential $\mu' = -(\mu - 4\mathrm{t}) \mp 2\mathrm{t}$ *for the chain.* We have investigated this paradigmatic system by analytic continuation in Sect. 3.1.5. If $\mu < 0$ or $\mu > 8\mathrm{t}$, both chains are in their topologically trivial regime. If $0 < \mu < 4\mathrm{t}$, then H_0 is in the non-trivial regime, but not $H_{-\pi}$. The opposite is true if $4\mathrm{t} < \mu < 8\mathrm{t}$. This analysis explains why is it that the fermionic parity of the ground state of the $p + ip$ superconductor is odd in the weak pairing phase [5], and suggests that one should expect surface bands crossing zero energy at $k_2 = 0$ ($k_2 = -\pi$) for $0 < \mu < 4$ ($4 < \mu < 8$). We already saw some of these bands in the previous section.

Let us focus here on the virtual Kitaev chain at $k_2 = 0$. Its effective chemical potential is $\mu' = \mu - 2\mathrm{t}$. Suppose we are in a parameter regime

$$4\Delta^2 = \mu(4\mathrm{t} - \mu), \quad 0 < \mu < 4\mathrm{t},$$

of the full 2D model. Then the $H_{\mathbf{k}_{\parallel}=\mathbf{0}}$ virtual Kitaev chain is in the topologically non-trivial parameter regime

$$\left(\frac{\mu'}{2\mathrm{t}}\right)^2 + \left(\frac{\Delta}{\mathrm{t}}\right)^2 = 1, \quad -2\mathrm{t} < \mu' < 2\mathrm{t}.$$

It is shown in Part I that the Majorana zero modes display an exotic power-law profile in this regime. For the $p + ip$ topological superconductor these remarks imply the following power-law zero-energy surface mode:

$$|\epsilon = 0, \mathbf{k}_{\parallel} = \mathbf{0}\rangle = \sum_{j=1}^{\infty} \sum_{j_1=1}^{N_1} j \left(\frac{-2(\mathrm{t}-\Delta)}{\mu - 2\mathrm{t}}\right)^j |j_1\rangle|j\rangle.$$

3.2.4 A Gapless s-Wave Topological Superconductor

A *gapless* SC is characterized by a vanishing single-particle excitation gap at particular k-points (or regions) of the Brillouin zone, whereas the superconducting order parameter remains non-vanishing. An example in $D = 2$ was analyzed in Ref. [38], where the nodeless character of the s-wave pairing in a two-band system was tuned to a gapless superconducting phase by introducing a suitable spin–orbit coupling. A remarkable feature of this system is the presence of zero-energy Majorana modes whose number grows with system size—a *continuum* in the thermodynamic limit, namely a MFB—as long as the system is subject to open BCs along one of the two spatial directions, but *not* the other. This anomalous bulk-boundary correspondence was attributed to an asymmetric (quadratic vs. linear) closing of the bulk excitation gap near the critical momenta. In this section, we revisit this phenomenon and show that the indicator of bulk-boundary correspondence we introduced in Chap. 2 captures it precisely. Furthermore, in the phase hosting a MFB, we demonstrate by combining our Bloch ansatz with numerical root evaluation that the characteristic length of the MFB wavefunctions diverges as we approach the critical values of momentum, similarly to what was observed in graphene (Eq. (3.31)). Finally, by comparing the equilibrium Josephson current in the gapless topological superconductor to the one of a corresponding gapped model, we show how, similar to the case of the local density of states at the surface [38], the presence of a MFB translates in principle into a substantial enhancement of the 4π-periodic supercurrent.

Analysis of Anomalous Bulk-Boundary Correspondence via Boundary Matrix

The relevant model Hamiltonian in real space is

$$\widehat{H} = \frac{1}{2}\sum_{\mathbf{j}}\left(\hat{\Psi}_{\mathbf{j}}^{\dagger} h_0 \hat{\Psi}_{\mathbf{j}} - 4\mu\right) + \frac{1}{2}\sum_{\mathbf{r}=\hat{x},\hat{z}}\left(\sum_{\mathbf{j}} \hat{\Psi}_{\mathbf{j}}^{\dagger} h_{\mathbf{r}} \hat{\Psi}_{\mathbf{j}+\mathbf{r}} + \text{H.c.}\right),$$

with respect to a local basis of fermionic operators given by

$$\hat{\Psi}_{\mathbf{j}}^{\dagger} \equiv \left[\hat{\Phi}_{\mathbf{j}1\uparrow}^{\dagger}\ \hat{\Phi}_{\mathbf{j}1\downarrow}^{\dagger}\ \hat{\Phi}_{\mathbf{j}2\uparrow}^{\dagger}\ \hat{\Phi}_{\mathbf{j}2\downarrow}^{\dagger}\ \hat{\Phi}_{\mathbf{j}1\uparrow}\ \hat{\Phi}_{\mathbf{j}1\downarrow}\ \hat{\Phi}_{\mathbf{j}2\uparrow}\ \hat{\Phi}_{\mathbf{j}2\downarrow}\right].$$

Here,

$$h_0 = -\mu\tau_z + \mathsf{u}\tau_z\nu_z - \Delta\tau_x\nu_y\sigma_x, \quad h_{\hat{x}(\hat{z})} = -\mathsf{t}\tau_z\nu_z + i\lambda\nu_x\sigma_{x(z)},$$

with Pauli matrices $\tau_v, \nu_v, \sigma_v,\ v = x, y, z$ for the Nambu, orbital, and spin space, respectively. This Hamiltonian can be verified to obey time-reversal and particle-hole symmetry, as well as a chiral symmetry $\mathcal{C} \equiv \tau_x\nu_z$. The topological response of the system was studied in Ref. [38] using a $\mathbb{Z}_2 \times \mathbb{Z}_2$ indicator $(Q_{k_\parallel=0}, Q_{k_\parallel=\pi})$, where $Q_{k_\parallel}$ stands for the parity of the partial Berry phase sum for the value of transverse momentum $k_\parallel$.[7] The bulk-boundary correspondence of the system was studied subject to two different configurations: BC1, in which the system is periodic along $\hat{z}$ and open along $\hat{x}$, and BC2, in which the system is periodic along $\hat{x}$ and open along $\hat{z}$. A MFB emerges along the open edges for BC1 in the phase characterized by $(Q_{k_z=0}, Q_{k_z=\pi}) = (1, 1)$. No MFB exists in the configuration BC2.

To shed light into this anomalous bulk-boundary correspondence using our generalized Bloch theorem framework, consider first the configuration BC1. Then, if N_x denotes the size of the lattice along the $\hat{x}$ direction, $\widehat{H}$ decouples into N_x virtual wires, parameterized by the transverse momentum k_z. These virtual $D = 1$ Hamiltonians have the form

$$H_{k_z,N_x} = \frac{1}{2}\sum_{j=1}^{N_x}\left(\hat{\Psi}_{j,k_z}^{\dagger} h_{k_z,0} \hat{\Psi}_{j,k_z} - 4\mu\right) + \frac{1}{2}\sum_{j=1}^{N_x-1}\left(\hat{\Psi}_{j,k_z}^{\dagger} h_{k_z,1} \hat{\Psi}_{j+1,k_z} + \text{H.c.}\right),$$

where $h_{k_z,0} \equiv h_0 + (e^{ik_z} h_{\hat{z}} + \text{H.c.})$ and $h_{k_z,1} \equiv h_{\hat{x}}$. The *total* number of Majorana modes hosted by each such chain (on its two ends) is given by the degeneracy indicator introduced in Sect. 2.3.2, namely $\mathscr{K}(0) \equiv \dim\ker[B_\infty(0)]$, where $B_\infty(0)$ is the boundary matrix in the large-N limit that we obtain after appropriately

[7] Reference [38] used the notation $P_{B,k_\parallel}$ ($\equiv Q_{k_\parallel}$), which however would be confusing in the present content.

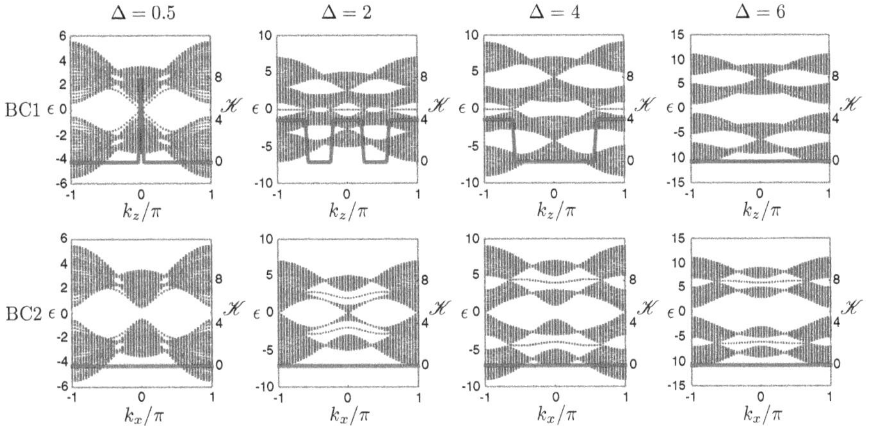

Fig. 3.9 Energy spectrum (blue scatter plot) and degeneracy indicator $\mathscr{K}(k_z)$ for the zero energy level (red solid line) in the large-N limit for BC1 (top panel) vs. BC2 (bottom panel) for various values of the superconducting pairing Δ. The other parameters are $\mu = 0$, $\mathsf{t} = \lambda = \mathsf{u} = 1$, $N_x = 120$, $N_y = 30$. Figure adapted with permission from Ref. [12]. Copyrighted by the American Physical Society

rescaling the extended bulk solutions corresponding to $|z_\ell| > 1$, and removing the un-normalizable extended solutions corresponding to $|z_\ell| = 1$. We calculate the above degeneracy indicator $\mathscr{K}(0) \equiv \mathscr{K}_{k_z}(0)$ for each wire parameterized by k_z, by evaluating the boundary matrix numerically. Representative results are shown in the top panel of Fig. 3.9. When the system is in a phase characterized by $(Q_{k_z=0}, Q_{k_z=\pi}) = (1, -1)$ $(\Delta = 2)$ and $(Q_{k_z=0}, Q_{k_z=\pi}) = (-1, -1)$ $(\Delta = 4)$ there are $\mathcal{O}(N)$ chains, each of them hosting four Majoranas (two pairs per edge). This is reflected in the fourfold degeneracy for a continuum of values of k_z. The values of k_z at which the excitation gap closes are also the points at which the indicator changes its nature.

The same analysis may be repeated for BC2, in which case periodic BCs are imposed along $\hat{x}$ instead. The resulting virtual $D = 1$ systems are now parameterized by k_x, with explicit expressions for the internal matrices given by $h_{k_x,0} = h_0 + (e^{ik_x} h_{\hat{x}} + \text{H.c.})$ and $h_{k_x,1} = h_{\hat{z}}$. In the BC2 configuration, the degeneracy indicator remains zero, showcasing the absence of MFBs, see bottom panel of Fig. 3.9.

Penetration Depth of Flat-Band Majorana Modes

Whether and how far the Majorana modes in the flat band penetrate in the bulk is important from the point of view of scattering. Our generalized Bloch theorem allows us to obtain a good estimate of the penetration depth without diagonalizing the system. In the large-N limit, the wavefunction corresponding to a Majorana mode for a single wire described by H_{k_z,N_x} must include left emergent solutions and decaying extended solutions, so that

$$|\epsilon = 0\rangle = \sum_{\substack{|z_\ell|<1,\\ \ell=0}} \sum_{s=1}^{s_\ell} \alpha_{\ell s} |\psi_{k_z \ell s}\rangle,$$

for complex amplitudes $\{\alpha_{\ell s}\}$. The emergent solutions are perfectly localized, and so the penetration depth is determined by the extended solutions only. The latter are labeled by the roots $\{z_\ell\}$, computed at $\epsilon = 0$, of the polynomial equation $z^{dR} \det(H_{k_z}(z) - \epsilon \mathbb{1}_8) = 0$, which is the dispersion relation. Each extended solution $|\psi_{k_z \ell s}\rangle$ corresponding to the root z_ℓ, $|z_\ell| < 1$ has penetration depth $(-\ln|z_\ell|)^{-1}$. A useful estimate of the penetration depth δ_p of a zero-energy mode may then be obtained by taking the maximum of the individual penetration depths of the bulk solutions [39], leading to the expression

$$\mathscr{L}_{\text{loc}} = -\frac{1}{\ln|z_p|}, \qquad |z_p| \equiv \max\{|z_\ell|,\ |z_\ell| < 1\}.$$

Since the roots $\{z_\ell\}$ depend on the value of the transverse momentum k_z, so does the penetration depth $\mathscr{L}_{\text{loc}}$. As seen in Fig. 3.10, the Majoranas penetrate more inside the bulk near the critical values of the transverse momentum, where the excitation gap closes. At these points, the penetration depth diverges, signifying that the corresponding Majorana excitations become part of the bulk bands.

Impact of a Majorana Flat Band on Josephson Current

Beside resulting in an enhanced local density of states at the surface [38], one expects that the MFB may impact the nature of the equilibrium (DC) Josephson

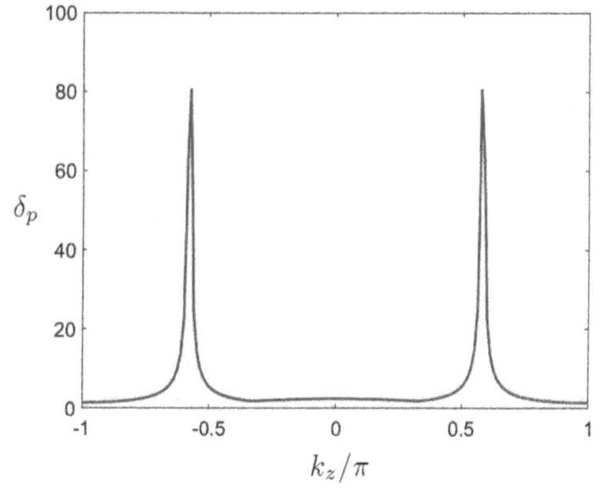

Fig. 3.10 Penetration depth (in units of the lattice constant) of flat-band Majoranas as a function of k_z. The parameters are $\mu = 0$, $\mathsf{u} = \mathsf{t} = \lambda = 1$, $\Delta = 4$. Figure adapted with permission from Ref. [12]. Copyrighted by the American Physical Society

current at zero temperature. We now show (numerically) that the Josephson current flowing through a strip of finite width is 4π-periodic, irrespective of the width of the strip. This is at variance with the behavior expected for a gapped $D = 2$ s-wave topological superconductor, in which case the 4π-periodic contribution resulting from a fixed number of Majorana modes is washed away once the strip width becomes large.

We model an SNS junction of the superconductor under investigation by connecting two superconducting planes along a metallic (non-superconducting) edge. The hopping and spin–orbit coupling between the two superconducting planes at this edge is assumed to be of the same type as each of the superconductors, but weaker by a factor of $\mathsf{w} = 0.2$. The magnetic flux ϕ is introduced by modifying the superconducting parameter of one of the planes according to $\Delta \mapsto \Delta e^{i\phi}$. The DC Josephson current can be calculated using the formula [26]

$$\mathrm{I}(\phi) = \frac{2e}{\hbar}\frac{\partial E_0}{\partial \phi} = -\frac{2e}{\hbar}\sum_{\epsilon_n > 0}\frac{\partial \epsilon_n}{\partial \phi},$$

where E_0 is the energy of the many-body ground state, ϵ_n are single-particle energy levels, and ϕ is the superconducting phase difference (or flux). As ϕ is varied, at the level crossings of low-lying energy levels with the many-body ground state associated with the 4π-periodic effect, the system continues in the state which respects fermionic parity and time-reversal symmetry in all the virtual wires.

The upper panels of Fig. 3.11 show the Josephson response $\mathrm{I}(\phi)$ of the gapless topological superconductor under the two BCs. While in the BC1 configuration the

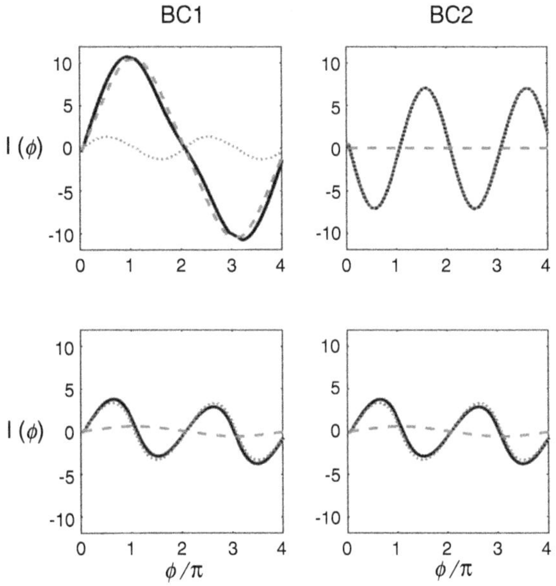

Fig. 3.11 Total Josephson current $\mathrm{I}(\phi)$ (black solid line), 2π-periodic component $\mathrm{I}_{2\pi}(\phi)$ (blue dotted line), and 4π-periodic component $\mathrm{I}_{4\pi}(\phi)$ (red dashed line) in units of $2e/\hbar$, as a function of flux ϕ. Top (bottom) panels correspond to the gapless (gapped) model of a $D = 2$ s-wave topological superconductor, whereas left (right) panels correspond to BC1 (BC2), respectively. The parameters used for both models are $\mu = 0$, $\mathsf{u} = \mathsf{t} = \lambda = 1$, $\Delta = 4$, $N_x = N_z = 60$. Figure adapted with permission from Ref. [12]. Copyrighted by the American Physical Society

behavior of the current $I(\phi)$ (solid black line) is 4π-periodic, the BC2 configuration displays standard 2π-periodicity, reflecting the presence of the MFB *only* under BC1. The lower panels of Fig. 3.11 show the Josephson response of the gapped s-wave topological superconductor model introduced and analyzed in Ref. [1, 2]. It can be seen that the Josephson current is now identical under BC1 and BC2, as expected from the fact that a standard bulk-boundary correspondence is in place.

Let us separate the total Josephson current $I(\phi)$ into 2π- and 4π-periodic components by letting $I(\phi) = I_{2\pi}(\phi) + I_{4\pi}(\phi)$, with

$$I_{2\pi}(\phi) \equiv \begin{cases} \frac{1}{2}[I(\phi) + I(\phi + 2\pi)] \text{ if } 0 \le \phi < 2\pi \\ \frac{1}{2}[I(\phi) + I(\phi - 2\pi)] \text{ if } 2\pi \le \phi < 4\pi \end{cases},$$

$$I_{4\pi}(\phi) \equiv \begin{cases} \frac{1}{2}[I(\phi) - I(\phi + 2\pi)] \text{ if } 0 \le \phi < 2\pi \\ \frac{1}{2}[I(\phi) - I(\phi - 2\pi)] \text{ if } 2\pi \le \phi < 4\pi \end{cases},$$

In the four panels of Fig. 3.11, the 2π- and 4π-periodic components are individually shown by (blue) dotted and (red) dashed lines, respectively. The nature of the supercurrent in the gapped topological superconductor (lower panels) is predominantly 2π-periodic, with only a small 4π-periodic component due to the presence of a finite number of Majoranas (two per edge). Further numerical simulations (data not shown) reveal that the amplitude of the 2π-periodic current relative to the 4π-periodic current increases linearly with the width of the strip, so that for large strip width, the Josephson current is essentially 2π-periodic. The origin of such a degradation of the 4π-periodicity lies in the fact that the number of Majorana modes is constant, irrespective of the width of the strip, as only one virtual wire hosts Majorana modes in this gapped model. Since only the Majorana modes can support 4π-periodic current, their contribution relative to the extensive 2π-periodic current arising from the bulk states diminishes as the strip width becomes large. In contrast, for the gapless topological superconductor in the MFB phase (top panels), the number of virtual wires hosting Majorana modes grows linearly with the width of the strip in the BC1 configuration. This leads to an *extensive contribution from the* 4π*-periodic component*, which may be easier to detect in experiments.

This concludes our discussion of topological edge states using generalized Bloch theorem. In the next section, we will employ a new tool, known as matrix Wiener–Hopf factorization, for the analysis of bulk-boundary correspondence in SPT phases.

References

1. S. Deng, L. Viola, G. Ortiz, Majorana modes in time-reversal invariant s-wave topological superconductors. Phys. Rev. Lett. **108**, 036803 (2012). https://link.aps.org/doi/10.1103/PhysRevLett.108.036803

2. S. Deng, G. Ortiz, L. Viola, Multiband s-wave topological superconductors: role of dimensionality and magnetic field response. Phys. Rev. B **87**, 205414 (2013). https://link.aps.org/doi/10.1103/PhysRevB.87.205414
3. C. Bena, Metamorphosis and taxonomy of Andreev bound states. Eur. Phys. J. B **85**, 196 (2012). https://doi.org/10.1140/epjb/e2012-30133-0
4. M. Kohmoto, Y. Hasegawa, Zero modes and edge states of the honeycomb lattice. Phys. Rev. B **76**, 205402 (2007). https://link.aps.org/doi/10.1103/PhysRevB.76.205402
5. N. Read, D. Green, Paired states of fermions in two dimensions with breaking of parity and time-reversal symmetries and the fractional quantum hall effect. Phys. Rev. B **61**, 10267–10297 (2000). https://link.aps.org/doi/10.1103/PhysRevB.61.10267
6. B.A. Bernevig, T.L. Hughes, *Topological Insulators and Topological Superconductors* (Princeton University Press, Princeton, 2013)
7. K. Kawabata, R. Kobayashi, N. Wu, H. Katsura, Exact zero modes in twisted Kitaev chains. Phys. Rev. B **95**, 195140 (2017). https://link.aps.org/doi/10.1103/PhysRevB.95.195140
8. D.-P. Liu, Topological phase boundary in a generalized Kitaev model. Chin. Phys. B **25**, 057101 (2016). https://doi.org/10.1088/1674-1056/25/5/057101
9. B.-Z. Zhou, B. Zhou, Topological phase transition in a ladder of the dimerized Kitaev superconductor chains. Chin. Phys. B **25**, 107401 (2016). https://doi.org/10.1088/1674-1056/25/10/107401
10. Y. He, K. Wright, S. Kouachi, C.-C. Chien, Topology, edge states, and zero-energy states of ultracold atoms in one-dimensional optical superlattices with alternating on-site potentials or hopping coefficients. Phys. Rev. A **97**, 023618 (2018). https://link.aps.org/doi/10.1103/PhysRevA.97.023618
11. A.A. Aligia, L. Arrachea, Entangled end states with fractionalized spin projection in a time-reversal-invariant topological superconducting wire. Phys. Rev. B **98**, 174507 (2018). https://link.aps.org/doi/10.1103/PhysRevB.98.174507
12. E. Cobanera, A. Alase, G. Ortiz, L. Viola, Generalization of Bloch's theorem for arbitrary boundary conditions: interfaces and topological surface band structure. Phys. Rev. B **98**, 245423 (2018). https://link.aps.org/doi/10.1103/PhysRevB.98.245423
13. A. Alase, E. Cobanera, G. Ortiz, L. Viola, Generalization of Bloch's theorem for arbitrary boundary conditions: theory. Phys. Rev. B **96**, 195133 (2017). https://link.aps.org/doi/10.1103/PhysRevB.96.195133
14. K. Tsutsui, Y. Ohta, R. Eder, S. Maekawa, E. Dagotto, J. Riera, Heavy quasiparticles in the Anderson lattice model. Phys. Rev. Lett. **76**, 279 (1996). https://link.aps.org/doi/10.1103/PhysRevLett.76.279
15. A.J. Heeger, S. Kivelson, J. Schrieffer, W.-P. Su, Solitons in conducting polymers. Rev. Mod. Phys. **60**, 781 (1988). https://link.aps.org/doi/10.1103/RevModPhys.60.781
16. D. Xiao, M.-C. Chang, Q. Niu, Berry phase effects on electronic properties. Rev. Mod. Phys. **82**, 1959 (2010). https://link.aps.org/doi/10.1103/RevModPhys.82.1959
17. E. Cobanera, G. Ortiz, Equivalence of topological insulators and superconductors. Phys. Rev. B **92**, 155125 (2015). https://doi.org/10.1103/PhysRevB.92.155125
18. E. Cobanera, G. Ortiz, Z. Nussinov, The bond-algebraic approach to dualities. Adv. Phys. **60**, 679–798 (2011). https://doi.org/10.1080/00018732.2011.619814
19. A.Y. Kitaev, Unpaired Majorana fermions in quantum wires. Phys.-Uspekhi **44**, 131–136 (2001). https://doi.org/10.1070/1063-7869/44/10s/s29
20. J. Alicea, New directions in the pursuit of Majorana fermions in solid state systems. Rep. Prog. Phys. **75**, 076501 (2012). https://doi.org/10.1088/0034-4885/75/7/076501
21. C. Beenakker, Search for Majorana fermions in superconductors. Annu. Rev. Condens. Matter Phys. **4**, 113–136 (2013). https://www.annualreviews.org/doi/full/10.1146/annurev-conmatphys-030212-184337
22. H.J. Mikeska, W. Pesch, Boundary effects on static spin correlation functions in the isotropic x–y chain at zero temperature. Z. Phys. B Condens. Matter **26**, 351–353 (1977). https://doi.org/10.1007/BF01570745

23. P. Pfeuty, The one-dimensional Ising model with a transverse field. Ann. Phys. **57**, 79–90 (1970). https://doi.org/10.1016/0003-4916(70)90270-8
24. S.S. Hegde, S. Vishveshwara, Majorana wave-function oscillations, fermion parity switches, and disorder in Kitaev chains. Phys. Rev. B **94**, 115166 (2016). https://link.aps.org/doi/10.1103/PhysRevB.94.115166
25. I.C. Fulga, A. Haim, A.R. Akhmerov, Y. Oreg, Adaptive tuning of Majorana fermions in a quantum dot chain. New J. Phys. **15**, 045020 (2013). https://doi.org/10.1088/1367-2630/15/4/045020
26. G.B. Lesovik, I.A. Sadovskyy, Scattering matrix approach to the description of quantum electron transport. Phys.-Uspekhi **54**, 1007 (2011). https://doi.org/10.3367/UFNe.0181.201110b.1041
27. G. Ortiz, J. Dukelsky, E. Cobanera, C. Esebbag, C. Beenakker, Many-body characterization of particle-conserving topological superfluids. Phys. Rev. Lett. **113**, 267002 (2014). https://link.aps.org/doi/10.1103/PhysRevLett.113.267002
28. K.Y. Arutyunov, D.S. Golubev, A.D. Zaikin, Superconductivity in one dimension. Phys. Rep. **464**, 1–70 (2008). https://doi.org/10.1016/j.physrep.2008.04.009
29. G. Ortiz, E. Cobanera, What is a particle-conserving topological superfluid? The fate of Majorana modes beyond mean-field theory. Ann. Phys. **372**, 357–374 (2016). https://doi.org/10.1016/j.aop.2016.05.020
30. A.C. Neto, F. Guinea, N.M. Peres, K.S. Novoselov, A.K. Geim, The electronic properties of graphene. Rev. Mod. Phys. **81**, 109 (2009). https://link.aps.org/doi/10.1103/RevModPhys.81.109
31. P. Delplace, D. Ullmo, G. Montambaux, Zak phase and the existence of edge states in graphene. Phys. Rev. B **84**, 195452 (2011). https://link.aps.org/doi/10.1103/PhysRevB.84.195452
32. S. Mao, Y. Kuramoto, K.-I. Imura, A. Yamakage, Analytic theory of edge modes in topological insulators. J. Phys. Soc. Jpn. **79**, 124709 (2010). https://doi.org/10.1143/JPSJ.79.124709
33. B. Dietz, F. Iachello, M. Macek, Algebraic theory of crystal vibrations: localization properties of wave functions in two-dimensional lattices. Crystals **7**, 246 (2017). https://doi.org/10.3390/cryst7080246
34. W. Yao, S.A. Yang, Q. Niu, Edge states in graphene: from gapped flat-band to gapless chiral modes. Phys. Rev. Lett. **102**, 096801 (2009). https://doi.org/10.1103/PhysRevLett.102.096801
35. F. Bechstedt, *Principles of Surface Physics*, 1st edn. (Springer, Berlin, 2012)
36. S.M. Rombouts, J. Dukelsky, G. Ortiz, Quantum phase diagram of the integrable px+ipy fermionic superfluid. Phys. Rev. B **82**, 224510 (2010). https://doi.org/10.1103/PhysRevB.82.224510
37. A.P. Mackenzie, Y. Maeno, The superconductivity of Sr2RuO4 and the physics of spin-triplet pairing. Rev. Mod. Phys. **75**, 657 (2003). https://doi.org/10.1103/RevModPhys.75.657
38. S. Deng, G. Ortiz, A. Poudel, L. Viola, Majorana flat bands in s-wave gapless topological superconductors. Phys. Rev. B **89**, 140507 (2014). https://link.aps.org/doi/10.1103/PhysRevB.89.140507
39. D.H. Lee, J.D. Joannopoulos, Simple scheme for surface-band calculations. I. Phys. Rev. B **23**, 4988–4996 (1981). https://link.aps.org/doi/10.1103/PhysRevB.23.4988

Chapter 4
Matrix Factorization Approach to Bulk-Boundary Correspondence

In statistical mechanics, the definition of thermodynamic phases effectively considers a system that extends infinitely in all directions, an assumption that underlies the "thermodynamic limit." This is also the limit in which the physics of clean free-fermionic systems is the most tractable, thanks to Bloch's theorem. Although SPT phases are, strictly speaking, defined in the thermodynamic limit, their distinctive physical features appear only in finite or semi-infinite systems, since only such systems can host boundary-localized states. The main goal of this chapter is to explore *exact* relations between the properties of the bulk of a system (defined in the thermodynamic limit) and the boundary states in possibly disordered settings. The emphasis is on investigating model-independent topological features of SPT phases that are determined by symmetries and spatial dimensionality alone [1].

We can broadly motivate the work in this chapter with a discussion of 2D topological insulators. As mentioned in the Introduction, a spin–orbit induced 2D topological insulator, also known as a quantum spin Hall insulator, hosts helical edge bands—the momentum of the electrons on each edge is correlated with their spin. These helical edge bands are known to be "symmetry-protected." In a weak sense, this means that they cannot be destroyed (removed altogether) by a perturbation, unless time-reversal symmetry is broken explicitly (e.g., by an external magnetic field), or the bulk energy gap is closed. This property has been verified in experiments [2], as well as by theoretical arguments at least in some cases [3, 4]. The presence of helical edge bands results in quantized values of the transverse conductance of the spin current. The fact that the edge bands can be removed by a time-reversal breaking perturbation reflects in the deviation of the transverse conductance from the quantized value as soon as an external magnetic field is applied [2].

The symmetry-based protection of the edge bands also guarantees that if the parameters of the bulk Hamiltonian are changed without closing the bulk band gap, the transverse conductance must remain quantized. Therefore, any system whose bulk (associated clean, periodic Hamiltonian) can be "adiabatically" connected,

A. Alase, *Boundary Physics and Bulk-Boundary Correspondence in Topological Phases of Matter*, Springer Theses, https://doi.org/10.1007/978-3-030-31960-1_4

while preserving the time-reversal symmetry, to that of a quantum spin Hall insulator must host the exotic edge bands that are distinctive of the quantum spin Hall phase. Similarly, a Hamiltonian whose bulk can be connected to an ordinary insulator cannot host such protected helical edge bands.[1] This leads us to the question of dividing the relevant bulk Hamiltonians into adiabatically connected components, and identifying those components which facilitate the presence of symmetry-protected edge bands. For time-reversal invariant 2D insulators, there are two adiabatically connected components of the bulk Hamiltonians, labeled by a $\mathbb{Z}_2$ topological invariant [3]. The latter is computed using the wavefunctions of the electrons in the filled bands. It is easy to identify which of these components corresponds to the quantum spin Hall insulators. More generally, we can identify three crucial questions that are fundamental to the discussion of SPT phases in arbitrary spatial dimensions:

1. What are the physical quantities that depend on the boundary physics, and are invariant under symmetry-preserving perturbations (such as the transverse conductance in the case of quantum spin Hall phase)?
2. Given the symmetries of a system, which quantity characterizes adiabatically connected "components" of periodic Hamiltonians? In other words, how do we identify whether two Hamiltonians satisfying the same set of symmetries can be transformed into each other without closing the bulk gap, and without breaking the symmetries?
3. Do distinct adiabatically connected components of periodic Hamiltonians *always* correspond to distinct signatures in the boundary physics?

The quantities that answer the first question are called "boundary invariants" [5]. By their definition, boundary invariants can be computed from the energy eigenvalues and wavefunctions of the boundary states. They remain unchanged in the presence of disorder, as long as the symmetries of the system are preserved and the bulk spectrum remains gapped.

The adiabatically connected "components" of bulk Hamiltonians, satisfying a certain set of discrete symmetries, can be characterized by the topology of the vector bundle formed by the wavefunctions of electronic states in the filled bands (at zero temperature) [6–8]. For example, consider a system with tight-binding Bloch Hamiltonian $H_{\mathbf{k}}$ with $\mathbf{k}$ taking values in the first Brillouin zone. Let us assume that only one band is filled, and let the corresponding eigenvalues and eigenstates be $\epsilon_{\mathbf{k}}$ and $|u_{\mathbf{k}}\rangle$, respectively, so that $H_{\mathbf{k}}|u_{\mathbf{k}}\rangle = \epsilon_{\mathbf{k}}|u_{\mathbf{k}}\rangle$. In this case, the vectors $\{|u_{\mathbf{k}}\rangle\}$ form a 1D vector bundle over the Brillouin zone. The topology of this vector bundle, which also depends on the symmetries that $H_{\mathbf{k}}$ satisfies, is characterized by integer-valued quantities known as "bulk topological invariants." The TKNN invariant [9] and the Chern number [10] are examples of bulk topological invariants. The topological invariant of a vector bundle remains unchanged under smooth symmetry-obeying deformations. Since the filled bands are ill-defined for a gapless

[1] Helical edge bands can exist for an ordinary insulator in principle, but they are not protected.

periodic Hamiltonian, the bulk topological invariant can change during a quantum phase transition, but not otherwise. Therefore, they provide a partial answer to the second question: if the bulk topological invariants of two periodic Hamiltonians differ in value, then those two Hamiltonians *cannot* be connected adiabatically without closing the energy gap. In most cases, the bulk topological invariants can also be expressed directly in terms of the Bloch Hamiltonian $H_{\mathbf{k}}$, instead of their eigenvectors. In this chapter, we will use this form of topological invariants for 1D SPT phases.

The third question motivates most of the work in this chapter. The bulk-boundary correspondence conjecture states that there is a one-to-one correspondence between the bulk and boundary topological invariants in SPT phases. Such correspondence has been rigorously proved for systems in all spatial dimensions, possibly containing disorder [5], in the absence of any anti-unitary (commuting or anticommuting) symmetries. For other symmetry combinations, the conjecture either remains unproven or established only for a restricted set of models [11, 12]. A significant amount of work in this direction uses mathematical tools from K-theory [5, 13–16]. As mentioned in the Introduction, a disadvantage of this approach is that no information about the boundary states, other than the boundary invariant, is obtained through such an analysis.

The main contribution of this chapter is to outline a unified approach, at least in 1D systems, to proving the bulk-boundary correspondence in SPT phases. We focus on physical interpretation of the intermediate and final results, rather than the detailed mathematical proofs, which will be presented elsewhere. We use a matrix factorization technique, known as the *matrix Wiener–Hopf factorization*, which allows a new way of "spectral flattening." The key step in our analysis is that this spectral flattening can be extended to systems with boundary, which is again made possible by the Wiener–Hopf factorization. It is this realization that allows us to shift our focus to bulk-boundary correspondence in much simpler flat-band Hamiltonian. We show that our analysis carries over to the case of interfaces, under the assumption that certain symmetry conditions are satisfied by the tunneling part of the Hamiltonian. In 1D SPT phases, the protected edge states always appear at zero energy.[2] It is interesting that we cannot deduce, in general, even the number of edge states from the topological invariant of the system. We discuss the conditions under which the number of edge states can be guaranteed to not change, and moreover, the zero-energy eigenspace is "stable" against symmetry-preserving perturbations. The Wiener–Hopf factorization makes the connection between the bulk Hamiltonian and the edge states transparent. This is the reason it also allows us to obtain a bound on the "sensitivity" of stable zero modes to symmetry-preserving perturbations. Finally, we apply the concept of stability to the 1D model of a s-wave topological superconductor considered in Sect. 3.1.6. Since the boundary states

[2]This is not true if we consider space group symmetries. In this chapter, we restrict our attention to symmetries that act locally in lattice space.

of this model are protected by multiple symmetries, this paves the way towards understanding SPT phases *beyond* the tenfold classification.

The outline of the rest of this chapter is as follows: Sect. 4.1 revisits the symmetry-based classification of topological phases and the structure of the 1D Hamiltonians in each of the symmetry classes along with their topological invariants. In Sect. 4.2, we outline and interpret physically a proof of the bulk-boundary correspondence in 1D SPT phases, using matrix Wiener–Hopf factorization. In Sect. 4.3, we define "stability" of zero-energy edge modes and "sensitivity to perturbations" of the edge states. We also provide an upper bound to the sensitivity to perturbations in terms of the generalized condition number of the single-particle Hamiltonian, and validate it numerically for the SSH Hamiltonian. Section 4.4 is devoted to the numerical analysis of the stability properties of a 1D s-wave topological superconductor. In Sect. 4.5, we take first steps towards establishing the bulk-boundary correspondence for interfaces by extending our results to interfaces between systems belonging to the same symmetry class.

4.1 Background

We first briefly review the five symmetry classes in the Altland–Zirnbauer classification scheme [17], which allow Hamiltonians to be topologically non-trivial in 1D. We discuss the topological invariants in each class and the exact relations between them as discussed in the literature. Next, we introduce the matrix Wiener–Hopf factorization, and describe how it can be used to determine the zero eigenspace of a block-Toeplitz operator.

4.1.1 *The Topologically Non-trivial Symmetry Classes in 1D*

In this chapter, we will focus on the SPT phases stabilized by unitary and anti-unitary symmetries acting locally in lattice space [18].[3] We are interested in the following three kinds of symmetries (see Table 4.1):

Table 4.1 Three kinds of discrete symmetries

	Commuting	Anticommuting
Unitary	–	Chiral/sublattice
Anti-unitary	Time-reversal	Particle-hole

[3] In particular, we do not discuss the phases stabilized by space group symmetries.

1. Unitary symmetries that anticommute with the Hamiltonians, known as "chiral" or "sublattice" symmetries;
2. Anti-unitary symmetries that commute with the Hamiltonians, known as "time-reversal" symmetries;
3. Anti-unitary symmetries that anticommute with the Hamiltonians, known as "particle-hole" symmetries.

We emphasize that a time-reversal symmetry need not be associated with the standard time-reversal operation, but to any anti-unitary operation that commutes with the set of Hamiltonians, irrespective of its origin. For instance, even in those cases where the time-reversal is broken, one may be able to find such a symmetry by composing time-reversal with another unitary commuting symmetry. The same is true about particle-hole and chiral symmetries. In some applications, a unitary anticommuting symmetry may arise as a sublattice symmetry (e.g., in graphene), while in others it may arise from a product of more fundamental time-reversal and particle-hole symmetries.

Unitary commuting symmetries are not considered in the classification of SPT phases [18]. The reason behind this omission is that, if some boundary states are protected by a unitary symmetry among others, then one can simultaneously block-diagonalize the Hamiltonian and the symmetry-preserving perturbations. It is therefore sufficient to analyze only the blocks of the Hamiltonian that protect the boundary states. We reach the same conclusion when studying the adiabatic components of the periodic Hamiltonians. If the Hamiltonians under consideration mutually obey a unitary symmetry, then they can all be block-diagonalized simultaneously. If this is the case, then the topology of the set of Hamiltonians can be inferred from the topology of each of the blocks. Therefore, one may instead focus attention on describing the topology of reduced blocks which do not have any common unitary symmetries. We will comment more on this aspect of classification in Sect. 4.4.

Notice that the product of any two symmetries of the same kind (sublattice/chiral, time-reversal, or particle-hole) leads to a commuting unitary symmetry. This would lead to a situation akin to what we just described, where the Hamiltonians can be simultaneously block-diagonalized. Therefore, in general it suffices to consider the blocks of Hamiltonians where at most one symmetry of each of the three kinds is present. Further, since the square of any symmetry of one of the three kinds is also a unitary commuting symmetry, in the cases where the square is different from the identity operator, it allows for further block-diagonalization. For the maximally reduced blocks, one must assume that the square of each of the symmetries is proportional to identity. From the normal form of any anti-unitary operator in terms of Wigner anti-unitaries [19], one can then deduce that the square of a time-reversal or a particle-hole symmetry must be either $+I$ or $-I$, where I denotes the identity operator.

We now review the structure of single-particle/BdG Hamiltonians in each of the five Altland–Zirnbauer classes which show non-trivial topology in one spatial dimension. We start with the discussion of Hamiltonians without assuming any spatial structure (effectively 0D systems). Therefore, the minimum structure we derive applies to systems in any spatial dimension, but in higher dimension the

Hamiltonian will be more constrained. We assume the form of time-reversal, particle-hole, and chiral symmetries pertaining to typical condensed matter systems. We arrive at specific structure for Hamiltonians in each class by choosing a basis in the internal space. We call this basis the "preferred" basis.

Throughout this paper, we will follow the order AIII, BDI, CII, D, and DIII in the descriptions, theorems, Remarks, etc. of the non-trivial symmetry classes in 1D. This is motivated from the observation that the first three are chiral classes whose topology arises from the analytic index of Fredholm operators, and therefore lead to a similar discussion. The topology of the latter two classes, as we shall describe later, arises from a different index.

- *Class AIII*: The typical systems in this class comprise insulators with sublattice (chiral) symmetry ν_z. In the basis of the chiral symmetry, the Hamiltonian takes the off-diagonal form

$$H = \begin{bmatrix} 0 & H_{21}^{\dagger} \\ H_{21} & 0 \end{bmatrix} \tag{4.1}$$

  The preferred basis is denoted by $\{|\nu\rangle|n\rangle,\ \nu = \pm 1,\ n = 1, \ldots, N\}$, where ν denotes the eigenvalue of the chiral symmetry ν_z.
- *Class BDI*: A superconductor with real parameters belongs to this class. The BdG Hamiltonian has particle-hole symmetry $\mathcal{P}_+ = \tau_x \mathrm{C}$, and spinless time-reversal symmetry $\mathcal{T}_+ = \mathrm{C}$, with τ_x being the Pauli X operator in the creation-annihilation basis, and C denoting complex conjugation. These symmetries force the structure of Hamiltonian to be

$$H = \begin{bmatrix} K & \Delta \\ -\Delta & -K \end{bmatrix},$$

  with K real symmetric and Δ real antisymmetric matrices, thanks to $\mathcal{T}_+$. After transforming to the eigenvector basis of τ_x (which is the chiral symmetry), we get

$$H = \begin{bmatrix} 0 & K + \Delta \\ K - \Delta & 0 \end{bmatrix} \equiv \begin{bmatrix} 0 & H_{21}^{\mathsf{T}} \\ H_{21} & 0 \end{bmatrix},$$

  with $H_{21} = K - \Delta$ having real entries. The new basis is closely related to the Majorana basis, and has its roots in the transform used by Lieb et al. in Ref. [20] to diagonalize a model of antiferromagnetic chain. We represent the preferred basis by $\{|\nu\rangle|n\rangle,\ \nu = \pm 1,\ n = 1, \ldots, N\}$, where ν denotes the eigenvalue of the chiral symmetry τ_x.
- *Class CII*: Superconductors with $SU(2)$ spin-invariance and a sublattice symmetry belong to this class. Consider the BdG Hamiltonian of a superconductor, which commutes with the generator $\tau_z\sigma_z$ of spin rotations about the z axis, where τ and σ act in the particle-hole and spin spaces, respectively. Such a Hamiltonian must be of the form

$$H = \begin{bmatrix} K_0 & 0 & 0 & \Delta_0 \\ 0 & K_1 & \Delta_1 & 0 \\ 0 & -\Delta_0^* & -K_0^* & 0 \\ -\Delta_1^* & 0 & 0 & -K_1^* \end{bmatrix}. \tag{4.2}$$

After further assuming commutation with the generator of spin rotations about the x axis, which is $\tau_z\sigma_x$, we find $\Delta_0 = -\Delta_1$ and $K_0 = K_1$. Each of the two diagonal blocks then satisfies the SU(2)-invariant particle-hole symmetry $\mathcal{P}_-$. We consider the $\tau_z\sigma_z = 1$ block of this Hamiltonian. When the order of tensor products of the combined Nambu-spin (τ,σ) label and other labels is interchanged, one sees that the $\tau_z\sigma_z = 1$ block of the Hamiltonian is a block matrix of 2 × 2 blocks, each taking values from $i\mathrm{H}$, where $\mathrm{H} \equiv \mathrm{span}_{\mathbb{R}}\{\mathbf{1}, \boldsymbol{i}, \boldsymbol{j}, \boldsymbol{k}\}$ is the matrix space of "quaternions," and

$$\mathbf{1} = \begin{bmatrix} 1 & 0 \\ 0 & 1 \end{bmatrix}, \ \boldsymbol{i} = \begin{bmatrix} 0 & i \\ i & o \end{bmatrix}, \ \boldsymbol{j} = \begin{bmatrix} 0 & 1 \\ -1 & 0 \end{bmatrix}, \ \boldsymbol{k} = \begin{bmatrix} i & 0 \\ 0 & -i \end{bmatrix}. \tag{4.3}$$

We refer to the above subspace of 2 × 2 complex matrices as the *quaternionic skew-field*. A further constraint of sublattice symmetry ν_z puts the Hamiltonian in off-diagonal form as in Eq. (4.1), where each off-diagonal block is a (possibly non-Hermitian) quaternionic matrix. We denote the preferred basis of single-particle space by $\{|\nu\rangle|n\rangle,\ \nu = \pm 1,\ n = 1, \ldots, N\}$, where ν denotes the eigenvalue of the chiral symmetry ν_z, and N is necessarily even, thanks to the quaternionic structure.

- *Class D*: A superconductor with possibly complex Hamiltonian parameters belongs to this class. The Hamiltonian is not assumed to have any other symmetry than $\mathcal{P}_+ = \tau_x \mathrm{C}$, and therefore can be expressed as

$$H = \begin{bmatrix} K & \Delta \\ -\Delta^* & -K^* \end{bmatrix},$$

with K complex Hermitian and Δ complex antisymmetric matrix. After transformation to the Majorana basis using

$$U = \frac{1}{\sqrt{2}} \begin{bmatrix} 1 & i \\ 1 & -i \end{bmatrix},$$

we find that H is i times a real antisymmetric matrix [21], namely

$$H = i \begin{bmatrix} K_{\mathrm{Im}} + \Delta_{\mathrm{Im}} & K_{\mathrm{Re}} - \Delta_{\mathrm{Re}} \\ -K_{\mathrm{Re}} - \Delta_{\mathrm{Re}} & K_{\mathrm{Im}} - \Delta_{\mathrm{Im}} \end{bmatrix},$$

where subscripts Re and Im denote real and imaginary parts, respectively. We denote the preferred basis by $\{|n\rangle,\ n = 1, \ldots, N\}$.

- *Class DIII*: A superconductor with spinful time-reversal symmetry $\mathcal{T}_- = i\sigma_y \mathrm{C}$ belongs to class DIII. The product of the particle-hole and time-reversal symmetry, $\mathcal{S} = \mathcal{T}_-\mathcal{P}_+$ is a chiral symmetry of the Hamiltonian. We first transform the Hamiltonian using a unitary (canonical) transformation in the internal space, that is,

$$U = \sqrt{12}\begin{bmatrix} 1 & i & 1 & -i \\ i & 1 & -i & 1 \\ 1 & -i & 1 & i \\ i & -1 & -i & -1 \end{bmatrix}.$$

In this new (fermionic) basis, the Hamiltonian is off-diagonal, which is to say, the Hamiltonian in the new basis has only pairing terms and no particle number-conserving terms,

$$H = \begin{bmatrix} 0 & -H_{21}^* \\ H_{21} & 0 \end{bmatrix}.$$

H_{21} is complex antisymmetric in this case. The preferred basis is denoted by $\{|\nu\rangle|n\rangle,\ \nu = \pm 1,\ n = 1, \ldots, N\}$, where ν is the eigenvalue of the chiral symmetry $\tau_x\sigma_y$.

We point out that we have chosen a convention such that for classes with chiral symmetry (AIII, BDI, CII, and DIII), we denote the relevant orthonormal basis of the single-particle Hilbert space $\mathcal{H}$ by $\{|\nu\rangle|n\rangle,\ \nu = \pm 1,\ n = 1, \ldots, N\}$, where ν is the eigenvalue of chiral symmetry ν_z, and $2N$ is the dimension of $\mathcal{H}$. For class D, since there is no chiral symmetry, we denote the relevant orthonormal basis by $\{|n\rangle,\ n = 1, \ldots, N\}$. The dimension of $\mathcal{H}$ in this case is N. A schematic diagram of the Hamiltonians in all five symmetry classes in their preferred basis is shown in Fig. 4.1.

In summary, regardless of spatial dimension, the Hamiltonians or their off-diagonal blocks of the five non-trivial symmetry classes are proportional to symmetric, antisymmetric, Hermitian, or anti-Hermitian matrices over real, complex, or quaternions fields/skew-field. These results are summarized in Table 4.2.

We now turn our attention specifically to the topology of 1D systems in each of these symmetry classes. We only consider the symmetries that act locally in the single-particle space, but the discussion can be extended to some other non-local symmetries. For a (infinite) tight-binding system in 1D, let $j \in \mathbb{Z} \equiv \{-\infty, \ldots, \infty\}$ label lattice sites. We label the internal degrees of freedom, such as spin, orbital, and creation-annihilation basis, by $\{(\nu, m),\ m = 1, \ldots, \bar{d},\ \nu = \pm 1\}$ for systems with chiral symmetry (classes AIII, BDI, CII, and DIII), and by $\{m,\ m = 1, \ldots, \bar{d}\}$ for those without chiral symmetry (class D). The dimension of internal space is then $d = 2\bar{d}$ for classes with chiral symmetry and $d = \bar{d}$ for those without chiral symmetry. The single-particle Hilbert space $\mathcal{H}$ is spanned by $\{|\nu\rangle|m\rangle|j\rangle,\ j \in \mathbb{Z},\ \nu = \pm 1,\ m = 1, \ldots, \bar{d}\}$ for chiral classes, and $\{|m\rangle|j\rangle,\ j \in \mathbb{Z},\ m = 1, \ldots, \bar{d}\}$

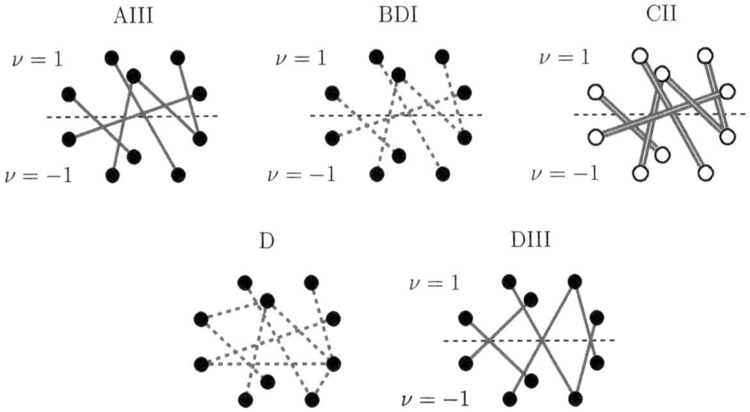

Fig. 4.1 Schematic diagram of Hamiltonians in the symmetry classes AIII, BDI, CII, D, and DIII assuming no spatial structure. Each filled black dot indicates an element of the preferred basis of the single-particle Hilbert space. For class CII, each hollow dot indicates a charge-conjugate pair of basis vectors. Blue solid, dashed, and double lines stand for complex, real, and quaternionic matrix entries. In classes with chiral symmetry (all except class D), the basis vectors with positive ($\nu = 1$) and negative chirality ($\nu = -1$) are separated by horizontal black dotted lines. In these classes, any basis vector with $\nu = 1$ can have non-zero element only with another basis vector with $\nu = -1$ and vice versa. For class DIII, the matrix elements satisfy $\langle \nu | \langle n | H | n' \rangle | -\nu \rangle = -\langle -\nu | \langle n | H | n' \rangle | \nu \rangle$, which is why the schematic looks symmetric about the black dotted line

Table 4.2 Symmetry classes and associated structure of Hamiltonian or off-diagonal blocks of Hamiltonian

Class	Structure of Hamiltonian/block
AIII	Complex
BDI	Real
CII	Quaternionic
D	Real antisymmetric
DIII	Complex antisymmetric

for class D. In this preferred basis, the action of an internal unitary (anti-unitary) operator $\mathcal{S} = U_{\text{int}}$ ($\mathcal{S} = U_{\text{int}}\,\mathrm{C}$) acts trivially on the lattice space spanned by $\{|j\rangle,\ j \in \mathbb{Z}\}$.

A given fully gapped Hamiltonian in a symmetry class can be placed in a topological phase by computing the bulk topological invariant of the class (e.g., the Berry phase, the Chern number, etc.). We consider finite-range, disorder-free single-particle/BdG Hamiltonians expressible in the form given in Eq. (2.9) in Chap. 2, which is

$$H = \sum_{r=1}^{R} h_r^\dagger V^{\dagger r} + h_0 I + \sum_{r=1}^{R} h_r V^r, \tag{4.4}$$

where $V = \sum_{j \in \mathbb{Z}} |j\rangle\langle j+1|$ is the periodic shift operator, the $d \times d$ matrices h_r describe hopping and pairing among fermions situated r sites apart in the bulk, and we assume the range R to be finite. Recall that the reduced bulk Hamiltonian is defined in Eq. (2.13) to be the matrix Laurent polynomial

$$H(z) = h_0 + \sum_{r-1}^{R} (z^r h_r + z^{-r} h_r^\dagger), \quad z \in \mathbb{C}, \quad z \neq 0,$$

which is the analytic continuation to the complex plane of the standard Bloch Hamiltonian, $H_k = H(z = e^{ik})$ for $k \in (-\pi, \pi]$. In the following, we provide explicit formulae for bulk invariants for the five non-trivial classes in 1D [22]. The Fermi energy level is assumed to be at zero for all Hamiltonians. For classes AIII, BDI, CII, and DIII, the Hamiltonian has a chiral symmetry and hence can be written in an off-diagonal form,

$$H_k = \begin{bmatrix} 0 & H_{k,21}^\dagger \\ H_{k,21} & 0 \end{bmatrix}.$$

The topological invariant for the first three of these classes, that is AIII, BDI, and CII, is the winding number of $\det H_{k,21}$, given by

$$Q^B(H) \equiv \frac{1}{2\pi i} \int_{k=-\pi}^{\pi} dk \frac{d}{dk} \log \det H_{k,21}. \tag{4.5}$$

For class D, the bulk invariant is [21]

$$Q^B(H) \equiv \operatorname{sgn}\left[\frac{\operatorname{Pf}(H_{k=0})}{\operatorname{Pf}(H_{k=\pi})}\right], \tag{4.6}$$

where Pf stands for "Pfaffian" [23] of an antisymmetric matrix.[4] Likewise, for class DIII, it is [24]

$$Q^B(H) \equiv \left[\frac{\operatorname{Pf}(H_{k=0,21})}{\operatorname{Pf}(H_{k=\pi,21})}\right] \exp\left(-\frac{1}{2}\int_{k=0}^{\pi} dk \frac{d}{dk} \log \det H_{k,21}\right), \tag{4.7}$$

[4]The Pfaffian of an even-dimensional antisymmetric matrix is a polynomial in the entries of the matrix, the square of which is equal to the determinant of the matrix. For an antisymmetric matrix A of size $2d \times 2d$ with entries $[A]_{ij} = a_{ij}$, the Pfaffian is given by

$$\operatorname{Pf}(A) = \frac{1}{2^d d!} \sum_{\sigma \in S_{2d}} \operatorname{sgn}(\sigma) \prod_{j=1}^{d} a_{\sigma(2j-1),\sigma(2j)},$$

where S_{2d} is the group of permutations of $2d$ indices, and $\operatorname{sgn}(\sigma)$ is the signature of the permutation σ.

To establish a bulk-boundary correspondence, we need to consider properties of systems terminated on one edge, which can host boundary-localized states. For a system on a semi-infinite lattice with lattice sites $\{|j\rangle,\ j \in \mathbb{Z}_+ \equiv \{0, \ldots, \infty\}\}$ labeled by non-negative integers, we consider Hamiltonians of the form $\widetilde{H} + \widetilde{W}$, with

$$\widetilde{H} = \sum_{r=1}^{R} h_r^\dagger S^{\dagger r} + h_0 I + \sum_{r=1}^{R} h_r S^r, \quad \widetilde{W} = \sum_{j,j'=0}^{R_w} w_{jj'} |j\rangle\langle j'|. \tag{4.8}$$

where $S = \sum_{j\in\mathbb{Z}_+} |j\rangle\langle j+1|$ is the "semi-finite (unilateral) shift" operator, the $\bar{d} \times \bar{d}$ matrices $w_{jj'}$ model a finite-strength perturbation that has support on finitely many sites R_w. In practice, this assumption of finite support is valid as long as R_w is much smaller than the total length of the system, but can be much larger than R. A clean system with open BCs is modeled by $\widetilde{W} = 0$. Notice that $\widetilde{H}$ is a block-Toeplitz operator. Capital letters with superscript tilde (e.g., $\widetilde{H}$) will be used to denote operators on semi-infinite 1D systems. Note that given a reduced bulk Hamiltonian $H(z)$, one can always construct the (block-Laurent) Hamiltonian H of the bi-infinite system, as well as the (block-Toeplitz) Hamiltonian $\widetilde{H}$ of the semi-infinite system subject to open BCs on the termination.

In Ref. [5], boundary invariants were defined for classes A and AIII using the K-groups of operator algebras of operators of relevance to the boundary of the system. Here, we refer to an integer-valued quantity as a boundary invariant if it depends only on the eigenvalues and eigenvectors of localized zero-energy states, and is provably invariant under compact perturbations obeying the symmetries of the class. This definition is consistent with the definition of Ref. [5] in the case of class AIII in 1D (class A is topologically trivial in 1D). Based on the conjectures for other symmetry classes in the literature, we identify

$$Q^\partial(\widetilde{H} + \widetilde{W}) \equiv \mathscr{N}_+ - \mathscr{N}_-$$

to be the boundary invariant for classes AIII, BDI, and CII, where $\mathscr{N}_\pm$ refers to the number of zero-energy edge states with chirality (eigenvalue of chiral operator) ± 1. For classes D and DIII, the boundary invariants are instead identified to be

$$Q^\partial(\widetilde{H} + \widetilde{W}) \equiv (-1)^{\mathscr{N}} \quad \text{and} \quad Q^\partial(\widetilde{H} + \widetilde{W}) \equiv (-1)^{\mathscr{N}/2},$$

respectively, where $\mathscr{N} = \dim\ker(\widetilde{H} + \widetilde{W})$ is the number of zero-energy edge states.

4.1.2 The Matrix Wiener–Hopf Factorization

We now describe the Wiener–Hopf factorization of matrix functions defined on the unit circle. While the factorization can be applied to a much broader class of matrix functions, we restrict our attention to matrix Laurent polynomials. The motivation behind such a factorization is to study the kernel and cokernel properties of a block-Toeplitz operator.[5] Consider a $d \times d$ matrix Laurent operator

$$A(z) = \sum_{r=R_{\min}}^{R_{\max}} z^r a_r,$$

which is taken to be invertible for all values of z on the unit circle. It can be associated with a parent banded block-Laurent operator

$$A = \sum_{r<0} a_r (V^\dagger)^r + a_0 I + \sum_{r>0} a_r V^r.$$

Since all eigenvalues (spectrum) of V lie on the unit circle, the invertibility of $A(z)$ on the unit circle immediately implies the invertibility of the block-Laurent operator A. This is in contrast with the block-Toeplitz operator

$$\widetilde{A} = \sum_{r<0} a_r (S^\dagger)^r + a_0 I + \sum_{r>0} a_r S^r,$$

which is *not* necessarily invertible even if $A(z)$ is invertible on the unit circle. This can be attributed to two main differences: (1) the spectra of S and $S^\dagger$ coincide with the unit disk, rather than the unit circle as in the case of V and $V^\dagger$, and (2) $S^\dagger$ is the right inverse of S, but not the left inverse, since $SS^\dagger = I,\ S^\dagger S = I - |0\rangle\langle 0|$. Because of these reasons, the usual functional calculus of normal operators breaks down. One must use the appropriate functional calculus for operators invertible on one side.

Let us review the definition of Wiener–Hopf factorization [25, 26] for matrix Laurent polynomials defined on the unit circle, in the absence of Hermiticity and other constraints.[6]

Definition 4.1 Let $A(z) = \sum_{r=R_{\min}}^{R_{\max}} z^r a_r$ be a $d \times d$ matrix Laurent polynomial that is invertible for all values of z on the unit circle. A factorization of the form

$$A(z) = A_+(z) D(z) A_-(z) \tag{4.9}$$

is called a "Wiener–Hopf factorization," if the factors $A_+(z)$, $A_-(z)$, and $D(z)$ satisfy the following properties:

[5] Recall that for an operator acting on a Hilbert space, its cokernel can be identified with the kernel of the adjoint operator.

[6] The domain on which the matrix Laurent polynomial is defined is irrelevant, since it gets completely defined everywhere in the complex plane by the values it takes on the unit circle.

1. $A_+(z) = \sum_{r \geq 0} a_{+,r} z^r$ is a $d \times d$ matrix polynomial in z that is invertible inside the unit circle ($|z| \leq 1$).
2. $A_-(z) = \sum_{r \geq 0} a_{-,r} z^{-r}$ is a $d \times d$ matrix polynomial in z^{-1} that is invertible outside the unit circle ($|z| \geq 1$), including at $z = \infty$.
3. $D(z)$ is a diagonal matrix function of the form $D(z) = \sum_{m=1}^{d} z^{\kappa_m} |m\rangle\langle m|$, where $\{\kappa_m,\ m = 1, \ldots, d\}$ are integers.

The powers of z in the diagonal factor, denoted by $\{\kappa_m\}$, are called the "partial indices" of $A(z)$. Such a factorization is always possible for matrix Laurent polynomials with non-vanishing determinant on the unit circle, see Ref. [26] for a constructive proof. Such a factorization of $A(z)$ is *not unique* in general; however, the partial indices and hence the diagonal factor $D(z)$ are unique. It is convenient to denote by $\mathcal{K}$, $\mathcal{K}_+$, and $\mathcal{K}_-$ the set of all, strictly positive and strictly negative *distinct* partial indices of $A(z)$, respectively. We also introduce the relabeling of the internal states on which A(z) acts, by using the partial indices $\{|\kappa, s\rangle,\ s = 1, \ldots, s_\kappa,\ \kappa \in \mathcal{K}\}$, such that

$$D(z)|\kappa, s\rangle = z^\kappa |\kappa, s\rangle \iff D(z) = \sum_\kappa \sum_{s=1}^{s_\kappa} z^\kappa |\kappa, s\rangle\langle\kappa, s|. \tag{4.10}$$

We denote the "multiplicity" of a partial index κ by s_κ, so that we may write $\sum_{\kappa \in \mathcal{K}} s_\kappa = d$.

Example 4.2 Let us consider the scalar polynomial $A(z) = 2 - 5z + 2z^2$. To find the Wiener–Hopf factorization, we first factorize the polynomial completely, $A(z) = 2(z-1/2)(z-2)$. We find that one of the roots is inside the unit circle, and the other is outside. We use the root outside the unit circle to form the factor A_+, and the other root to form A_-, the remaining power is absorbed in D:

$$A(z) = \underbrace{-(2-z)}_{A_+} z \underbrace{(2-z^{-1})}_{A_-}.$$

One may verify that A_+ and A_- satisfy the required properties. In this case we find $D(z) = z$, and hence $\kappa = 1$. Since the internal space is one-dimensional, the relabeling is $|m = 1\rangle \mapsto |\kappa = 1, s = 1\rangle$.

Back to the general case, we note since $V^\dagger = V^{-1}$, the factorization of $A(z)$ leads to the factorization of the block-Laurent operator A, namely

$$A(z) = A_+(z) D(z) A_-(z)\ \forall z,\ z \neq 0 \implies A = A_+ D A_-,$$

where

$$A_\pm = \sum_{r \geq 0} a_{\pm,r} V^{\pm r}, \quad D = \sum_\kappa \sum_{s=1}^{s_\kappa} V^\kappa |\kappa, s\rangle\langle\kappa, s|.$$

However, we are interested in whether this factorization carries over to the block-Toeplitz case. This is not trivial, since the product of two Toeplitz operators in general yields a sum of a Toeplitz and a compact operator, which reflects the presence of a boundary. However, post-multiplying S^r by $(S^\dagger)^{r'}$, for $r, r' \in \mathbb{Z}_+$, *always* yields an operator that is purely Toeplitz with no compact part, that is,

$$S^r S^{\dagger r'} = \begin{cases} S^{r-r'} & \text{if } r > r' \\ S^{\dagger r'-r} & \text{if } r < r' \\ I & \text{if } r = r' \end{cases} .$$

From this observation, it is immediate to see that the factorization of $A(z)$ in Eq. (4.9) leads to the factorization $\widetilde{A} = \widetilde{A}_+ \widetilde{D} \widetilde{A}_-$ of the block-Toeplitz operator $\widetilde{A}$, where

$$\widetilde{A}_+ = \sum_{r \geq 0} a_{+,r} S^r, \quad \widetilde{A}_- = \sum_{r \geq 0} a_{-,r} S^{\dagger r},$$

and

$$\widetilde{D} = \sum_{\kappa \in \mathcal{K}_-} \sum_{s=1}^{s_\kappa} |\kappa, s\rangle\langle \kappa, s| (S^\dagger)^{-\kappa} + \sum_{s=1}^{s_0} |0, s\rangle\langle 0, s| I + \sum_{\kappa \in \mathcal{K}_+} \sum_{s=1}^{s_\kappa} |\kappa, s\rangle\langle \kappa, s| S^\kappa .$$

Further, the conditions (1) and (2) in the definition of Wiener–Hopf factorization ensure the invertibility of the operators $\widetilde{A}_+$ and $\widetilde{A}_-$.[7] The kernel and cokernel of powers of shift operators can be deduced from

$$S^r |j\rangle = \begin{cases} |j-r\rangle & \text{if } j \geq r \\ 0 & \text{if } j < r \end{cases}, \quad \langle j'|(S^\dagger)^r = \begin{cases} \langle j'-r| & \text{if } j' \geq r \\ 0 & \text{if } j' < r \end{cases},$$

so that we can express the kernel and cokernel of $\widetilde{D}$ as

$$\begin{aligned} \operatorname{Ker} \widetilde{D} &= \operatorname{span}\{|\kappa, s\rangle |j\rangle, \ j = 0, \ldots, \kappa - 1, \quad s = 1, \ldots, s_\kappa, \ \kappa \in \mathcal{K}_+\}, \\ \operatorname{Coker} \widetilde{D} &= \operatorname{span}\{\langle \kappa, s| \langle j|, \ j = 0, \ldots, -\kappa - 1, \quad s = 1, \ldots, s_\kappa, \ \kappa \in \mathcal{K}_-\}. \end{aligned}$$

The kernel and cokernel of $\widetilde{A}$ are then given by

$$\operatorname{Ker} \widetilde{A} = \widetilde{A}_-^{-1} \operatorname{Ker} \widetilde{D}, \quad \operatorname{Coker} \widetilde{A} = (\operatorname{Coker} \widetilde{D}) \widetilde{A}_+^{-1}. \tag{4.11}$$

[7] The functions $A_+(z)$ and $A_-(z)$ belong to the *Wiener algebra*, which has the property that the $A_+^{-1}(z)$ and $A_-^{-1}(z)$ have an absolutely convergent Fourier series inside and outside the unit circle, respectively. Since the spectrum of T and $T^\dagger$ is contained in the unit disk, $\widetilde{A}_\pm$ are invertible.

Equation (4.11) provides an analytic way of computing the kernel of BBT operators, whose symbol is invertible on the unit circle. It follows that the dimensions of the kernel and the cokernel of $\widetilde{A}$, also called its "defect numbers," are given, respectively, by

$$\dim \mathrm{Ker}\, \widetilde{A} = \dim \mathrm{Ker}\, \widetilde{D} = \sum_{\kappa \in \mathcal{K}_+} \kappa s_\kappa, \quad \dim \mathrm{Coker}\, \widetilde{A} = \dim \mathrm{Coker}\, \widetilde{D} = -\sum_{\kappa \in \mathcal{K}_-} \kappa s_\kappa,$$

which are completely determined by the partial indices. Further, these formulas for defect numbers provide the expression

$$\mathrm{index}(\widetilde{A}) \equiv \dim \mathrm{Ker}\, \widetilde{A} - \dim \mathrm{Coker}\, \widetilde{A} = \sum_{\mathcal{K}} \kappa s_\kappa \tag{4.12}$$

for the "analytic index" [27] of $\widetilde{A}$. We will also make use of the "secondary index" [27, 28]

$$\mathrm{index}_2(\widetilde{A}) \equiv \dim \mathrm{Ker}\, \widetilde{A} \bmod 2 = \left(\sum_{\mathcal{K}_+} \kappa s_\kappa \right) \bmod 2.$$

4.2 Establishing the Bulk-Boundary Correspondence in 1D

4.2.1 Wiener–Hopf Spectral Flattening of Translation-Invariant Systems

It was mentioned at the beginning of this chapter that the bulk topological invariants are computed using the wavefunctions of the filled electronic states. Since the eigenvalues of the filled band do not affect the value of the topological invariant, it is often more convenient to work with "spectrally flattened" Hamiltonians [1]. The standard way of spectral flattening is to assign energy eigenvalues -1 ($+1$) in arbitrary units to all the filled (empty) states, without changing their wavefunctions. Assuming that zero energy lies in the band gap, mathematically this operation can be expressed as $H \mapsto \mathrm{sgn}(H)$.

We now move on to describe a non-conventional spectral flattening of Hamiltonians in all five symmetry classes under consideration, using a modification of the Wiener–Hopf factorization, that we call "symmetric Wiener–Hopf factorization" (see the Appendix). Our method of spectral flattening ensures that, in all symmetry classes, the resultant flat-band system is in the *same symmetry class* as the original one. Further, the flat-band system shares the same value of the topological bulk invariant as the original system, so that it remains in the same topological phase as the original system. An advantage of this new way of spectral flattening is that it leads to finite-range flat-band Hamiltonians, at the cost of deforming the eigenstates of the original Hamiltonian.

In the presence of additional symmetry constraints on H, the factors in the Wiener–Hopf factorization can be chosen to have more structure. Various works in the mathematics literature have been devoted to determining the structure of the factors [29–32] and factorization methods [33, 34] for matrix polynomials satisfying various constraints [35–38]. The symmetries can be exploited to make sure that the middle factor satisfies the same symmetry constraints as the original Bloch Hamiltonian. In view of this requirement, the middle factor in the symmetric factorization can be viewed as new Hamiltonian that is related to the original Hamiltonian. A common feature of such new Hamiltonians is that their energy bands are *non-dispersive*. Therefore, we refer to the symmetric factorization as the "Wiener–Hopf spectral flattening." Similar to the standard spectral flattening [1], the flat-band Hamiltonian belongs to the same symmetry class as the original Hamiltonian, and shares the same value of topological invariant.

The exact structure of the Hamiltonians and the three factors for each symmetry class is listed in Table 4.3. That such kind of factorization is possible for the symmetry requirements of each class under consideration is proved in the Appendix. Since we always assume Hermiticity for the Hamiltonians, the side factors can always be chosen to satisfy $H_-(z) = H_+^\dagger(z)$. We caution the reader about our notation here: $H_+^\dagger(z)$ is the symbol of the block-Laurent operator $H_+^\dagger$, and is different from $H_+(z)^\dagger = H_+^\dagger(z^*)$ in general. We use the letter $H_F(z)$ to denote the middle factor, which we interpret as the reduced bulk Hamiltonian of the resulting flat-band system (see Fig. 4.2). For the chirally symmetric classes AIII, BDI, DIII, and CII, we exploit the off-diagonal structure of the Hamiltonian (recall Eq. (4.1)) to write the factors as

$$H_+ = \begin{bmatrix} H_{21,-}^\dagger & 0 \\ 0 & H_{21,+} \end{bmatrix}, \quad H_F = \begin{bmatrix} 0 & D^\dagger \\ D & 0 \end{bmatrix},$$

Table 4.3 The form of the three factors in the factorization of H (H_{21}) for the non-trivial symmetry classes without (with) chiral symmetry. The arguments (z, z^{-1}) are dropped for brevity. $\mathbb{F}_d$ denotes the set of $d \times d$ matrices with entries in field $\mathbb{F} = \mathbb{R}, \mathbb{C}$ or skew-field $\mathbb{F} = \mathbb{H}$

Class	Side factors	Middle factor
AIII	$H_{21,+}, H_{21,-} \in \mathbb{C}_{\tilde{d}}[z]$	$D = \sum_{\kappa\in\mathcal{K}} \sum_{s=1}^{s_\kappa} z^\kappa \lvert\kappa, s\rangle\langle\kappa, s\rvert$
BDI	$H_{21,+}, H_{21,-} \in \mathbb{R}_{\tilde{d}}[z]$	$D = \sum_{\kappa\in\mathcal{K}} \sum_{s=1}^{s_\kappa} z^\kappa \lvert\kappa, s\rangle\langle\kappa, s\rvert$
CII	$H_{21,+}, H_{21,-} \in \mathbb{H}_{\tilde{d}}[z]$	$D = \sum_{\kappa\in\mathcal{K}} \sum_{s=1}^{s_\kappa} z^\kappa \lvert\kappa, s\rangle\langle\kappa, s\rvert, \quad s_\kappa \in 2\mathbb{Z}$
D	$H_+, H_- \in \mathbb{R}_{\tilde{d}}[z]$	$H_F = i(\sum_{\kappa\in\mathcal{K}_+} \sum_{s=1}^{s_\kappa} (z^\kappa \lvert\kappa, s\rangle\langle-\kappa, s\rvert - z^{-\kappa}\lvert-\kappa, s\rangle\langle\kappa, s\rvert)$ $+i(\sum_{s=1}^{s_0/2} \lvert 0, 2s\rangle\langle 0, 2s-1\rvert - \lvert 0, 2s-1\rangle\langle 0, 2s\rvert)$
DIII	$H_{21,+} = H_{21,-}^{\mathsf{T}} \in \mathbb{C}_{\tilde{d}}[z]$	$D = i(\sum_{\kappa\in\mathcal{K}_+} \sum_{s=1}^{s_\kappa} (z^\kappa \lvert\kappa, s\rangle\langle-\kappa, s\rvert - z^{-\kappa}\lvert-\kappa, s\rangle\langle\kappa, s\rvert)$ $+i(\sum_{s=1}^{s_0/2} \lvert 0, 2s\rangle\langle 0, 2s-1\rvert - \lvert 0, 2s-1\rangle\langle 0, 2s\rvert)$

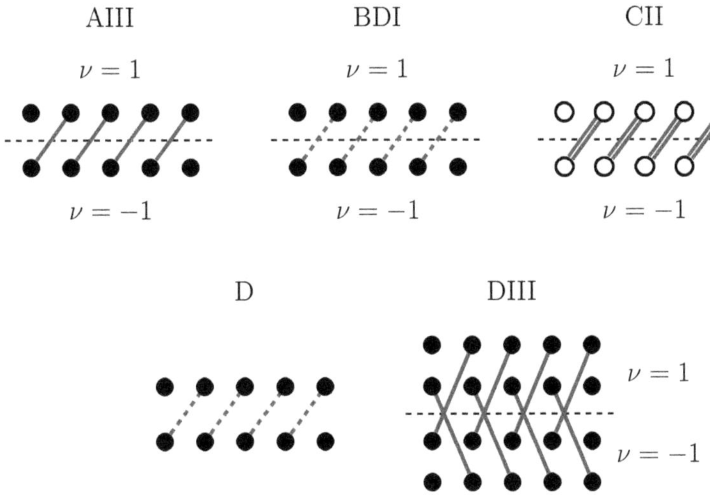

Fig. 4.2 Schematic diagram of the smallest flat-band Hamiltonian of a topologically non-trivial Hamiltonian in each symmetry class in 1D. For systems shown in classes AIII, BDI, and CII in the figure, the only partial index of $D(z)$ is $\kappa = 1$ with multiplicity 1, 1, and 2, respectively. For class D, the partial indices of $H_F(z)$ are $\{1, -1\}$, each with multiplicity 1. For class DIII, the partial indices of $D(z)$ are $\{1, -1\}$, each with multiplicity 1. For gapped systems in symmetry classes CII and DIII, $\bar{d} = d/2$ is necessarily even

where $H_{21,+}$ and $H_{21,-}$ are factors in the factorization $H_{21} = H_{21,+}DH_{21,-}$. For these classes, we relabel the internal states $\{|\nu\rangle|m\rangle,\ \nu = \pm 1,\ m = 1, \ldots, \bar{d}\}$ by $\{|\nu\rangle|\kappa, s\rangle\}$, where $\nu = \pm 1$ is the chirality, and

$$\nu_z|\nu\rangle|\kappa, s\rangle \equiv \nu|\nu\rangle|\kappa, s\rangle, \quad H_F(z)|\nu\rangle|\kappa, s\rangle \equiv z^{\nu\kappa}|-\nu\rangle|\kappa, s\rangle.$$

Here, $\{\kappa, s_\kappa\}$ are partial indices and multiplicities of $H_{21}(z)$, and therefore $\sum_{\kappa\in\mathcal{K}} s_\kappa = d$ equals half the internal dimension of $H(z)$.

Example 4.3 We will elucidate the meaning of Table 4.3 with the example of spectral flattening of the SSH Hamiltonian (see Sect. 3.1.4). Note that for this particularly simple model, a completely closed-form expression for the Wiener–Hopf factorization can be obtained. The reduced bulk Hamiltonian is

$$H(z) = -\begin{bmatrix} 0 & \mathrm{t}_1 + \mathrm{t}_2 z \\ \mathrm{t}_1 + \mathrm{t}_2 z^{-1} & 0 \end{bmatrix},$$

The factorization of $H(z)$ according to the relevant class BDI is

$$H(z) = \begin{cases} \begin{bmatrix} 1 & 0 \\ 0 & -(\mathrm{t}_1 z + \mathrm{t}_2) \end{bmatrix} \begin{bmatrix} 0 & z \\ z^{-1} & 0 \end{bmatrix} \begin{bmatrix} 1 & 0 \\ 0 & -(\mathrm{t}_1 z^{-1} + \mathrm{t}_2) \end{bmatrix} & \text{if } \mathrm{t}_1 < \mathrm{t}_2, \\ \begin{bmatrix} -(\mathrm{t}_1 + \mathrm{t}_2 z) & 0 \\ 0 & 1 \end{bmatrix} \begin{bmatrix} 0 & 1 \\ 1 & 0 \end{bmatrix} \begin{bmatrix} -(\mathrm{t}_1 + \mathrm{t}_2 z^{-1}) & 0 \\ 0 & 1 \end{bmatrix} & \text{if } \mathrm{t}_1 > \mathrm{t}_2. \end{cases} \tag{4.13}$$

Notice that both $H(z)$ and $H_F(z)$ are Hermitian and obey the same chiral symmetry. That H_F is indeed a flat-band Hamiltonian is seen from its dispersion relation,

$$\det[H_F(e^{ik}) - \epsilon] = 0 \implies \epsilon^2 = 1.$$

It is immediate to check that $H_-(z) = H_+^\dagger(z)$ is satisfied in both the parameter regimes.

The factorization in Eq. (4.13) immediately allows us to compute the zero-energy edge state for the case $\mathrm{t}_1 < \mathrm{t}_2$. Notice that the unnormalized kernel of $\widetilde{H}$ is given by

$$\ker \widetilde{H} = \widetilde{H}_-^{-1} \ker \widetilde{D} = \begin{bmatrix} 1 & 0 \\ 0 & -(\mathrm{t}_1 S^\dagger + \mathrm{t}_2) \end{bmatrix}^{-1} \begin{bmatrix} 0 \\ 1 \end{bmatrix}.$$

Some algebra reveals that the zero-energy edge state $|\psi\rangle$ has a wavefunction given by

$$|\psi\rangle = \sqrt{1 - (\mathrm{t}_1/\mathrm{t}_2)^2} \begin{bmatrix} 0 \\ 1 \end{bmatrix} \sum_{j \in \mathrm{Z}_+} (-\mathrm{t}_1/\mathrm{t}_2)^j |j\rangle.$$

Lemma 4.4 *The Wiener–Hopf spectral flattening leaves the symmetries of the Hamiltonian and the value of the topological (bulk) invariant unchanged. That is, for a symmetric factorization of the Hamiltonian* $H = H_+ H_F H_+^\dagger$, H_F *satisfies the constraints of the symmetry class of* H, *and*

$$Q^B(H) = Q^B(H_F). \tag{4.14}$$

One way to prove this result is to observe that the symmetric Wiener–Hopf factorization in each class is equivalent to an explicit homotopy between H and H_F, up to an inconsequential basis change in the internal space.[8] In other words, $H(z)$ can be transformed (up to change in internal basis) to $H_F(z)$ adiabatically while preserving all the symmetries of the class. Recall that the bulk topological invariant

---

[8]There are another ways of establishing this crucial result by checking directly that the bulk topological invariants yield the same value for both H and H_F. The details of this proof will be provided in Ref. [39].

for two Hamiltonians which can be connected adiabatically without breaking the symmetries must take the same values. It is not very difficult to construct a family of Hamiltonians $H(z,t)$, such that $H(z,0) = H(z)$ and $H(z,T) = H_F(z)$, and $H(z,t)$ remains invertible for $0 < t < T$. For instance, we can choose

$$H(z,t) = H_+(z(1-2t/T))H_F(z)H_+^\dagger(z/(1-2t/T)), \quad 0 \leq t < T/2,$$

so that the roots of $\det H_+(z,t) = 0$ are $\{z_\ell/(1-t/T)\}$, in terms of roots $\{z_\ell\}$ of $\det H_+(z) = 0$. Therefore, they remain outside the unit circle throughout the transformation, so that $H_+(z,t)$ remains invertible inside the unit circle. It is easy to check that $H(z,t)$ satisfies the symmetries of the class of $H(z)$. Notice that for $t = T/2$, the side factor $H_+(0)$ is a constant, invertible matrix. The symmetries of the class are preserved during this transformation, thanks to the property that G, G' defined by $G(z) \equiv H_+(\alpha z)$ and $G'(z) \equiv H_+^\star(\alpha^{-1}z)$ for $\star = \dagger, \mathsf{T}, \alpha \in \mathbb{R}$ satisfy $G'(z) = G^\star(z)$. In the second step, we transform $H_+(0)$ to a norm-preserving matrix of the class (unitary for AIII, DIII, orthogonal for D, BDI, and quaternionic unitary for CII). Any real, complex, or quaternionic invertible matrix M admits a polar decomposition [40–42] of the form $M = S_M U$, where S_M is a positive definite matrix, and U is a norm-preserving matrix. Let $H_+(0) = S_+U_+$ be a polar decomposition of $H_+(0)$, such that S_+ and U_+ share the same diagonal block structure as $H_+(0)$. $H(z,T) = H_+(0)H_F(z)H_+^\dagger(0)$ can be adiabatically transformed to $U_+H_F(z)U_+^\dagger$ via

$$H(z,t) = S_+^{(2-2t/T)}U_+H_FU_+^\dagger S_+^{(2-2t/T)}, \quad T/2 \leq t \leq T.$$

Notice that $H(z,T) = U_+H_FU_+^\dagger$ is H_F in a different internal basis.

4.2.2 Wiener–Hopf Spectral Flattening of Systems with Boundary

We now show that the Wiener–Hopf spectral flattening described in the previous section can be naturally extended to semi-infinite systems, and that, in addition, the boundary invariant is preserved. We consider the Hamiltonian $\widetilde{H} + \widetilde{W}$ defined in Eq. (4.8), where both $\widetilde{H}$ and $\widetilde{W}$ satisfy the constraints of the symmetry class. Let $H(z) = H_+(z)H_F(z)H_F^\dagger(z)$ be the symmetric factorization of $H(z)$. We can then write

$$\widetilde{H} + \widetilde{W} \equiv \widetilde{H}_+(\widetilde{H}_F + \widetilde{W}_F)\widetilde{H}_+^\dagger, \quad \widetilde{W}_F = (\widetilde{H}_+)^{-1}\widetilde{W}(\widetilde{H}_+^\dagger)^{-1}.$$

We are now interested in the Hamiltonian $(\widetilde{H}_F + \widetilde{W}_F)$. First, we will show that $\widetilde{W}_F$ is a finite-range disorder satisfying the constraints of the symmetry class. Using the form of $\widetilde{W}$ in Eq. (4.8), it is easy to deduce that $\widetilde{W}_F$ has support at most on the first R_w lattice sites. In the case of chiral classes, $\widetilde{H}_+$ commutes with the chiral symmetry, which means that $\widetilde{W}_F$ anticommutes and hence obeys the chiral symmetry. In classes BDI and D, the entries of $\widetilde{H}_+$ are real; therefore, the entries of $\widetilde{W}_F$ are imaginary according to the requirement of these classes. Similarly, in the case of class CII, the (block-) entries of $\widetilde{H}_+$, and hence of $\widetilde{W}_F$, are quaternionic. Therefore, we conclude that the entries of $\widetilde{H}_+$ are real, and thus the entries of $\widetilde{W}_F$ always satisfy the symmetries of the class.

Having established the symmetry properties of $\widetilde{W}_F$, we are now in a position to prove the following Lemma:

Lemma 4.5 *The boundary invariant for systems with finite range disorder remains unchanged under spectral flattening, that is, we have*

$$Q^\partial(\widetilde{H} + \widetilde{W}) = Q^\partial(\widetilde{H}_F + \widetilde{W}_F). \tag{4.15}$$

For classes AIII, BDI, and CII, the boundary invariant is $Q^\partial = \mathcal{N}_+ - \mathcal{N}_-$, with $\mathcal{N}_+ = \dim\ker(\widetilde{H}_{21} + \widetilde{W}_{21})$ and $\mathcal{N}_- = \dim\ker(\widetilde{H}_{21}^\dagger + \widetilde{W}_{21}^\dagger)$. Because $\widetilde{H}_{21,+}$ and $\widetilde{H}_{21,-}$ are invertible, we find that $\dim\ker(\widetilde{H}_{21} + \widetilde{W}_{21}) = \dim\ker(\widetilde{D} + \widetilde{W}_{F,21})$ and $\dim\ker(\widetilde{H}_{21}^\dagger + \widetilde{W}_{21}^\dagger) = \dim\ker(\widetilde{D}^\dagger + \widetilde{W}_{F,21}^\dagger)$. Therefore, the boundary invariant remains unchanged under spectral flattening. For classes D and DIII, since $\widetilde{H}_+$ is invariant, we get $\dim\ker(\widetilde{H} + \widetilde{W}) = \dim\ker(\widetilde{D} + \widetilde{W}_F) = \mathcal{N}$. Since the boundary invariant depends only on $\mathcal{N}$ ($(-1)^{\mathcal{N}}$ and $(-1)^{\mathcal{N}/2}$ for classes D and DIII, respectively), the boundary invariant is seen to remain unchanged under spectral flattening.

4.2.3 *Equality of Bulk and Boundary Invariants*

Building on the results of Sects. 4.2.1 and 4.2.2, we can now present the main result of this chapter, namely the bulk-boundary correspondence in one spatial dimension.

Theorem 4.6 (Bulk-Boundary Correspondence) *The following equalities hold for 1D systems in all the symmetry classes:*

1. *The bulk invariant of a bi-infinite system is equal to the boundary invariant of the semi-infinite system under open BCs, that is,*

$$Q^B(H) = Q^\partial(\widetilde{H}).$$

2. *The boundary invariant of a semi-infinite system under open boundary conditions remains unchanged under addition of finite range disorder that satisfies the symmetries of the class. That is,*

$$Q^\partial(\widetilde{H}) = Q^\partial(\widetilde{H} + \widetilde{W}).$$

A detailed mathematical proof of these statements is outside the scope of the present chapter. The key idea behind the proof is to utilize the Wiener–Hopf spectral flattening of the semi-infinite system, so that one may instead focus on a much simpler class of "dimerized" flat-band Hamiltonians. Such a system can be exactly separated into two parts—one near the boundary which supports the boundary disorder $\widetilde{W}$, and one sufficiently away from it, and hence trivial in nature. This separation depends on the partial indices of the reduced bulk Hamiltonian, and hence indirectly on the bulk topological invariant. Finally, it allows us to focus only on the former finite-dimensional space, where equalities of bulk and boundary invariants can be shown using elementary matrix identities.

With advanced mathematical tools, more general statement regarding the boundary invariant can be proved for all classes. For classes AIII, BDI, and CII, the boundary invariant is the analytic index of the off-diagonal block H_{21}, which is known to be a continuous (hence constant on connected components) function on the set of Fredholm operators [43]. Similarly, for classes D and DIII, the boundary invariant is the secondary index of iH and H_{21}, respectively, which is constant on connected components of antisymmetric Fredholm operators [28]. It follows that

$$Q^{\partial}(\widetilde{H}+\widetilde{V}) = Q^{\partial}(\widetilde{H}),$$

as long as $||\widetilde{V}|| < \Delta E/2$, where ΔE is the single-particle bulk energy gap, $\widetilde{V}$ satisfies the symmetries of the class, and the operator norm is considered in the left-hand side. Furthermore, the analytic (secondary) index is known to be invariant under the addition of a (antisymmetric) compact operator of arbitrary norm. Therefore, one may replace the finite-range $\widetilde{W}$ by a *compact* $\widetilde{W}$ in the second part of the theorem. This result also follows from the one for finite-range operators, since compact operators are limits of finite-range operators in the operator-norm topology, and the limit of each boundary invariant is well-defined, as it depends on the spectral projector of the discrete spectrum.

4.3 Protection of Zero Modes

Symmetry conditions guarantee protection of bulk and boundary invariants against compact perturbations of arbitrary strength, as well as against bulk disorder that is sufficiently weak for the gap not to close. However, for all five non-trivial classes in 1D, the total dimension of the zero-energy single-particle eigenspace, and hence of the many-body ground manifold, is *not* guaranteed to be invariant under such perturbations [44]. The aim of this section is to establish necessary and sufficient conditions for the stability of the ground manifold. This can be especially useful in assessing the performance of symmetry-protected quantum memories. A similar notion was discussed in Sec. 7.4 of Ref. [5] in the context of the integer quantum Hall effect.

4.3.1 A Criterion for the Stability of Zero Modes

The stability of the ground manifold depends entirely on the stability of the zero-energy eigenspace of the single-particle Hamiltonian, both in the case of topological insulators as well as topological superconductors. In our analysis, we will assume that the system is long enough to ignore finite size effects, so that we can focus on the zero-energy eigenspace of the semi-infinite system. We say that the zero-energy eigenspace of a Hamiltonian $\widetilde{H}_0$ is stable against perturbations in the symmetry class, if the zero-energy eigenspace is *nearly* unchanged under any such perturbation $\widetilde{V} = \lambda\widetilde{\boldsymbol{v}}$ for small enough strength $\lambda \in \mathbb{R}$. Since we are only interested in the strength of the perturbation $\widetilde{V}$ relative to that of $\widetilde{H}_0$, we will assume $\|\widetilde{\boldsymbol{v}}\| = \|\widetilde{H}_0\|$ to have unit norm, so that $\lambda = \|\widetilde{V}\|/\|\widetilde{H}_0\|$ is the dimensionless parameter of interest. Let $P_{\text{Ker}\,(\cdot)}$ denote the orthogonal projector on Ker $(\cdot)$.

Definition 4.7 Ker $(\widetilde{H}_0)$ is said to be "stable against a perturbation $\widetilde{\boldsymbol{v}}$" of unit norm if

$$\lim_{\lambda\to 0} P_{\text{Ker}\,(\widetilde{H}_0+\lambda\widetilde{\boldsymbol{v}})} = P_{\text{Ker}\,(\widetilde{H}_0)}, \quad \lambda \in R.$$

Ker $(\widetilde{H}_0)$ is said to be "stable in the symmetry class of $\widetilde{H}_0$" if it is stable against every perturbation $\widetilde{\boldsymbol{v}}$ of unit norm in the symmetry class of $\widetilde{H}_0$.

We will next show that a necessary and sufficient condition for the stability of the zero-energy eigenspace is that its dimension is invariant for small but arbitrary perturbations in the class. Before formally proving this statement, some additional tools are needed. Let $\widetilde{H}_0^{(-1)}$ be a "reflexive generalized inverse" of $\widetilde{H}_0$, defined by the properties

$$\widetilde{H}_0^{(-1)}\widetilde{H}_0\widetilde{H}_0^{(-1)} = \widetilde{H}_0^{(-1)}, \quad \widetilde{H}_0\widetilde{H}_0^{(-1)}\widetilde{H}_0 = \widetilde{H}_0.$$

We define the "generalized condition number" [45] of $\widetilde{H}_0$ with respect to $\widetilde{H}_0^{(-1)}$ by

$$\varkappa(\widetilde{H}_0, \widetilde{H}_0^{(-1)}) = \|\widetilde{H}_0\|\|\widetilde{H}_0^{(-1)}\|.$$

Generally, a large value of the generalized condition number implies that the system of equations is ill-conditioned. We will see in the next lemma that it also means that the kernel of the operator is very sensitive to perturbations.

Lemma 4.8 *For* $\lambda < 1/\varkappa(\widetilde{H}_0, \widetilde{H}_0^{(-1)})$, *we have* $\dim \text{Ker}\,(\widetilde{H}_0 + \widetilde{V}) \leq \dim \text{Ker}\,(\widetilde{H}_0) \equiv \mathcal{N}$. *If, in addition, the equality holds for some* $\widetilde{V}$, *then*

$$\text{Ker}\,(\widetilde{H}_0 + \widetilde{V}) = (\widetilde{I} + \widetilde{H}_0^{(-1)}\widetilde{V})^{-1}\text{Ker}\,(\widetilde{H}_0). \tag{4.16}$$

The proof of the lemma follows from a simple observation, namely $\widetilde{H}_0^{(-1)}(\widetilde{H}_0 + \widetilde{V}) = \widetilde{H}_0^{(-1)}\widetilde{H}_0(\widetilde{I} + \widetilde{H}_0^{(-1)}\widetilde{V})$. Due to the restriction on the norm of $\widetilde{V}$, the term inside the bracket on the right-hand side is invertible, and hence

$$\mathrm{Ker}\, \widetilde{H}_0^{(-1)}(\widetilde{H}_0 + \widetilde{V}) = (\widetilde{I} + \widetilde{H}_0^{(-1)}\widetilde{V})^{-1}\mathrm{Ker}\, (\widetilde{H}_0^{(-1)}\widetilde{H}_0) = (\widetilde{I} + \widetilde{H}_0^{(-1)}\widetilde{V})^{-1}\mathrm{Ker}\, \widetilde{H}_0.$$

Since $\dim \mathrm{Ker}\, \widetilde{H}_0^{(-1)}(\widetilde{H}_0 + \widetilde{V}) \geq \dim \mathrm{Ker}\, (\widetilde{H}_0 + \widetilde{V})$, we have proved the first statement. In the case of equality of dimensions, we have $\mathrm{Ker}\, \widetilde{H}_0^{(-1)}(\widetilde{H}_0 + \widetilde{V}) = \mathrm{Ker}\, (\widetilde{H}_0 + \widetilde{V})$, from which the second statement of the lemma follows.
We are now in a position to characterize the stability of the zero-energy eigenspace:

Lemma 4.9 *A necessary and sufficient condition for the stability of* $\mathrm{Ker}\, (\widetilde{H}_0)$ *is that the number of zero modes is invariant against small but arbitrary perturbations in the symmetry class, that is,*

$$\lim_{\lambda \mapsto 0} \dim \mathrm{Ker}\, (\widetilde{H}_0 + \lambda \widetilde{\boldsymbol{v}}) = \dim \mathrm{Ker}\, (\widetilde{H}_0), \qquad \lambda \in \mathbb{R},$$

for every $\widetilde{\boldsymbol{v}}$ *of unit norm in the symmetry class.*

The proof of this lemma follows from a straightforward application of Lemmas 4.8 and 4.9. What is remarkable is that the stability of zero eigenspace is determined completely by the stability of its dimension!

Our next step, naturally, is to determine the conditions under which the dimension of the zero-energy eigenspace is stable for each of the symmetry classes. This condition can be written in terms of dimension of the zero-energy subspace and the topological invariants.

Theorem 4.10 (Stability of Zero Modes) *The zero-energy eigenspace of* $\widetilde{H}_0$ *is stable with respect to symmetry-preserving perturbations if and only if*

$$\mathcal{N} = \begin{cases} |Q^\partial(\widetilde{H}_0)| & \text{for } \widetilde{H}_0 \text{ in chiral classes} \\ (1 - Q^\partial(\widetilde{H}_0))/2 & \text{for } \widetilde{H}_0 \text{ in class D} \\ 1 - Q^\partial(\widetilde{H}_0) & \text{for } \widetilde{H}_0 \text{ in class DIII} \end{cases} . \tag{4.17}$$

The proof of this theorem rests on the simple observation that one can always infer from the value of the boundary invariant a lower bound on the dimension of zero-energy eigenspace. Further, if this lower bound is tight (that is, the dimension of zero-energy eigenspace is equal to the lower bound), then by Lemma 4.8 the zero-energy eigenspace is stable. For example, in the case of chiral classes, we have $\mathcal{N}_+ - \mathcal{N}_- = Q^\partial(\widetilde{H}_0)$. Then $\mathcal{N} = \mathcal{N}_+ + \mathcal{N}_- \geq |Q^\partial(\widetilde{H}_0)|$. Because $\widetilde{V}$ satisfies the symmetries of the class, the boundary invariant is continuous. Therefore, we have $\dim \mathrm{Ker}\, (\widetilde{H}_0 + \widetilde{V}) \geq \dim \mathrm{Ker}\, \widetilde{H}_0$. However, for small enough $\widetilde{V}$, Lemma 4.8 states that $\dim \mathrm{Ker}\, (\widetilde{H}_0 + \widetilde{V}) \leq \dim \mathrm{Ker}\, \widetilde{H}_0$. These two statements lead to the conclusion that $\dim \mathrm{Ker}\, (\widetilde{H}_0 + \widetilde{V}) = \dim \mathrm{Ker}\, \widetilde{H}_0$, which is a sufficient condition for the stability of $\mathrm{Ker}\, \widetilde{H}_0$. The strategy to prove the necessity of the condition (4.17) is to construct explicitly a perturbation in the symmetry class of the system which changes the dimension of the zero-energy eigenspace.

Remark 4.11 Note that the condition in Eq. (4.17) is also a necessary and sufficient condition for protecting the *degeneracy* of the many-body ground state, which depends entirely on the number of zero-energy boundary states.

We now discuss the stability of zero modes of some representative systems in light of our definition and criterion based on dimension. We have already discussed the chiral zero modes of the SSH Hamiltonian in Sect. 4.2. For the parameter regime $\mathrm{t}_1 < \mathrm{t}_2$, we have $Q^\partial(\widetilde{H}) = -1$ and $\mathscr{N} = 1$, which means the condition for stability in Eq. (4.17) is satisfied. Therefore, the edge state of the SSH Hamiltonian subject to open BCs is stable. The same can be said about the Majorana zero mode at the end of Kitaev's Majorana chain under open BCs, discussed in Sect. 3.1.5.

We emphasize that the condition in Eq. (4.17) is *not* a consequence of the non-trivial topology of the Hamiltonian. Rather, it should be regarded as an additional constraint, on top of non-trivial topology, that a system's Hamiltonian must meet in order to host *stable* zero modes. This becomes evident from the fact that Eq. (4.17) is explicitly violated by some Hamiltonians, despite the latter having non trivial topology. For instance, if $\widetilde{H}_0$ represents a stack of three Kitaev chains in non-trivial regime subject to open BCs, then the system as a whole hosts three Majorana modes on each end. This system in class D does not satisfy Eq. (4.17). The fact that the zero-energy eigenspace is not stable can be seen from the effect of particle-hole symmetric perturbations. If $\hat{\eta}_1$ and $\hat{\eta}_2$ are two of these Majoranas (the hat indicates operators in Fock space), then a perturbation of arbitrarily small strength of the form $\widehat{V} = i\lambda\hat{\eta}_1\hat{\eta}_2$ splits the two Majoranas away from zero energy while satisfying particle-hole symmetry.

Finally, our definition of stability necessarily implies that Andreev bound states on a junction that appear at zero energy accidentally are necessarily unstable. Since Andreev bound states obey fermionic statistics, their anti-excitations must also appear at zero energy. This means that $\widetilde{H}_0$ in this case will have $\mathscr{N} \geq 2$, which is unstable in class D.

4.3.2 Sensitivity of Stable Modes to Perturbations

Next, we will make some quantitative remarks on the sensitivity to symmetry-preserving perturbations of zero-energy localized states. If the zero-energy eigenspace is not stable, then it is extremely sensitive to some perturbations, in particular, the ones that change its dimension. Therefore, we will only consider the zero-energy eigenspaces of 1D SPT phases that are stable according to Definition 4.7.

We will use the "maximal angle" [46] between the subspaces $\mathrm{Ker}\,(\widetilde{H}_0 + \lambda\widetilde{\mathfrak{v}})$ and $\mathrm{Ker}\,\widetilde{H}_0$, defined by

$$\sin\theta_{\max}(\widetilde{\mathfrak{v}},\lambda) = \|P_{\text{Ker}\,(\widetilde{H}_0+\lambda\widetilde{\mathfrak{v}})} - P_{\text{Ker}\,(\widetilde{H}_0)}\|, \quad \theta_{\max}(\widetilde{\mathfrak{v}},\lambda) \in [0,\pi/2],$$

as a measure of "distortion" of the zero-energy eigenspace.

Definition 4.12 The "sensitivity $\mathscr{X}(\widetilde{\mathfrak{v}})$ of ker $\widetilde{H}_0$ to a perturbation $\widetilde{\mathfrak{v}}$," where $\widetilde{\mathfrak{v}}$ is an operator of unit norm, is given by

$$\mathscr{X}(\widetilde{\mathfrak{v}}) \equiv \frac{d}{d\lambda}\theta_{\max}(\widetilde{\mathfrak{v}},\lambda)\Big|_{\lambda=0^+},$$

where "+" indicates the right derivative at $\lambda = 0$. The "sensitivity $\mathscr{X}$ of ker $\widetilde{H}_0$ to perturbations in the symmetry class of $\widetilde{H}_0$" is then defined as

$$\mathscr{X} \equiv \sup_{\|\widetilde{\mathfrak{v}}\|=1} \mathscr{X}(\widetilde{\mathfrak{v}}),$$

with $\widetilde{\mathfrak{v}}$ being in the symmetry class of $\widetilde{H}_0$.

Theorem 4.13 (Sensitivity to Perturbations) *For $\lambda < 1/2\varkappa(\widetilde{H}_0, \widetilde{H}_0^{(-1)})$, the sine of the maximal angle is upper-bounded by*

$$\sin\theta_{\max}(\widetilde{\mathfrak{v}},\lambda) \leq \frac{\lambda\varkappa(\widetilde{H}_0,\widetilde{H}_0^{(-1)})}{\sqrt{1-2\lambda\varkappa(\widetilde{H}_0,\widetilde{H}_0^{(-1)})}}. \tag{4.18}$$

The sensitivity to symmetry-preserving perturbations is upper-bounded by

$$\mathscr{X} \leq \varkappa(\widetilde{H}_0,\widetilde{H}_0^{(-1)}).$$

The proof relies on simple algebraic inequalities, and will be presented in Ref. [39].

Finally we turn our attention to computing a bound on the generalized condition number $\varkappa$. Computing the condition number for clean systems with open BCs is made possible by the Wiener–Hopf factorization, thanks to the following proposition:

Proposition 4.14 *Let $\widetilde{A} = \widetilde{A}_+\widetilde{D}\widetilde{A}_-$ be the Wiener–Hopf factorization of the block-Toeplitz operator $\widetilde{A}$. Its pseudo-inverse is given by $\widetilde{A}^{(-1)} = \widetilde{A}_-^{-1}\widetilde{D}^\dagger\widetilde{A}_+^{-1}$.*

The statement follows from the fact that $\widetilde{A}_\pm$ are invertible, and S, $S^\dagger$ are pseudo-inverses of each other.

One can now bound the generalized condition number of $\widetilde{A}$ by

$$\varkappa(\widetilde{A},\widetilde{A}^{(-1)}) \leq \varkappa(\widetilde{A}_+)\varkappa(\widetilde{A}_-).$$

A similar analysis for the spectral flattening (symmetric factorization) of a Hamiltonian subject to open BCs leads to the inequality $\varkappa(\widetilde{H},\widetilde{H}^{(-1)}) \leq \varkappa(\widetilde{H}_+)^2$. The quantity on the right-hand side is easily computable, since [47]

$$\|\widetilde{H}_+\| = \max_{k\in[0,2\pi)} (\|H_+(e^{ik})\|),$$

$$\|\widetilde{H}_+^{-1}\| = \max_{k\in[0,2\pi)} (\|H_+^{-1}(e^{ik})\|).$$

We can use this bound on the generalized condition number to derive a new bound on the maximal angle,

$$\sin\theta_{\max}(\widetilde{\boldsymbol{v}}, \lambda) \le \frac{\lambda\varkappa(\widetilde{H}_+)^2}{\sqrt{1-2\lambda\varkappa(\widetilde{H}_+)^2}} \tag{4.19}$$

which follows from Eq. (4.18). Since the Wiener–Hopf factorization of $\widetilde{H}$ is not unique, two different factorizations can lead to two different values of the bound on $\varkappa(\widetilde{H})$.

Example 4.15 Let us compute a bound on the stability of the zero mode of the SSH Hamiltonian under BCs for $t_1 < t_2$ using the Wiener–Hopf factorization in Sect. 4.2.1. Since

$$H_+ = \begin{bmatrix} 1 & 0 \\ 0 & -(t_1 z + t_2) \end{bmatrix},$$

we can easily deduce that

$$\|\widetilde{H}_+\| = \|H_+(e^{ik})\|\,|_{k=0} = t_2 + t_1, \quad \|\widetilde{H}_+^{-1}\| = \|H_+^{-1}(e^{ik})\|\,|_{k=\pi} = 1/(t_2 - t_1),$$

so that (see also Fig. 4.3)

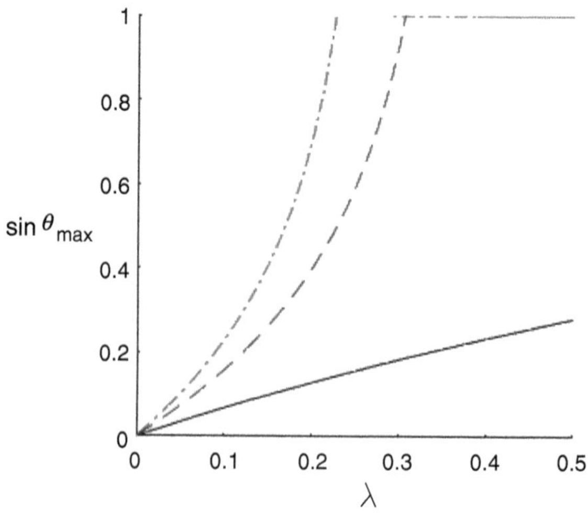

Fig. 4.3 The sine of the maximal angle (solid black line), the upper bound in Eq. (4.18) using full diagonalization (dashed blue line) and in Eq. (4.19) using the Wiener–Hopf factorization (dotted red line) as a function of the perturbation strength λ for an arbitrarily chosen $\widetilde{\boldsymbol{v}}$. The SSH Hamiltonian parameters take values $t_1 = 0.15$, $t_2 = 1$

$$\mathscr{X} \leq \varkappa(\widetilde{H}, \widetilde{H}^{(-1)}) \leq \left(\frac{\mathsf{t}_1 + \mathsf{t}_2}{\mathsf{t}_1 - \mathsf{t}_2}\right)^2.$$

The bound approaches its best possible value 1 as $\mathsf{t}_1 \to 0$. Near the phase boundary $\mathsf{t}_1 = \mathsf{t}_2$, the value of this bound diverges, indicating instability of the zero mode.

4.4 Beyond the Tenfold Way: Case Study of an s-Wave Topological Superconductor

The tenfold classification scheme of topological phases focuses on systems in which the boundary states are protected only by "extremely generic" symmetries. For instance, Kitaev's p-wave superconductor enjoys *only* the particle-hole symmetry for generic (complex) parameter values; hence, the latter is the only symmetry that provides protection to the Majorana zero modes. However, especially from a quantum engineering perspective, it may be important to investigate cases where multiple sets of symmetries provide protection to the same boundary states. From the point of view of the tenfold classification, such a system will be designated a symmetry class based on the most *generic* symmetries that protect the boundary states, with a generic symmetry being one that is unlikely to be broken by the disorder in the system. Here, we investigate the same s-wave topological superconductor model we considered in Sect. 3.1.6. The Majorana modes in this model are protected by multiple sets of symmetries. We check the stability of the Majorana modes against various perturbations, and explain the role of various topological invariants in this context.

4.4.1 Model Hamiltonian and Perturbations Under Consideration

The BdG Hamiltonian of a semi-infinite wire of the s-wave topological superconductor described in Sect. 3.1.6 can be expressed as $\widetilde{H} = h_1^\dagger S^\dagger + h_0 I + h_1 S$ in terms of the matrices h_0 and h_1 describing hopping and pairing, where

$$h_0 = -\mu\tau_z + \mathsf{u}\tau_z\nu_x + \Delta\tau_y\nu_z\sigma_y, \quad h_1 = -t\tau_z\nu_x + i\lambda\nu_z\sigma_x.$$

Recall that the real parameters $\mu, \mathsf{u}, \mathsf{t}, \lambda, \Delta$ denote the chemical potential, the interband hybridization, hopping, spin–orbit coupling, and pairing potential strengths, respectively, and $\tau_\alpha, \nu_\alpha, \sigma_\alpha, \alpha = \{x, y, z\}$, are Pauli matrices in Nambu, orbital, and spin spaces. We consider three types of perturbation based on their support in lattice space:

1. Boundary disorder with support only on the first few sites;
2. Bulk disorder with support on a few bulk sites; and
3. Uniform field throughout the length of the chain.

The third type of perturbation can be modeled as a change in hopping matrices h_0, h_1.

This model is time-reversal invariant and therefore, according to the tenfold classification, it belongs to the symmetry class DIII. The characteristic of the topologically non-trivial phase of class DIII is the presence of an odd number of Kramer's pairs of Majorana modes at its open end. In agreement, in the non-trivial phase of this model, we find one pair of Majorana modes at the open end. However, the Majorana modes are protected by more symmetries, rather than just the time-reversal and particle-hole symmetries as predicted by the tenfold classification. To perform the stability analysis with respect to various types of perturbations, we pick parameters μ, u, t, λ, Δ generically from the non-trivial topological phase. In Table 4.4, we have listed the stability of zero modes against product perturbations of the form $\widetilde{\boldsymbol{v}} = \widetilde{\boldsymbol{v}}_I \widetilde{\boldsymbol{v}}_L$, where $\widetilde{\boldsymbol{v}}_I (\widetilde{\boldsymbol{v}}_L)$ act non-trivially on the internal (lattice) space only.

4.4.2 *Analysis of the Stability Results*

From the results described in Table 4.4 and similar simulations, we conclude that the zero modes are stable against the perturbation provided it obeys *at least one* of the four symmetries considered in the table. The table also suggests that if none of these four symmetries is obeyed, then the perturbation will split the zero modes away from zero energy. However, in order to conclusively establish this numerically, one would need to check results of a much larger class of simulations.

We conclude this section by discussing the topological invariants that are responsible for the stability of zero modes in the presence of one of the four symmetries.

- In the presence of time-reversal symmetry ($\mathcal{T}_- = i\sigma_y \mathrm{C}$), the topological invariant of class DIII (Eq. (4.7)) takes the non-trivial value $Q^B = -1$, which leads to stability of a Kramer's pair of Majorana modes on the edge.
- In the presence of chiral symmetry $\mathcal{C}_3 = \tau_y \sigma_z$, the topological invariant of class AIII (Eq. (4.5)) yields non-trivial value $Q^B = 2$, which leads to stability of two Majorana modes (not Kramer's pair in general) on the edge.
- In the presence of unitary symmetries $\mathcal{S}_1 = \nu_x \sigma_z$ or $\mathcal{S}_2 = \tau_z \nu_x \sigma_y$, the Hamiltonian can be block-diagonalized in two 4×4 blocks in the internal space. Because $\mathcal{S}_1$ and $\mathcal{S}_2$ commute with the particle-hole symmetry ($\mathcal{P}_+ = \tau_x \mathrm{C}$), the resulting blocks are invariant under particle-hole symmetry, and hence belong to class D. The topological invariant (Eq. (4.6)) for each of the blocks is $Q^B = -1$, because of which one Majorana mode is stable in each of the blocks.

Table 4.4 Stability against perturbations of the Majorana modes of the s-wave topological superconductor. The symmetries of the model in the top column are $\mathcal{T}_- = \sigma_y\kappa$, $\mathcal{C}_3 = \tau_y\sigma_z$, $\mathcal{S}_1 = \nu_x\sigma_z$, $\mathcal{S}_2 = \tau_z\nu_x\sigma_y$. ✓and × indicate whether the perturbations under consideration (left column) satisfy or break, respectively, the symmetry (top row)

No.	Internal factor ($\widetilde{\boldsymbol{v}}_I$)	$\mathcal{T}_-$	$\mathcal{C}_3$	$\mathcal{S}_1$	$\mathcal{S}_2$	Stability
1	$\nu_y\sigma_x$, $i\nu_z\sigma_x$, $i\tau_x\sigma_z$, $i\tau_x\nu_x\sigma_z$, $i\tau_y\nu_y\sigma_y$, $\tau_y\nu_z\sigma_y$, τ_z, $\tau_z\nu_x$	✓	✓	✓	✓	Stable
2	$i\sigma_x$, $i\nu_x\sigma_x$, $\tau_x\nu_y\sigma_z$, $i\tau_x\nu_z\sigma_z$, $\tau_y\sigma_y$, $\tau_y\nu_x\sigma_y$, $i\tau_z\nu_y$, $\tau_z\nu_z$	✓	✓	×	×	Stable
3	$i\sigma_z$, $i\nu_x\sigma_z$, $\tau_x\nu_y\sigma_x$, $i\tau_x\nu_z\sigma_x$, $i\tau_y$, $i\tau_y\nu_x$, $\tau_z\nu_y\sigma_y$, $i\tau_z\nu_z\sigma_y$	✓	×	✓	×	Stable
4	$\nu_y\sigma_z$, $i\nu_z\sigma_z$, $i\tau_x\sigma_x$, $i\tau_x\nu_x\sigma_x$, $\tau_y\nu_y$, $i\tau_y\nu_z$, $i\tau_z\sigma_y$, $i\tau_z\nu_x\sigma_y$	✓	×	×	✓	Stable
5	$i\nu_y\sigma_y$, $\nu_z\sigma_y$, $i\tau_x$, $i\tau_x\nu_x$, $\tau_y\nu_y\sigma_x$, $i\tau_y\nu_z\sigma_x$, $\tau_z\sigma_z$, $\tau_z\nu_x\sigma_z$	×	✓	✓	×	Stable
6	σ_y, $\nu_x\sigma_y$, $\tau_x\nu_y$, $i\tau_x\nu_z$, $i\tau_y\sigma_x$, $i\tau_y\nu_x\sigma_x$, $i\tau_z\nu_y\sigma_z$, $\tau_z\nu_z\sigma_z$	×	✓	×	✓	Stable
7	$i\mathbb{1}_8$, $i\nu_x$, $i\tau_x\nu_y\sigma_y$, $\tau_x\nu_z\sigma_y$, $i\tau_y\sigma_z$, $i\tau_y\nu_x\sigma_z$, $i\tau_z\nu_y\sigma_x$, $\tau_z\nu_z\sigma_x$	×	×	✓	✓	Stable
8	ν_y, $i\nu_z$, $\tau_x\sigma_y$, $\tau_x\nu_x\sigma_y$, $\tau_y\nu_y\sigma_z$, $i\tau_y\nu_z\sigma_z$, $\tau_z\sigma_x$, $\tau_z\nu_x\sigma_x$	×	×	×	×	Unstable

In summary, at least three different topological invariants are applicable to this model, in such a way that each of them guarantees stability of Majorana modes against a specific set of perturbations.

4.5 Towards a Bulk-Boundary Correspondence for Interfaces

So far, we have discussed the bulk-boundary correspondence in 1D systems with one open end. We conclude this chapter by showing how our analysis may be extended to interfaces formed by two bulks belonging to the *same* symmetry class.

We consider the simplest case of two 1D systems, both belonging to a given symmetry class, forming a bridge (see Fig. 4.4). Both systems are assumed to extend to infinity in the direction away from the 0D interface. Let us label the fermionic degrees in the first system by $\mathcal{H}_1 = \text{Span}\,\{|j\rangle|m_1\rangle,\ j = 0, \ldots, \infty,\ m_1 = 1, \ldots, d_1\}$, and the second system by $\mathcal{H}_2 = \text{Span}\,\{|j\rangle|m_2\rangle,\ j = 0, \ldots, \infty,\ m_2 = 1, \ldots, d_2\}$, so that the degrees labeled by $j = 0$ in the two systems are adjacent. Let $S_\text{i} : \mathcal{H}_\text{i} \mapsto \mathcal{H}_\text{i},\ \text{i} = 1, 2$ denote the two shift operators with action

$$S_\text{i}|j\rangle|m_\text{i}\rangle = \begin{cases} |j-1\rangle|m_\text{i}\rangle & \text{if } j > 0 \\ 0 & \text{if } j = 0 \end{cases}.$$

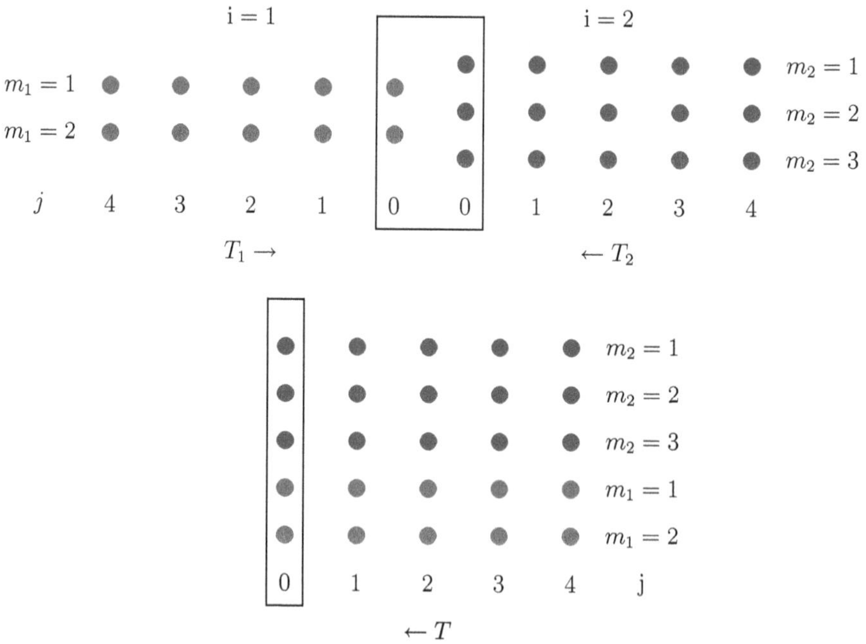

Fig. 4.4 The filled and empty circles stand for single-particle Hilbert space basis for two different 1D systems in the same symmetry class. The top panel shows the sketch of an interface between them. Visualizing this system in a different way as shown in the lower panel allows us to describe the system using a single block-Toeplitz operator

We define S to be the new shift operator on $\mathcal{H} = \mathcal{H}_1 \oplus \mathcal{H}_2$ with action $S|j\rangle|m_i\rangle = S_i|j\rangle|m_b\rangle$ on the above-specified basis. In this representation, the system in the bridge configuration can be described by one block-Toeplitz operator, with reduced bulk Hamiltonian

$$H(z) = \begin{bmatrix} H_1(z) & 0 \\ 0 & H_2(z) \end{bmatrix},$$

where $H_1(z)$ and $H_2(z)$ are reduced bulk Hamiltonians of systems 1 and 2. The bulk invariant of this system can be then computed from $H(z)$ in the same way as the bulk invariants of $H_1(z)$ and $H_2(z)$ are computed. Similarly, the boundary invariant of $\widetilde{H} + \widetilde{W}$ follows from the appropriate expression for the symmetry class. Here, $\widetilde{W}$ denotes the finite-range disorder at the interface. Note that even though the two systems forming the interface are assumed to be in the same symmetry class, we have not made any assumption on the operators of the two systems. In fact, the two systems are allowed to have different number of energy bands in general. However, for our results to hold, the operator $\widetilde{W}$ must satisfy the symmetry $\mathcal{S}$ defined by

$$\mathcal{S}\mathcal{H}_{\rm i} = \mathcal{S}_{\rm i}\mathcal{H}_{\rm i}, \quad {\rm i} = 1, 2, \tag{4.20}$$

if the individual bulks satisfy the symmetries $\{\mathcal{S}_{\rm i}\}$ of same kind.

We first provide a definition of bulk and boundary topological invariants for interfaces in all symmetry classes, which has not been made explicit in the literature to the best of our knowledge. In particular, we associate the bulk invariant

$$Q_{\rm I}^B = \sum_{\rm i=1}^{\rm n} Q_{\rm i}^B \quad \text{(AIII, BDI, CII)}, \tag{4.21}$$

$$Q_{\rm I}^B = \prod_{\rm i=1}^{\rm n} Q_{\rm i}^B \quad \text{(D, DIII)}, \tag{4.22}$$

to the junction where n bulk wires meet. Here, $Q_{\rm i}^B$ is the bulk invariant of the i-th wire. The boundary invariant for the junction can be defined similar to individual wires, namely

$$Q_{\rm I}^\partial = \begin{cases} \mathscr{N}_+ - \mathscr{N}_- & \text{AIII, BDI, CII} \\ (-1)^{\mathscr{N}} & \text{D} \\ (-1)^{\mathscr{N}/2} & \text{DIII} \end{cases}. \tag{4.23}$$

With these definitions, the bulk-boundary correspondence for interfaces follows as a corollary of Theorem 4.6. Note that Eq. (4.21) implies that the invariants of individual systems combine according to the group operation (addition/multiplication) of the group of homotopy invariant ($\mathbb{Z}/\mathbb{Z}_2$).

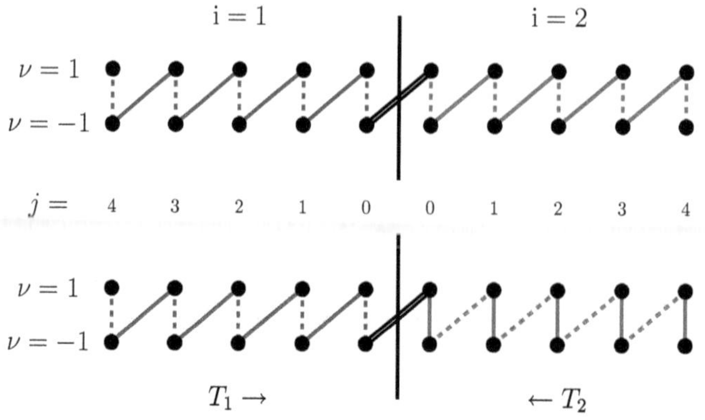

Fig. 4.5 Schematic of interfaces formed by two wires individually described by SSH Hamiltonian. On both sides of the interface (solid black lines), the dotted lines indicate weaker hopping strength compared to the solid lines. The double lines indicate hopping between the two wires. Notice the exchange of strong and weak hopping strengths for $b = 2$ wire in the interface shown in the bottom figure, compared to the one shown in the top. The shift operator S_1 implements right shift on the wire $b = 1$, whereas S_2 implements left shift on the wire $b = 2$

The same arguments as described above allow us to establish bulk-boundary correspondence in a 3-way bridge, where one end of three 1D systems in the same symmetry class is connected together. Further, if the bulk-boundary correspondence holds for a single bulk in higher dimension, the same approach can be used to extend it to interfaces formed by bulks in the same symmetry class. Consider a bridge formed by two 2D systems forming a 1D interface. In this case, we label the single-particle bases of the two systems by $\{|\mathbf{j}_\parallel\rangle|j\rangle|m_b\rangle,\ \mathbf{j}_\parallel \in \mathbb{Z},\ j = 1, \ldots, \infty,\ m_\mathrm{i} = 1, \ldots, d_\mathrm{i}\}$, $\mathrm{i} = 1, 2$, and proceed as above.

Example 4.16 Let us look at the bulk-boundary correspondence of two wires, each described by a SSH Hamiltonian and connected together at one end. In the interfaces shown in Fig. 4.5, assuming $\{\mathrm{t}_{1,\mathrm{i}}, \mathrm{t}_{2,\mathrm{i}},\ \mathrm{i} = 1, 2\}$ to be the strengths of intra-cell and inter-cell hoppings, the reduced bulk Hamiltonians of the individual wires are

$$H_\mathrm{i}(z) = -\begin{bmatrix} 0 & \mathrm{t}_{1,\mathrm{i}} + \mathrm{t}_{2,\mathrm{i}}z \\ \mathrm{t}_{1,\mathrm{i}} + \mathrm{t}_{2,\mathrm{i}}z^{-1} & 0 \end{bmatrix}, \quad \mathrm{i} = 1, 2.$$

Notice the interchange between z and z^{-1}, which is a result of the convention that S_1 implements a right shift on $\mathrm{i} = 1$, whereas S_2 implements a left shift on $\mathrm{i} = 2$. The hopping between the two wires is such that it respects the global chiral symmetry $\nu_{z,1} \oplus \nu_{z,2}$. For the interface shown in the top panel, we have $\mathrm{t}_{0,\mathrm{i}} < \mathrm{t}_{1,\mathrm{i}}$ for $\mathrm{i} = 1, 2$, which leads to $Q_1^B = -1$ and $Q_2^B = 1$. Therefore, we have $Q_\mathrm{I}^B = Q_1^B + Q_2^B = 0$ as expected, and boundary invariant $Q_\mathrm{I}^\partial = 0$. For the interface shown in the lower panel, we have $\mathrm{t}_{1,1} < \mathrm{t}_{2,1}$ and $\mathrm{t}_{1,2} > \mathrm{t}_{2,2}$. This leads to $Q_\mathrm{I}^B = Q_1^B + Q_2^B = -1+0 = -1$, which must equal the boundary invariant as well. In this case, the interface must host at least one edge state of chirality -1.

References

1. A. Kitaev, Periodic table for topological insulators and superconductors, in *AIP Conference Proceedings*, vol. 1134, no. 1 (AIP, 2009), pp. 22–30. https://doi.org/10.1063/1.3149495
2. M. König, S. Wiedmann, C. Brüne, A. Roth, H. Buhmann, L.W. Molenkamp, X.-L. Qi, S.-C. Zhang, Quantum spin hall insulator state in HgTe quantum wells. Science **318**, 766–770 (2007). https://science.sciencemag.org/content/318/5851/766
3. C.L. Kane, E.J. Mele, Quantum spin hall effect in graphene. Phys. Rev. Lett. **95**, 226801 (2005). https://link.aps.org/doi/10.1103/PhysRevLett.95.226801
4. B. Wu, J. Song, J. Zhou, H. Jiang, Disorder effects in topological states: brief review of the recent developments. Chin. Phys. B **25**, 117311 (2016). https://doi.org/10.1088/1674-1056/25/11/117311
5. E. Prodan, H. Schulz-Baldes, *Bulk and Boundary Invariants for Complex Topological Insulators: From k-theory to Physics*, 1st edn., vol. 117 (Springer International Publishing AG, Cham, 2016)
6. J. Zak, Berry's phase for energy bands in solids. Phys. Rev. Lett. **62**, 2747 (1989). https://doi.org/10.1103/PhysRevLett.62.2747
7. G. Ortiz, R.M. Martin, Macroscopic polarization as a geometric quantum phase: many-body formulation. Phys. Rev. B **49**, 14202 (1994). https://doi.org/10.1103/physrevb.49.14202
8. G. Ortiz, P. Ordejón, R.M. Martin, G. Chiappe, Quantum phase transitions involving a change in polarization. Phys. Rev. B **54**, 13515 (1996). https://doi.org/10.1103/PhysRevB.54.13515
9. D.J. Thouless, M. Kohmoto, M.P. Nightingale, M. den Nijs, Quantized hall conductance in a two-dimensional periodic potential. Phys. Rev. Lett. **49**, 405–408 (1982). https://link.aps.org/doi/10.1103/PhysRevLett.49.405
10. F.D.M. Haldane, Model for a quantum hall effect without landau levels: condensed-matter realization of the "parity anomaly". Phys. Rev. Lett. **61**, 2015–2018 (1988). https://link.aps.org/doi/10.1103/PhysRevLett.61.2015
11. R.S.K. Mong, V. Shivamoggi, Edge states and the bulk-boundary correspondence in Dirac Hamiltonians. Phys. Rev. B **83**, 125109 (2011). https://link.aps.org/doi/10.1103/PhysRevB.83.125109
12. G.M. Graf, M. Porta, Bulk-edge correspondence for two-dimensional topological insulators. Commun. Math. Phys. **324**, 851–895 (2013). https://doi.org/10.1007/s00023-018-0657-7
13. V. Mathai, G.C. Thiang, T-duality of topological insulators. J. Phys. A Math. Theor. **48**, 42FT02 (2015). https://doi.org/10.10882F1751-8113/48/42/42ft02
14. K.C. Hannabuss, T-duality and the bulk-boundary correspondence. J. Geom. Phys. **124**, 421–435 (2018). http://www.sciencedirect.com/science/article/pii/S0393044017302966
15. J.C. Avila, H. Schulz-Baldes, C. Villegas-Blas, Topological invariants of edge states for periodic two-dimensional models. Math. Phys. Anal. Geom. **16**, 137–170 (2013). https://doi.org/10.1007/s11040-012-9123-9
16. C. Bourne, J. Kellendonk, A. Rennie, The k-theoretic bulk–edge correspondence for topological insulators. Ann. Henri Poincaré **18**, 1833–1866 (2017). https://doi.org/10.1007/s00023-016-0541-2
17. A. Altland, M.R. Zirnbauer, Nonstandard symmetry classes in mesoscopic normal-superconducting hybrid structures. Phys. Rev. B **55**, 1142–1161 (1997). https://link.aps.org/doi/10.1103/PhysRevB.55.1142
18. S. Ryu, A.P. Schnyder, A. Furusaki, A.W.W. Ludwig, Topological insulators and superconductors: tenfold way and dimensional hierarchy. New J. Phys. **12**, 065010 (2010). https://doi.org/10.1088/1367-2630/12/6/065010
19. E.P. Wigner, Normal form of antiunitary operators, in *The Collected Works of Eugene Paul Wigner* (Springer, Berlin, 1993), pp. 551–555. https://doi.org/10.1063/1.1703672
20. E. Lieb, T. Schultz, D. Mattis, Two soluble models of an antiferromagnetic chain. Ann. Phys. **16**, 407–466 (1961). https://doi.org/10.1016/0003-4916(61)90115-4

21. A.Y. Kitaev, Unpaired Majorana fermions in quantum wires. Phys.-Uspekhi **44**, 131–136 (2001). https://doi.org/10.1070/1063-7869/44/10s/s29
22. C.-K. Chiu, J.C. Teo, A.P. Schnyder, S. Ryu, Classification of topological quantum matter with symmetries. Rev. Mod. Phys. **88**, 035005 (2016). https://doi.org/10.1103/RevModPhys.88.035005
23. D. Serre, *Matrices: Theory and Applications*, 2nd edn. (Springer, New York, 2010)
24. X.-L. Qi, T.L. Hughes, S.-C. Zhang, Topological invariants for the Fermi surface of a time-reversal-invariant superconductor. Phys. Rev. B **81**, 134508 (2010). https://doi.org/10.1103/PhysRevB.81.134508
25. I. Gohberg, I.A. Fel_dman, *Convolution Equations and Projection Methods for their Solution*, vol. 41 (American Mathematical Society, Providence, 2005)
26. I. Gohberg, M.A. Kaashoek, I.M. Spitkovsky, An overview of matrix factorization theory and operator applications, in *Factorization and Integrable Systems* (Springer, Basel, 2003), pp. 1–102
27. M.F. Atiyah, I.M. Singer, The index of elliptic operators: I. Ann. Math. **87**, 484 (1968). https://www.jstor.org/stable/1970715
28. H. Schulz-Baldes, Z2-indices and factorization properties of odd symmetric Fredholm operators. Doc. Math. **20**, 1481–1500 (2015)
29. A.C. Ran, L. Rodman, Factorization of matrix polynomials with symmetries. SIAM J. Matrix Anal. Appl. **15**, 845–864 (1994). https://doi.org/10.1137/S0895479892235502
30. L. Rodman, I.M. Spitkovsky, Factorization of matrices with symmetries over function algebras. Integr. Equ. Oper. Theory **80**, 469–510 (2014). https://doi.org/10.1007/s00020-014-2155-8
31. Y. Shelah, Quaternionic Wiener algebras, factorization and applications. Adv. Appl. Clifford Algebras **27**, 2805–2840 (2017). https://doi.org/10.1007/s00006-016-0750-2
32. P. Lancaster, L. Rodman, Minimal symmetric factorizations of symmetric real and complex rational matrix functions. Linear Algebra Appl. **220**, 249–282 (1995). https://doi.org/10.1016/0024-3795(94)00151-3
33. A.F. Voronin, A method for determining the partial indices of symmetric matrix functions. Sib. Math. J. **52**, 41–53 (2011). https://doi.org/10.1134/S0037446606010058
34. T.-Y. Guo, B.-W. Lin, C. Hwan, A new method for factoring matrix polynomials relative to the unit circle. J. Chin. Inst. Eng. **21**, 87–92 (1998). https://doi.org/10.1109/ACC.1997.609013
35. V.G. Kravchenko, A.B. Lebre, J.S. Rodriguez, Matrix functions consimilar to the identity and singular integral operators. Complex Anal. Oper. Theory **2**, 593–615 (2008). https://doi.org/10.1007/s11785-008-0068-8
36. T. Ehrhardt, Invertibility theory for Toeplitz plus Hankel operators and singular integral operators with flip. J. Funct. Anal. **208**, 64–106 (2004). https://doi.org/10.1016/S0022-1236(03)00113-7
37. O. Iftime, H. Zwart, J-spectral factorization and equalizing vectors. Syst. Control Lett. **43**, 321–327 (2001). https://doi.org/10.1016/0167-6911(94)00077-9
38. D. Youla, N. Kazanjian, Bauer-type factorization of positive matrices and the theory of matrix polynomials orthogonal on the unit circle. IEEE Trans. Circuits Syst. **25**, 57–69 (1978). https://doi.org/10.1109/TCS.1978.1084443
39. A. Alase, E. Cobanera, G. Ortiz, L. Viola, Matrix factorization approach to bulk-boundary correspondence and stability of zero modes (in preparation)
40. R. Bhatia, *Matrix Analysis* (Springer Science & Business Media, New York, 2013)
41. T.A. Loring, Factorization of matrices of quaternions. Exp. Math. **30**, 250–267 (2012). https://doi.org/10.1016/j.exmath.2012.08.006
42. L. Feng, Decompositions of some types of quaternionic matrices. Linear Multilinear Algebra **58**, 431–444 (2010). https://doi.org/10.1080/03081080802632735
43. I. Gohberg, N. Krupnik, *One-dimensional Linear Singular Integral Equations: I. Introduction*, vol. 53 (Birkhäuser, Basel, 2012)

44. L. Isaev, Y. Moon, G. Ortiz, Bulk-boundary correspondence in three-dimensional topological insulators. Phys. Rev. B **84**, 075444 (2011). https://doi.org/10.1103/PhysRevB.84.075444
45. G. Chen, Y. Wei, Y. Xue, The generalized condition numbers of bounded linear operators in Banach spaces. J. Aust. Math. Soc. **76**, 281–290 (2004)
46. C.D. Meyer, *Matrix Analysis and Applied Linear Algebra*, vol. 71 (SIAM, Philadelphia, 2000)
47. P.R. Halmos, *A Hilbert Space Problem Book*, vol. 19 (Springer Science & Business Media, New York, 2012)

Chapter 5
Mathematical Foundations to the Generalized Bloch Theorem

In this chapter, we provide a detailed proof of the mathematical results that were used in deriving the generalized Bloch theorem in Chap. 2. Following a review of the required linear-algebraic concepts and tools in Sect. 5.1, our core results are presented in Sect. 5.2. In particular, we show how our main task is equivalent to solving an appropriate linear system, consisting of a *bulk and boundary equations*, and prove how the abovementioned auxiliary tasks are achieved. Section 5.3 focuses on developing a general eigenvalue-dependent ansatz that our approach provides for the eigenvectors of a corner-modified BBT matrix, as well as *infinite* corner-modified BBT operators on Hilbert space, as relevant to semi-infinite systems.

5.1 Preliminaries

Let ϵ denote an eigenvalue of M. Recall that $|\psi\rangle \in \mathcal{V}$ is a "generalized eigenvector" of M, of rank $\upsilon \in \mathbb{N}$, if

$$|\psi\rangle \in \text{Ker}\,(M-\epsilon)^{\upsilon}, \quad |\psi\rangle \notin \text{Ker}\,(M-\epsilon)^{\upsilon-1}.$$

A generalized eigenvector of rank $\upsilon = 1$ is an eigenvector in the usual sense. Since $\text{Ker}\,(M-\epsilon)^{\upsilon} \subseteq \text{Ker}\,(M-\epsilon)^{\upsilon'}$, $\forall \upsilon' \geq \upsilon$, any generalized eigenvector $|\psi\rangle$ of rank υ also belongs to $\text{Ker}\,(M-\epsilon)^{\upsilon'}$. Hence, one may define the "generalized eigenspace" of M corresponding to its eigenvalue ϵ as

$$\mathcal{N}_{M,\epsilon} = \bigcup_{\upsilon \in \mathbb{N}} \text{Ker}\,(M-\epsilon)^{\upsilon}.$$

If $\mathcal{V}$ is a finite-dimensional complex vector space, then for every eigenvalue ϵ of M, there exists a $\upsilon_{\max} \in \mathbb{N}$, such that $\text{Ker}\,(M-\epsilon)^{\upsilon_{\max}} = \text{Ker}\,(M-\epsilon)^{\upsilon_{\max}+1} = \mathcal{N}_{M,\epsilon}$.

A. Alase, *Boundary Physics and Bulk-Boundary Correspondence in Topological Phases of Matter*, Springer Theses, https://doi.org/10.1007/978-3-030-31960-1_5

Since $\mathcal{V} = \bigoplus_\epsilon \mathcal{N}_{M,\epsilon}$ [1], where the direct sum runs over all the eigenvalues of M, it follows that a maximal linearly independent subset of the set of all generalized eigenvectors of M yields a basis of generalized eigenvectors of $\mathcal{V}$. The "Jordan basis" is one example of a basis of generalized eigenvectors.

5.1.1 *Corner-Modified Banded Block-Toeplitz Matrices*

A matrix $A_N \in \mathbb{C}_{dN}$ of size $dN \times dN$ is a "block-Toeplitz matrix" if there exists a sequence $\{a_j \in \mathbb{C}_d\}_{j=-N+1}^{N-1}$ of $d \times d$ matrices such that $[A_N]_{ij} \equiv a_{j-i}$, $1 \le i, j \le N$ as an array of $d \times d$ blocks. Graphically, A_N has the structure

$$A_N = \begin{bmatrix} a_0 & a_1 & \cdots & a_{N-2} & a_{N-1} \\ a_{-1} & a_0 & \ddots & & a_{N-2} \\ \vdots & \ddots & \ddots & \ddots & \vdots \\ a_{-N+2} & & \ddots & a_0 & a_1 \\ a_{-N+1} & a_{-N+2} & \cdots & a_{-1} & a_0 \end{bmatrix}, \quad a_j \in \mathbb{C}_d. \tag{5.1}$$

A block-Toeplitz matrix is "banded" if there exist "bandwidth parameters" $p, q \in \mathbb{Z}$, with $-N+1 < p \le q < N-1$, such that $a_p, a_q \neq 0$ and $a_r = 0$ if $r < p$ or $r > q$. Accordingly, the graphical representation of a BBT matrix is

$$A_N = \begin{bmatrix} a_0 & \dots & a_q & 0 & \cdots & 0 \\ \vdots & \ddots & & \ddots & \ddots & \vdots \\ a_p & & \ddots & & \ddots & 0 \\ 0 & \ddots & & \ddots & & a_q \\ \vdots & \ddots & \ddots & & \ddots & \vdots \\ 0 & \cdots & 0 & a_p & \cdots & a_0 \end{bmatrix}.$$

The entries a_p, a_q are the "leading coefficients" of A_N, and the pair (p, q) determines the bandwidth $q - p + 1$. The adjoint or transpose of a BBT matrix of bandwidth (p, q) is a BBT matrix of bandwidth $(-q, -p)$. Let $p' \equiv \min(p, 0)$, $q' \equiv \max(0, q)$. The "principal coefficients" $a_{p'}, a_{q'}$ of A_N are defined by

$$a_{p'} \equiv \begin{cases} a_p & \text{if } p \le 0 \\ 0 & \text{if } p > 0 \end{cases}, \quad a_{q'} \equiv \begin{cases} a_q & \text{if } q \ge 0 \\ 0 & \text{if } q < 0 \end{cases}.$$

Leading and principal coefficients differ only if A_N is strictly upper or lower triangular.

Block matrices of size $dN \times dN$ induce linear transformations of $\mathcal{V} = \mathbb{C}^N \otimes \mathbb{C}^d$. Let $\{|j\rangle,\ j = 1, 2, \ldots, N\}$ and $\{|m\rangle, m = 1, 2, \ldots, d\}$ denote the canonical basis of $\mathbb{C}^N$ and $\mathbb{C}^d$, respectively. Then,

$$A_N|\psi\rangle = \sum_{i,j=1}^{N} |i\rangle[A_N]_{ij}|\psi_j\rangle.$$

For fixed N, d and bandwidth parameters p, q, the following projectors will play a key role in our discussion.

Definition 5.1 The "right bulk projector" and "right boundary projector" are given by

$$P_B|j\rangle|m\rangle \equiv \begin{cases} |j\rangle|m\rangle & 1 - p' \leq j \leq N - q',\ 1 \leq m \leq d \\ 0 & \text{otherwise} \end{cases}, \qquad P_\partial \equiv I - P_B. \tag{5.2}$$

The "left bulk projector" and "left boundary projector" are similarly defined by $Q_B \equiv P_B^{(-q,-p)}$, $Q_\partial \equiv I - Q_B$.

Since $j = 1, \ldots, N$ and $p' = \min(p, 0)$, the condition $1 - p' \leq j$ is trivially satisfied if $p \geq 0$. This situation corresponds to having an upper-triangular BBT matrix. Similarly, $j \leq N - q'$ is trivially satisfied if $q \leq 0$, which corresponds to a lower-triangular BBT matrix. It is immediate to check that $\dim \operatorname{Range} P_\partial = d(q' - p')$, and so the bulk of a BBT matrix is non-empty, that is, $P_B \neq 0$ only if $N > q' - p'$. This makes $q' - p'$ one of the key "length scales" of the problem, and so we will introduce a special symbol for it,

$$\tau \equiv q' - p' \geq q - p.$$

The condition for a non-empty bulk, $N > \tau$, is similar-looking to the relationship $2(N - 1) > p - q$ that we found as part of the definition of BBT matrix. It is quite possible for a BBT matrix to have an empty bulk, especially for small N. Such matrices are outside the scope of our methods. Similarly, one may think of the generic block Toeplitz matrix of Eq. (5.1) as having $P_B = I$, which also invalidates our methods.

Definition 5.2 A "corner modification" for bandwidth parameters (p, q) is any block matrix W such that $P_B W = 0$. A corner modification is *symmetrical* if, in addition, $W Q_B = 0$. A "corner-modified BBT matrix" A_{tot} is any block matrix of the form $A_{\text{tot}} = A_N + W$, with A_N as a BBT matrix of size $dN \times dN$ and bandwidth (p, q).

If W is a symmetrical corner modification for bandwidth (p, q), then W^{T} and $W^\dagger$ are both symmetrical corner modifications for bandwidth $(-q, -p)$, and the other way around. Moreover, symmetrical corner modifications do indeed look like "corner modifications" graphically, whereas this is not necessarily the case for non-symmetrical ones.

5.1.2 Banded Block-Laurent Matrices

Let $\{a_j \in \mathbb{C}_d\}_{j\in Z}$ denote a doubly infinite sequence of $d \times d$ square matrices. A "block-Laurent matrix" $\boldsymbol{A}$ is a doubly infinite matrix with entries $[\boldsymbol{A}]_{ij} = a_{j-i} \in \mathbb{C}_d,\ i, j \in \mathbb{Z}$. A block-Laurent matrix $\boldsymbol{A}$ is "banded" if there exist integers p, q, with $p \leq q$, such that $a_j = 0$ if $j < p$ or $j > q$, and $a_p, a_q \neq 0$. The graphical representation of a "banded block-Laurent" (BBL) matrix is

$$\boldsymbol{A} = \begin{bmatrix} \ddots & & \ddots & \ddots & & & \\ & a_0 & \cdots & a_q & 0 & & \\ \ddots & \vdots & \ddots & & \ddots & \ddots & \\ \ddots & a_p & & \ddots & & \ddots & 0 \\ & 0 & \ddots & & \ddots & & a_q & \ddots \\ & & \ddots & \ddots & & \ddots & \vdots & \ddots \\ & & & 0 & a_p & \cdots & a_0 & \\ & & & & \ddots & \ddots & & \ddots \end{bmatrix}. \tag{5.3}$$

The bandwidth (p, q), as well as the leading and principal coefficients, of a BBL matrix is defined just as for BBT matrices.

A BBL matrix induces a linear transformation of the space

$$\mathcal{V}_d^S \equiv \left\{ \{|\psi_j\rangle\}_{j\in Z} \,\middle|\, |\psi_j\rangle \in \mathbb{C}^d,\ \forall j \right\} \tag{5.4}$$

of vector-valued, doubly infinite sequences. Let us write $\Psi \equiv \{|\psi_j\rangle\}_{j\in Z}$ for compactness and, since $\mathcal{V}_d^S$ is *not* a Hilbert space, we do not case Ψ in a ket. Then, for a BBL matrix of bandwidth (p, q),

$$\boldsymbol{A}\Psi = \{a_p|\psi_{j+p}\rangle + \cdots + a_q|\psi_{j+q}\rangle\}_{j\in Z} = \left\{ \sum_{r=p}^{q} a_r|\psi_{j+r}\rangle \right\}_{j\in Z} \in \mathcal{V}_d^S \tag{5.5}$$

is the associated BBL transformation. If one pictures Ψ as a doubly infinite block-column vector, one can think of this equation as matrix–vector multiplication. The "support" of a sequence $\{|\psi_j\rangle\}_{j\in Z}$ is finite if the sequence vanishes but for finitely many values of j. Otherwise, it is infinite.

In the special non-block case where $d = 1$, $\mathcal{V}_d^S$ becomes the space of scalar sequences $\mathcal{V}_1^S$. There is a natural identification $\mathcal{V}_d^S \simeq \bigoplus_{m=1}^{d} \mathcal{V}_1^S$ that we will use often. It is established by noticing that

$$\{|\psi_j\rangle\}_{j\in Z} = \left\{ \sum_{m=1}^{d} \psi_{jm}|m\rangle \right\}_{j\in Z} = \sum_{m=1}^{d} \{\psi_{jm}|m\rangle\}_{j\in Z} \cong \bigoplus_{m=1}^{d} \{\psi_{jm}\}_{j\in Z}, \tag{5.6}$$

with respect to the canonical basis of $\mathbb{C}^d$. Let us define a multiplication of scalar sequences by vectors in terms of the bilinear mapping

$$\mathcal{V}_1^S \times \mathbb{C}^d \ni (\Phi, |\psi\rangle) \mapsto \Phi|\psi\rangle = |\psi\rangle\Phi = \{\phi_j|\psi\rangle\}_{j\in\mathbb{Z}} \in \mathcal{V}_d^S.$$

Combining this definition with Eq. (5.6), we finally obtain the most convenient representation of vector sequences, namely

$$\{|\psi_j\rangle\}_{j\in\mathbb{Z}} = \sum_{m=1}^{d}\{\psi_{jm}\}_{j\in\mathbb{Z}}|m\rangle, \quad \boldsymbol{A}\Psi = \sum_{m',m=1}^{d}|m'\rangle\Big\{\sum_{r=p}^{q}\psi_{j+r,m}\langle m'|a_r|m\rangle\Big\}_{j\in\mathbb{Z}}.$$

Unlike BBT transformations, BBL transformations close an associative algebra with identity **1**, isomorphic to the algebra of "matrix Laurent polynomials" [2]. The algebra of complex Laurent polynomials $\mathbb{C}[w, w^{-1}]$ consists of complex polynomials in two variables, an indeterminate w and its inverse w^{-1}, with coefficients in $\mathbb{C}$. Matrix Laurent polynomials are described similarly, but with coefficients in $\mathbb{C}_d$. We will denote a concrete but arbitrary matrix Laurent polynomial as

$$A(w, w^{-1}) = \sum_{r=p}^{q} w^r a_r, \quad a_r \in \mathbb{C}_d, \quad a_p, a_q \neq 0. \tag{5.7}$$

In order to map a matrix Laurent polynomial to a BBL matrix $\boldsymbol{A}$, it is convenient to write sequences as formal power series,

$$\Psi = \{|\psi_j\rangle\}_{j\in\mathbb{Z}} = \sum_{j\in\mathbb{Z}} w^{-j}|\psi_j\rangle,$$

which, together with Eq. (5.7), yields

$$A(w, w^{-1})\sum_{j\in\mathbb{Z}} w^{-j}|\psi_j\rangle = \sum_{j\in\mathbb{Z}} w^{-j}\big(a_p|\psi_{j+p}\rangle + \cdots + a_q|\phi_{j+q}\rangle\big).$$

Comparing this equation with Eq. (5.5), it follows that one may regard the matrix Laurent polynomial $A(w, w^{-1})$ as inducing a BBL transformation via the following algebra isomorphism:

$$A(w, w^{-1}) \mapsto \boldsymbol{A} \equiv \rho_d(A(w, w^{-1})). \tag{5.8}$$

Since the algebra of matrix Laurent polynomials is generated by $w\mathbb{1}$, $w^{-1}\mathbb{1}$, and $w^0 a = a \in \mathbb{C}_d$, the algebra of BBL transformations is generated by the corresponding linear transformations of $\mathcal{V}_d^S$, which we denote by $\boldsymbol{a} \equiv \rho_d(w^0 a)$, the left shift $V \equiv \rho_d(w\mathbb{1})$, and the right shift $V^{-1} = \rho_d(w^{-1}\mathbb{1})$. Explicitly, the effect of these BBL matrices on sequences is[1]

$$\begin{aligned} V\Psi &= w\mathbb{1}\sum_{j\in Z} w^{-j}|\psi_j\rangle = \sum_{j\in Z} w^{-j}|\psi_{j+1}\rangle, \\ a\Psi &= w^0 a\sum_{j\in Z} w^{-j}|\psi_j\rangle = \sum_{j\in Z} w^{-j}a|\psi_j\rangle, \\ V^{-1}\Psi &= w^{-1}\mathbb{1}\sum_{j\in Z} w^{-j}|\psi_j\rangle = \sum_{j\in Z} w^{-j}|\psi_{j-1}\rangle. \end{aligned}$$

In general, $\rho_d\left(\sum_{r=p}^{q} a_r w^r\right) = \sum_{m=p}^{q} a_m V^m$ (notice that $[a, V] = 0 = [a, V^{-1}]$ and $\rho_d(\mathbb{1}) = \mathbf{1}$). The relationship between BBT and BBL transformations can be formalized in terms of projectors. For integers $-\infty \leq L \leq R \leq \infty$, let

$$\mathcal{V}_{L,R} \equiv \left\{\{|\psi_j\rangle\}_{j\in Z} \in \mathcal{V}_d^S \,\middle|\, |\psi_j\rangle = 0 \text{ if } j < L \text{ or } j > R\right\}.s \tag{5.9}$$

In particular, $\mathcal{V}_{-\infty,\infty} = \mathcal{V}_d^S$, as defined in Eq. (5.4). One may think of $\mathcal{V}_{L,R}$ as the range of the projector

$$\boldsymbol{P}_{L,R}\{|\psi_j\rangle\}_{j\in Z} \equiv \{|\chi_j\rangle\}_{j\in Z}, \quad |\chi_j\rangle = \begin{cases} 0 & \text{if } j < L \\ |\psi_j\rangle & \text{if } j = L, \ldots, R \\ 0 & \text{if } j > R \end{cases}.$$

The linear transformation $\boldsymbol{P}_{L,R}\boldsymbol{A}|_{\mathcal{V}_{L,R}}$ of $\mathcal{V}_{L,R}$ is induced by a block-Toeplitz matrix A_N, with $N = R - L + 1$. To see this, write $\boldsymbol{A}\Psi = \sum_{i,j\in Z} w^{-i}[\boldsymbol{A}]_{ij}|\psi_j\rangle = \sum_{i,j\in Z} w^{-i}a_{j-i}|\psi_j\rangle$. The sum over i is formal, and the sum over j exists because $a_{j-i} = 0$ if $j - i < p$ or $j - i > q$. If $\Psi \in \mathcal{V}_{L,R}$,

$$\boldsymbol{P}_{L,R}\boldsymbol{A}\Psi = \sum_{i,j=L}^{R} w^{-i}a_{j-i}|\psi_j\rangle = \sum_{i,j=1}^{N} w^{-(i+L-1)}[A_N]_{ij}|\psi_{j+L-1}\rangle,$$

with $N = R - L + 1$ and $[A_N]_{ij} = a_{j-i}$ a $N \times N$ block-Toeplitz matrix. A similar calculation shows that, if $R - L \geq \tau \equiv q' - p'$, then the linear transformation $\boldsymbol{P}_{L-p',R-q'}\boldsymbol{A}|_{\mathcal{V}_{L,R}}$ is induced by the corner-modified BBT matrix $P_B A_N$. In this sense, we will write

$$A_N = \boldsymbol{P}_{L,R}\boldsymbol{A}|_{\mathcal{V}_{L,R}}, \quad P_B A_N = \boldsymbol{P}_{L-p',R-q'}\boldsymbol{A}|_{\mathcal{V}_{L,R}}.$$

[1] By our conventions, we have implicitly agreed to use the same symbols V, V^{-1} to denote the left and right shifts of scalar ($d = 1$) and vector ($d > 1$) sequences. As a consequence, e.g., $V(\Phi|\psi\rangle) = (V\Phi)|\psi\rangle$, illustrating how V may appear in multiples places of an equation with meanings determined by its use.

The matrix A_N is banded only if $2(R-L) > q-p$, and its bulk is non-empty if $R-L \geq \tau$. In the following it is always implicitly assumed that these conditions are met.

5.1.3 Regularity and the Smith Normal Form

A Laurent matrix polynomial may be associated with a standard matrix polynomial, involving only non-negative powers. In particular, the matrix polynomial (in the variable w) of the matrix Laurent polynomial $A(w, w^{-1})$ in Eq. (5.7) is

$$G(w) \equiv w^{-p} A(w, w^{-1}) = \sum_{s=0}^{q-p} w^s a_{s+p}. \tag{5.10}$$

A matrix polynomial is called "regular" if its determinant is not the zero polynomial. Otherwise, it is "singular." For example, direct calculation shows that $G(w)$ in

$$A(w, w^{-1}) = w^{-1} G(w) \equiv \begin{bmatrix} w + w^{-1} - \epsilon & w - w^{-1} \\ -w + w^{-1} & -w - w^{-1} - \epsilon \end{bmatrix}$$

is regular unless the parameter $\epsilon = \pm 2$. By extension, a matrix Laurent polynomial $A(w, w^{-1})$ and associated BBL matrix $\rho(A(w, w^{-1}))$ are regular or singular according to whether the polynomial factor of $A(w, w^{-1})$ is regular or singular. Finally, a BBT matrix $A_N = \boldsymbol{P}_{L,R} \boldsymbol{A}|_{\mathcal{V}_{L,R}}$ is regular (singular) if $\boldsymbol{A}$ is.

A useful fact from the theory of matrix polynomials is that they can be put in "Smith normal form" by Gaussian elimination [3]. That is, there exist $d \times d$ matrix polynomials $E(w)$, $D(w)$, $F(w)$ such that the "Smith factorization" holds:

$$G(w) = E(w) D(w) F(w). \tag{5.11}$$

Here, $E(w)$, $F(w)$ are non-unique, invertible matrix polynomials, and

$$D(w) \equiv \begin{bmatrix} g_1(w) & & & & & \\ & \ddots & & & & \\ & & g_{d_0}(w) & & & \\ & & & 0 & & \\ & & & & \ddots & \\ & & & & & 0 \end{bmatrix} \tag{5.12}$$

is the unique diagonal matrix polynomial with the property that each $g_m(w)$ is monic (i.e., has unit leading coefficient) and $g_m(w)$ divides $g_{m'}(w)$ if $1 \leq m < m' \leq d_0 \leq d$. $D(w)$ is the "Smith normal form" of $G(w)$. The matrix polynomial $G(w)$ is singular if and only if $d_0 < d$, so that its Smith normal form $D(w)$ has zeroes on the main diagonal. Since, from Eq. (5.10), $G(0) = a_p$ and $[w^{q-p}G(w^{-1})]_{w=0} = a_q$, it is immediate to check that a BBL (or BBT) matrix with at least one invertible leading coefficient is necessarily regular.

The Smith factorization of $G(w)$ immediately implies the factorization $A(w, w^{-1}) = w^p E(w)D(w)F(w)$. By combining this result with the representation defined in Eq. (5.8), one obtains what one might reasonably call the Smith factorization

$$\boldsymbol{A} = \boldsymbol{V}^p \boldsymbol{E}\boldsymbol{D}\boldsymbol{F}, \quad \text{with} \quad \boldsymbol{E} = \rho_d(E(w)), \quad \boldsymbol{D} = \rho_d(D(w)), \quad \text{and} \quad \boldsymbol{F} = \rho_d(F(w)), \tag{5.13}$$

of a BBL matrix. By construction, the linear transformations $\boldsymbol{E}$, $\boldsymbol{F}$ of $\mathcal{V}_d^S$ are invertible BBL matrices. The BBL matrix $\boldsymbol{D}$ is the Smith normal form of $\boldsymbol{A}$. In array form,

$$\boldsymbol{D} = \rho_d(D(w)) = \begin{bmatrix} \boldsymbol{g}_1 & & & & \\ & \ddots & & & \\ & & \boldsymbol{g}_{d_0} & & \\ & & & 0 & \\ & & & & \ddots & \\ & & & & & 0 \end{bmatrix}, \quad \text{with} \quad \boldsymbol{g}_m = \rho_{d=1}(g_m(w)). \tag{5.14}$$

The $\boldsymbol{g}_m$ are banded, non-block Laurent matrices: they are linear transformations of the space $\mathcal{V}_1^S$ of scalar sequences.

5.2 Structural Characterization of Kernel Properties

5.2.1 *The Bulk-Boundary System of Equations*

Our algorithm for computing the kernel of a corner-modified BBT matrix builds on an indirect method for determining the kernel of a linear transformation that we term "kernel determination by projectors." The starting point is the following:

Definition 5.3 Let M be a linear transformation of $\mathcal{V}$, and P_1, $P_2 \equiv I - P_1$ non-trivial projectors, that is, neither the zero nor the identity map. The "compatibility map" of the pair (M, P_1) is the linear transformation $B \equiv P_1 M|_{\text{Ker } P_2 M}$.

The kernel condition $M|\psi\rangle = 0$ is equivalent to the system of equations

$$P_1 M|\psi\rangle = 0, \quad P_2 M|\psi\rangle = 0.$$

In view of the above definition, this means that $\text{Ker}\, M = \text{Ker}\, P_2 M \cap \text{Ker}\, P_1 M = \text{Ker}\, B$. Roughly speaking, the subspaces $\text{Ker}\, P_2 M$ and $\text{Range}\, P_1$ may be "much smaller" than $\mathcal{V}$, and so it may be advantageous to determine the kernel of M indirectly, by way of its compatibility map. One can make these ideas more precise if $\dim \mathcal{V} < \infty$.

Lemma 5.4 *Let M be a linear transformation acting on a finite-dimensional vector space $\mathcal{V}$. Then $\dim \text{Range}\, P_1 \leq \dim \text{Ker}\, P_2 M \leq \dim \text{Ker}\, M + \dim \text{Range}\, P_1$.*

Proof The dimension of kernel of $M^{\mathsf{T}} P_2 = (P_2 M)^{\mathsf{T}}$ is bounded below by the dimension of the kernel of P_2, which is precisely dim Range P_1. This establishes the first inequality, because, in finite dimensions, the dimension of the kernel of a matrix coincides with that of its transpose. For the second inequality, notice that the solutions of $P_2 M|\psi\rangle = 0$ are of two types. There are the kernel vectors of M, and then there are the vectors that are mapped by M into the kernel of P_2. Hence, the number of linearly independent solutions of $P_2 M|\psi\rangle = 0$ is bounded above by dim Ker M + dim Range P_1. □

It is instructive to notice that, if dim Ker $P_2 M >$ dim Range P_1, then Ker M is necessarily non-trivial. Since dim Range $P_1 = n_{P_1}$ determines the number of rows of the matrix of the compatibility map (the "compatibility matrix" from now on), and dim Ker $P_2 M \equiv n_{P_2}$ determines the number of columns, this condition implies that the compatibility matrix is an $n_{P_1} \times n_{P_2}$ rectangular matrix with more columns than rows. As a consequence, its kernel is necessarily non-trivial. The following general property is also worth noting, for later use. Suppose that $M' = M + W$ and $P_2 W = 0$. Then, $W = P_1 W$, and $\text{Ker}\, P_2 M' = \text{Ker}\, P_2(M + W) = \text{Ker}\, P_2 M$. It follows that

$$B' = P_1 M'|_{\text{Ker}\, P_2 M'} = P_1 M|_{\text{Ker}\, P_2 M} + W|_{\text{Ker}\, P_2 M} = B + W|_{\text{Ker}\, P_2 M}.$$

Let now $A_{\text{tot}} = A_N + W$ denote a $dN \times dN$ corner-modified BBT matrix, with associated boundary projector $P_1 \equiv P_\partial$, and bulk projector $P_2 \equiv P_B = I - P_\partial$. The task is to compute Ker A_{tot}, which coincides with the kernel of the compatibility map

$$B = P_\partial A_{\text{tot}}|_{\text{Ker}\, P_B A_{\text{tot}}} = (P_\partial A_N + W)|_{\text{Ker}\, P_B A_N}. \tag{5.15}$$

In Chap. 2 this approach was understood as solving a bulk/boundary system of equations. In order to explain this connection more fully, let us introduce an index b such that

$$\text{Range}\, P_\partial \equiv \text{Span}\,\{|b\rangle|m\rangle \mid b = 1, \ldots, -p', N - q' + 1, \ldots, N;\ m = 1, \ldots, d\}. \tag{5.16}$$

If $p' = 0$ or $q' = 0$, then the corresponding list of vectors is empty, and so $n_\partial \equiv \dim \text{Range}\, P_\partial = d\tau$. In addition, let

$$\mathcal{B} \equiv \{|\psi_s\rangle \mid s = 1, \ldots, n_B\} \tag{5.17}$$

denote a fixed but arbitrary basis of Ker $P_B A_N$, so that $n_B = \dim \text{Ker}\, P_B A_N$.

Definition 5.5 The "bulk equation" is the kernel equation $P_B A_N |\psi\rangle = 0$. The "boundary matrix" is the $n_\partial \times n_B$ block matrix $[B]_{bs} = \langle b|B|\psi_s\rangle$, of block-size $d \times 1$. The "boundary equation" is the (right) kernel equation for B.

Thus, one can set up the boundary equation only *after* solving the bulk equation. Let us show explicitly how a solution of the boundary equation determines a basis of Ker A_{tot}. With respect to the bases described in Eqs. (5.16) and (5.17), the boundary matrix is related to the compatibility map of Eq. (5.15) as $B|\psi_s\rangle = \sum_b |b\rangle [B]_{bs}$. Let $|\epsilon_k\rangle = \sum_{s=1}^{n_B} \alpha_{ks} |\psi_s\rangle$. It then follows that

$$B|\epsilon_k\rangle = \sum_b |b\rangle \sum_{s=1}^{n_B} [B]_{bs}\alpha_{ks} = 0 \quad \text{if and only if} \quad \sum_{s=1}^{n_B} [B]_{bs}\alpha_{ks} = 0.$$

We conclude that $\{|\epsilon_k\rangle\}_{k=1}^{\mathcal{K}}$ is a basis of Ker A_{tot} if and only if the column vectors of complex coefficients $\boldsymbol{\alpha}_k = \left[\alpha_{k1} \ldots \alpha_{kn_B}\right]^{\mathsf{T}}$ constitute a basis of the right kernel of the boundary matrix.

In summary, Ker A_{tot} is encoded in two pieces of information: a basis for the solution space of the bulk equation and the boundary matrix B. Given these two pieces of information, one may determine Ker A_{tot} completely and unambiguously. Is this encoding advantageous from a computational perspective? There are three factors to consider:

(i) How difficult is it to solve the bulk equation and store the solution?
(ii) How difficult is it to multiply the vectors in this basis by the matrix $P_\partial A_{\text{tot}}$?
(iii) How hard is it to solve the boundary equation?

The remarkable answer to (i) is that it is easy to compute and store a basis of Ker $P_B A_N$. More precisely, the complexity of this task is *independent of N*. The reason is that most (or even all) of the solutions of the bulk equation $P_B A_N |\psi\rangle = 0$ are obtained by determining the kernel for the associated BBL transformation $\mathbf{A}$ such that $A_N = \mathbf{P}_{L,R}\mathbf{A}|_{\mathcal{V}_{L,R}}$. The latter can be described exactly, in terms of elementary functions and the roots of a polynomial of degree *at most* $d(q-p)$, as we prove formally below (Theorem 5.7). The answer to (ii) depends on $W = A_{\text{tot}} - A_N$. In order to compute the boundary matrix, it is necessary to multiply a basis of the solution space to the bulk equation by the matrix $P_\partial A_{\text{tot}} = P_\partial A_N + W$. The cost of this task is independent of N if W is a symmetrical corner modification—which, fortunately, is most often the case in applications. Otherwise, the cost of computing the boundary matrix is roughly $O(dN)$. Lastly, the boundary matrix is not a structured matrix; thus, the answer to (iii) boils down to whether B is small enough to handle efficiently with standard routines of kernel determination. Generically, if the target BBT matrix is regular, we shall prove later (see Theorem 5.16) that the boundary matrix is necessarily *square*, of size $n_\partial \times n_\partial$. In particular, its size does not grow with N.

Next we will state and prove the main technical results of this section, both concerning the bulk equation. Let us begin with an important definition.

Definition 5.6 Let $A_N = \boldsymbol{P}_{L,R}\boldsymbol{A}|_{\mathcal{V}_{L,R}}$ denote a BBT transformation of bandwidth (p, q) and non-empty bulk (that is, $R - L \geq \tau$). Its "bulk solution space" is

$$\mathcal{M}_{L,R}(\boldsymbol{A}) \equiv \text{Ker}\, \boldsymbol{P}_{L-p',R-q'}\boldsymbol{A}|_{\mathcal{V}_{L,R}} = \text{Ker}\, P_B A_N. \tag{5.18}$$

Notice that $\boldsymbol{V}^{-n}\mathcal{M}_{L,R}(\boldsymbol{A}) = \mathcal{M}_{L+n,R+n}(\boldsymbol{A})$. The spaces $\mathcal{M}_{L,\infty}(\boldsymbol{A}) = \text{Ker}\, \boldsymbol{P}_{L-p',\infty}\boldsymbol{A}$ and $\mathcal{M}_{-\infty,R}(\boldsymbol{A}) = \text{Ker}\, \boldsymbol{P}_{-\infty,R-q'}\boldsymbol{A}$ should not be regarded as "limits" of $\mathcal{M}_{L,R}(\boldsymbol{A})$. We will usually write simply $\mathcal{M}_{L,R}$ if $\boldsymbol{A}$ is fixed.

Theorem 5.7

1. $\boldsymbol{P}_{L,R}\,\text{Ker}\, \boldsymbol{A} \subseteq \mathcal{M}_{L,R}$.
2. *If the principal coefficients of* A_N *are invertible, then* $\boldsymbol{P}_{L,R}\,\text{Ker}\,\boldsymbol{A} = \mathcal{M}_{L,R}$.

Proof

(i) We will prove a stronger result that highlights the usefulness of the notion of bulk solution space. For any $\widetilde{L}$, $\widetilde{R}$ such that $-\infty \leq \widetilde{L} \leq L$ and $R \leq \widetilde{R} \leq \infty$, the "nesting property"

$$\boldsymbol{P}_{L,R}\mathcal{M}_{\widetilde{L},\widetilde{R}} \subseteq \mathcal{M}_{L,R} \tag{5.19}$$

holds. In particular, $\boldsymbol{P}_{L,R}\mathcal{M}_{-\infty,\infty} = \boldsymbol{P}_{L,R}\text{Ker}\,\boldsymbol{A} \subset \mathcal{M}_{L,R} = \text{Ker}\, P_B A_N$ establishes our claim. By definition, the sequences in $\mathcal{M}_{L,R}$ are sequences in $\mathcal{V}_{L,R}$ annihilated by $\boldsymbol{P}_{L-p',R-q'}\boldsymbol{A}$. Hence, we can prove Eq. (5.19) by showing that $\boldsymbol{P}_{L-p',R-q'}\boldsymbol{A}\boldsymbol{P}_{L,R}$ annihilates $\mathcal{M}_{\tilde{L},\tilde{R}}$. To begin with,

$$\begin{aligned}\boldsymbol{P}_{L-p',R-q'}\boldsymbol{A}\boldsymbol{P}_{L,R} &= \boldsymbol{P}_{L-p',R-q'}\sum_{r=p}^{q} a_r \boldsymbol{V}^r \boldsymbol{P}_{L,R} \\ &= \sum_{r=p}^{q} \boldsymbol{P}_{L-p',R-q'}\boldsymbol{P}_{L-r,R-r} a_r \boldsymbol{V}^r.\end{aligned}$$

Since $p' = \min(p, 0)$ and $q' = \max(0, q)$, it follows that $L-r \leq L-p \leq L-p'$, and $R-r \geq R-q \geq R-q'$, $\forall r = p, \ldots, q$, hence $\boldsymbol{P}_{L-p',R-q'}\boldsymbol{P}_{L-r,R-r} = \boldsymbol{P}_{L-p',R-q'}$. We conclude that $\boldsymbol{P}_{L-p',R-q'}\boldsymbol{A}\boldsymbol{P}_{L,R} = \boldsymbol{P}_{L-p',R-q'}\boldsymbol{A}$, and, in particular, $\boldsymbol{P}_{L-p',R-q'}\boldsymbol{A}\boldsymbol{P}_{L,R} = \boldsymbol{P}_{L-p',R-q'}(\boldsymbol{P}_{\tilde{L}-p',\tilde{R}-q'}\boldsymbol{A})$. This last identity makes it explicit that $\boldsymbol{P}_{L-p',R-q'}\boldsymbol{A}\boldsymbol{P}_{L,R}$ annihilates $\mathcal{M}_{\widetilde{L},\widetilde{R}}$.

(ii) It suffices to show that Ker $P_B A_N \subseteq \boldsymbol{P}_{L,R}\text{Ker}\,\boldsymbol{A}|_{\mathcal{V}_{L,R}}$ if the principal coefficients are invertible. The principal coefficients are invertible only if $p \leq 0 \leq q$ and the leading coefficients a_p, a_q are invertible. Then, since in this case

$$P_B A_N = \begin{bmatrix} 0 & \cdots & & \cdots & 0 & \cdots & 0 \\ \vdots & & & & \vdots & & \vdots \\ 0 & \cdots & & \cdots & 0 & \cdots & 0 \\ a_p & \cdots & a_0 & \cdots & a_q & 0 & \cdots & 0 \\ 0 & \ddots & & \ddots & & \ddots & \ddots & \vdots \\ \vdots & \ddots & \ddots & & \ddots & & \ddots & 0 \\ 0 & \cdots & 0 & a_p & \cdots & a_0 & \cdots & a_q \\ 0 & \cdots & 0 & \cdots & & & \cdots & 0 \\ \vdots & & \vdots & & & & & \vdots \\ 0 & \cdots & 0 & \cdots & & & \cdots & 0 \end{bmatrix},$$

it follows that a state $\Psi \in$ Ker $P_B A_N$ satisfies

$$P_B A_N \begin{bmatrix} |\psi_L\rangle \\ \vdots \\ |\psi_R\rangle \end{bmatrix} = \begin{bmatrix} 0 \\ \vdots \\ 0 \\ a_p|\psi_L\rangle + \cdots + a_q|\psi_{L+q-p}\rangle \\ \vdots \\ a_p|\psi_{R-q+p}\rangle + \cdots + a_q|\psi_R\rangle \\ 0 \\ \vdots \\ 0 \end{bmatrix} = 0\,,$$

in terms of the notation $\Psi = \boldsymbol{P}_{L,R}\Psi = \{|\psi_j\rangle\}_{j=L}^{R} = \big[|\psi_L\rangle \ldots |\psi_R\rangle\big]^{\mathsf{T}}$. It follows that Ψ can be uniquely extended in order to obtain a sequence $\Psi' \in$ Ker $\boldsymbol{A}$. Compute $|\psi_{L-1}\rangle$ (the $L-1$ entry of Ψ') as $|\psi_{L-1}\rangle = -a_p^{-1}(a_{p+1}|\psi_L\rangle + \cdots + a_q|\psi_{L+q-p-1}\rangle)$, and repeat the process in order to obtain $|\psi_{L-j}\rangle$ for all $j \geq 1$. Similarly, compute $|\psi_{R+1}\rangle$ as $|\psi_{R+1}\rangle = -a_q^{-1}(a_{q-1}|\psi_{R-(q-p)+1}\rangle + \cdots + a_p|\psi_R\rangle)$, and repeat in order to compute $|\psi_{R+j}\rangle$ for all $j \geq 1$. □

Theorem 5.8 *If* $A_N = \boldsymbol{P}_{L,R}\boldsymbol{A}|_{\mathcal{V}_{L,R}}$ *is regular with non-empty bulk, then* $\dim \mathcal{M}_N = d(q' - p') \equiv d\tau$.

Proof Since $\dim \mathcal{M}_N = \dim \text{Ker}\, P_B A_N = \dim \text{Ker}\, A_N^\dagger P_B$, we can focus on keeping track of the linearly independent solutions of $\langle\phi|P_B A_N = 0$. First, there are the boundary vectors $\langle\phi|P_\partial = \langle\phi|$. From the definition of P_∂, there are precisely $d\tau$ such solutions, showing that $\dim \mathcal{M}_N \geq d\tau$. Suppose that $\dim \mathcal{M}_N > d\tau$. We will show that $\boldsymbol{A}$ must then be singular. By assumption, there exists a non-zero $\langle\phi| = \langle\phi|P_B = \sum_{j=-p'+1}^{N-q'} \langle j|\langle\phi_j| \neq 0$, such that $\langle\phi|P_B A_N = \langle\phi|A_N = 0$. Let $T^r = \boldsymbol{P}_N \boldsymbol{V}^r|_{\mathcal{V}_N}$. Then,

$$0 = \langle\phi|A_N = \sum_{j=-p'+1}^{N-q'} \langle j|\langle\phi_j| \sum_{r=p}^{q} T^r \otimes a_r = \sum_{j=-p'+1}^{N-q'} \sum_{r=p}^{q} \langle j+r|\langle\phi_j|a_r.$$

It is useful to rearrange the above equation as

$$\begin{aligned} 0 = \langle N-q'+q|\langle\phi_{N-q'}|a_q &+ \langle N-q'-1|\big(\langle\phi_{N-q'-1}|a_q + \langle\phi_{N-q'}|a_{q-1}\big) \\ &+ \cdots + \langle 2|\big(\langle\phi_{1-p'+p}|a_{p+1} + \langle\phi_{1-p'+p+1}|a_p\big) \\ &+ \langle 1|\langle\phi_{1-p'+p}|a_p, \end{aligned} \tag{5.20}$$

where all the labels are consistent because $-N+1 < p' \leq p \leq q \leq q' < N+1$. Since, by assumption, $\{\langle\phi_j|\}_{j=1-p'}^{N-q'}$ is not the zero sequence, the vector polynomial $\langle\phi(w)| \equiv \sum_{j=-p'+1}^{N-q'} w^j \langle\phi_j|$ is not the zero vector polynomial. It is immediate to check that $\langle\phi(w)|A(w, w^{-1}) = 0$, because this equation induces precisely the same relations among the $\{\langle\phi_j\}_{j=1-p'}^{N-q'}$ as the bulk equation does, Eq. (5.20). The result follows if we can show that this implies $\det A(w, w^{-1}) = 0$. Consider the Smith decomposition $A(w, w^{-1}) = w^p E(w)D(w)F(w)$. A non-zero vector polynomial cannot be annihilated by an invertible matrix polynomial. Hence, $\langle\psi(w)| \equiv \langle\phi(w)|E(w)^{-1}$ is a non-zero vector polynomial annihilated by $D(w)$, $\langle\psi(w)|D(w) = 0$. Since $D(w)$ is diagonal, this is only possible if at least one of the entries on its main diagonal vanishes, in contradiction with the assumption that $\boldsymbol{A}$ is regular. □

5.2.2 *Exact Solution of the Bulk Equation*

In this section we focus on solving the bulk equation, $P_B A_N|\psi\rangle = 0$. There are two types of solutions. Solutions with *extended support* are associated with kernel vectors of the BBL matrix $\boldsymbol{A}$ such that $A_N = \boldsymbol{P}_N \boldsymbol{A}|_{\mathcal{V}_N}$, as implied by Theorem 5.7. From a physical standpoint, it is interesting to observe that these solutions can be constructed to be translation-invariant, since $[\boldsymbol{A}, \boldsymbol{V}] = 0$. Any other solution, necessarily of *finite support* as we will show, may be thought of as emergent. They exist only because of the translation-symmetry-breaking projection that leads from the infinite system $\boldsymbol{A}$ to the finite system A_N, and *only if* the principal coefficients are *not* invertible.

Extended-Support Solutions

In order to determine the kernel of an arbitrary BBL transformation, it is convenient to first establish a few results concerning the kernel of a special subclass of BBL transformations. Given a non-negative integer v and $j \in \mathbb{Z}$, let

$$j^{(v)} = \begin{cases} 1 & \text{if } v = 0, \\ (j - v + 1)(j - v + 2) \dots j & \text{if } v = 1, 2, \dots \end{cases} \tag{5.21}$$

and, for any complex number $z \neq 0$ and $v \in N$, introduce the family of scalar sequences defined by

$$\Phi_{z,0} = 0, \quad \Phi_{z,1} = \{z^j\}_{j \in \mathbb{Z}}, \quad \text{and} \quad \Phi_{z,v} = \frac{1}{(v-1)!} \frac{d^{v-1} \Phi_{z,1}}{dz^{v-1}}$$
$$= \{j^{(v-1)} z^{j-v+1}\}_{j \in \mathbb{Z}}. \tag{5.22}$$

Moreover, we will write $\mathcal{S}_{z,s} \equiv \text{Span}\{\Phi_{z,v}\}_{v=1}^{s}$.

Lemma 5.9

1. *The scalar sequences $\Phi_{z,v}$ satisfy the following properties:*

$$\boldsymbol{V} \Phi_{z,v} = z \Phi_{z,v} + \Phi_{z,v-1}, \quad \textit{and} \quad \text{Ker}\,(\boldsymbol{V} - z)^s = \mathcal{S}_{z,s}.$$

2. *For any $s_1, s_2 \in N$, if $z_1 \neq z_2$, the intersection* $\text{Ker}\,(\boldsymbol{V} - z_1)^{s_1} \cap \text{Ker}\,(\boldsymbol{V} - z_2)^{s_2} = \{0\}$.

Proof

(i) It is immediate to check that $\boldsymbol{V}\Phi_{z,1} = z\Phi_{z,1}$. Hence,

$$\boldsymbol{V}\Phi_{z,v} = \frac{1}{(v-1)!} \frac{d^{v-1}}{dz^{v-1}} \boldsymbol{V}\Phi_{z,1} = \frac{1}{(v-1)!} \frac{d^{v-1}}{dz^{v-1}} (z\Phi_{z,1}) = z\Phi_{z,v} + \Phi_{z,v-1}.$$

Next, let $K_z \Phi = \{z^j \phi_j\}_{j \in \mathbb{Z}}$ for $z \neq 0$. Then,

$$K_z(\boldsymbol{V} - \mathbf{1})\Phi = \sum_{j \in \mathbb{Z}} w^{-j} z^j (\phi_{j+1} - \phi_j) = (z^{-1}\boldsymbol{V} - \mathbf{1}) K_z \Phi.$$

In other words, $zK_z(\boldsymbol{V} - \mathbf{1}) = (\boldsymbol{V} - z)K_z$, and so $(\boldsymbol{V} - z)^s K_z = z^s K_z (\boldsymbol{V} - \mathbf{1})^s$. As a consequence, $\text{Ker}\,(\boldsymbol{V} - z)^s = K_z\, \text{Ker}\,(\boldsymbol{V} - \mathbf{1})^s$, and we can zoom in on the special case $z = 1$. Since $(\boldsymbol{V} - \mathbf{1})\Phi = \sum_{j \in \mathbb{Z}} w^{-j}(\phi_{j+1} - \phi_j)$, the kernel of $\boldsymbol{V} - \mathbf{1}$ is spanned by the constant sequence $\Phi_{1,1} = \{1\}_{j \in \mathbb{Z}}$. Suppose $\mathcal{S}_{1,s} = \text{Ker}\,(\boldsymbol{V} - \mathbf{1})^s$. We will prove that, as a consequence, $\mathcal{S}_{1,s+1} = \text{Ker}\,(\boldsymbol{V} - \mathbf{1})^{s+1}$. Every $\Psi \in \text{Ker}\,(\boldsymbol{V} - \mathbf{1})^{s+1}$ satisfies $(\boldsymbol{V} - 1)^s(\boldsymbol{V} - 1)\Psi = 0$. Then, by the induction hypothesis, $(\boldsymbol{V} - \mathbf{1})\Psi \in \mathcal{S}_{1,s}$ and so there exist numbers α_v such that $(V-1)\Psi = \sum_{v=1}^{s} \alpha_v \Phi_{1,v}$. Since $(V-1)\Phi_{1,v} = \Phi_{1,v-1}$, it follows that $\Psi = \Psi' + \sum_{v=1}^{s} \alpha_v \Phi_{1,v+1}$, with $\Psi' \in \mathrm{Ker}\,(V-1)$. This shows that Ψ belongs to $\mathcal{S}_{1,s+1}$, and so $\mathrm{Ker}\,(V-1)^{s+1} \subseteq \mathcal{S}_{1,s+1}$. The reverse inclusion $\mathcal{S}_{1,s+1} \subseteq \mathrm{Ker}\,(V-1)^{s+1}$ holds because $\mathcal{S}_{1,s} = \mathrm{Ker}\,(V-1)^s \subset \mathrm{Ker}\,(V-1)^{s+1}$ by the induction hypothesis, and

$$(V-1)^{s+1}\Phi_{1,s+1} = (V-1)^s\Phi_{1,s} = \cdots = (V-1)\Phi_{1,1} = 0.$$

(ii) We will prove a slightly more general result: Let $Y_i = \rho(y_i(w))$, $i = 1, 2$, denote two upper-triangular banded Laurent transformations of $\mathcal{V}_1^S$. If $\gcd(y_1, y_2) = 1$, then $\mathrm{Ker}\,Y_1 \cap \mathrm{Ker}\,Y_2 = \{0\}$. In order to prove this, let $\deg y_2 \geq \deg y_1$ without loss of generality, and suppose $\Phi \in \mathrm{Ker}\,Y_1 \cap \mathrm{Ker}\,Y_2$. The banded Laurent transformation $\rho(r)$ induced by the remainder of dividing y_2 by y_1 also annihilates Φ, since $y_2 = cy_1 + r$. Continuing this process, we conclude that $\gcd(y_1, y_2)$ annihilates Φ. Hence, if $\gcd(y_1, y_2) = 1$, then $\Phi = 0$. □

Lemma 5.10 *The span of vector sequences* $\mathcal{T}_{z,s} \equiv \mathrm{Span}\{\Phi_{z,v}|m\rangle \mid v = 1, \ldots, s;\ m = 1, \ldots, d\}$ *is an invariant subspace of the algebra of BBL matrices. The mapping*

$$A(w, w^{-1}) \mapsto A|_{\mathcal{T}_{z,s}} \equiv [(A, z, s)] \tag{5.23}$$

defines a ds*-dimensional representation of the algebra of matrix Laurent polynomials, and the block matrix*

$$[(A, z, s)]_{xv} \equiv \begin{cases} \frac{1}{(v-x)!}A^{(v-x)}(z, z^{-1}) & \text{if } 1 \leq x \leq v \leq s \\ 0 & \text{if } 1 \leq v < x \leq s \end{cases}, \qquad A^{(v-x)} \equiv \frac{d^{v-x}A}{dz^{v-x}}, \tag{5.24}$$

is the matrix of $[(A, z, s)]$ *relative to the defining basis of* $\mathcal{T}_{z,s}$, *with* $A^{(0)} = A$.

Proof Since $V\Psi = \sum_{m=1}^{d} |m\rangle (V\{\psi_{jm}\}_{j\in\mathbb{Z}})$, it is immediate to check that $\mathcal{T}_{z,s} = \mathrm{Ker}\,(V-z)^s$. Moreover, $(V-z)^s A\mathcal{T}_{z,s} = A(V-z)^s\mathcal{T}_{z,s} = \{0\}$. This proves that $A\mathcal{T}_{z,s} \subseteq \mathcal{T}_{z,s}$. The matrix of $A|_{\mathcal{T}_{z,s}}$ is computed as follows. On the one hand, by definition,

$$A\Phi_{z,v}|m\rangle = [(A, z, s)]\Phi_{z,v}|m\rangle = \sum_{x=1}^{s}\sum_{m'=1}^{d} \Phi_{z,x}|m'\rangle\langle m'|[(A, z, s)]_{xv}|m\rangle.$$

On the other hand,

$$A\Phi_{z,v}|m\rangle = \frac{1}{(v-1)!}\frac{d^{v-1}}{dz^{v-1}}\Phi_{z,1}A(z, z^{-1})|m\rangle$$

$$= \sum_{x=1}^{v} \sum_{m'=1}^{d} \Phi_{z,x}|m'\rangle \frac{1}{(v-x)!} \langle m'|A^{(v-x)}(z, z^{-1})|m\rangle,$$

where we have taken advantage of $\boldsymbol{A}\Phi_{z,1}|m\rangle = \sum_{r=p}^{q}(\boldsymbol{V}^r \Phi_{z,1})a_r|m\rangle = A(z, z^{-1})|m\rangle$. □

We call the finite dimensional representation $A(w, w^{-1}) \mapsto [(A, z, s)]$ ($\boldsymbol{A} \mapsto [(A, z, s)]$) of the algebra of Laurent polynomials (BBL matrices) the "generalized evaluation map at z of degree s." In block array form,

$$[(A, z, s)] = \begin{bmatrix} A & A^{(1)} & \frac{1}{2}A^{(2)} & \dots & \frac{1}{(s-1)!}A^{(s-1)} \\ 0 & A & A^{(1)} & \ddots & \vdots \\ \vdots & \ddots & \ddots & \ddots & \frac{1}{2}A^{(2)} \\ \vdots & & \ddots & \ddots & A^{(1)} \\ 0 & \dots & \dots & 0 & A \end{bmatrix},$$

and $[(A, z, 1)] = A(z, z^{-1})$ is in particular the usual evaluation map of the theory of polynomials. Notice also that

$$\boldsymbol{A}\sum_{v=1}^{s} \Phi_{z,v}|\psi_v\rangle = \begin{bmatrix}\Phi_{z,1} & \Phi_{z,2} & \dots & \Phi_{z,s-1} & \Phi_{z,s}\end{bmatrix}[(A, z, s)] \begin{bmatrix} |\psi_1\rangle \\ |\psi_2\rangle \\ \vdots \\ |\psi_{s-1}\rangle \\ |\psi_s\rangle \end{bmatrix}.$$

Finally, it is instructive to check the representation property explicitly in an example. Direct calculation shows that

$$[(w\mathbb{1}, z, 3)] = \begin{bmatrix} z\mathbb{1} & \mathbb{1} & 0 \\ 0 & z\mathbb{1} & 2\mathbb{1} \\ 0 & 0 & z\mathbb{1} \end{bmatrix} \quad \text{and} \quad [(w^{-1}\mathbb{1}, z, 3)] = \begin{bmatrix} \mathbb{1}/z & -\mathbb{1}/z^2 & 2\mathbb{1}/z^3 \\ 0 & \mathbb{1}/z & -2\mathbb{1}/z^2 \\ 0 & 0 & \mathbb{1}/z \end{bmatrix}.$$

Now it is immediate to check that

$$[(w\mathbb{1}, z, 3)][(w^{-1}\mathbb{1}, z, 3)] = [(w^{-1}\mathbb{1}, z, 3)][(w\mathbb{1}, z, 3)] = [(\mathbb{1}, z, 3)],$$

where $[(\mathbb{1}, z, 3)]$ is the $3d$ dimensional identity matrix.

The Smith decomposition of $\boldsymbol{A}$, Eq. (5.13), implies that $\text{Ker}\,\boldsymbol{A} = \boldsymbol{F}^{-1}\text{Ker}\,\boldsymbol{D}$, where $\boldsymbol{F}$ is a nonsingular BBL matrix and the Smith normal form $\boldsymbol{D}$ of $\boldsymbol{A}$ has the simple structure shown in Eq. (5.14). In this way, the problem of determining a basis of $\text{Ker}\,\boldsymbol{A}$ reduces to two independent tasks, namely determining a basis of

Ker $\boldsymbol{D}$, and the change of basis $\boldsymbol{F}^{-1}$. With regard to Ker $\boldsymbol{D}$, its structure may be characterized quite simply after introducing some notation. Let $\{z_\ell\}_{\ell=0}^n$ denote the *distinct* roots of $g_{d_0}(w) = D_{d_0 d_0}(w)$. By convention, $z_0 = 0$. Hence, if $g_{d_0}(w)$ does not have a vanishing root, then we just drop z_0 from the picture. Since every $g_m(w) = D_{mm}(w)$ divides $g_{d_0}(w)$, it must be that the roots of the $g_m(w)$ are also roots of $g_{d_0}(w)$. We need a unified notation that organizes these structural facts. Hence, in the following we will write

$$g_m(w) = \prod_{\ell=0}^{n} (w - z_\ell)^{s_{m\ell}}, \quad m = 1, \ldots, d_0, \tag{5.25}$$

where $s_{d_0\ell}$ is the multiplicity of z_ℓ as a root of $g_{d_0}(w)$, and $s_{d_0 0} = 0$ if $z_0 = 0$ is not a root. For $1 \leq m < d_0$, $s_{m\ell}$ is *either* the multiplicity of z_ℓ as a root of $g_m(w)$ *or* it vanishes if z_ℓ is not a root of g_m to begin with. The projector $\pi \equiv \sum_{m=d_0+1}^{d} |m\rangle\langle m|$ keeps track of the vanishing entries on the main diagonal of $D(w)$, $(\mathbb{1} - \pi)D(w) = D(w)(\mathbb{1} - \pi) = D(w)$.

Lemma 5.11 *Let* $\boldsymbol{D}$ *be the Smith normal form of* $\boldsymbol{A}$, *and* $\rho(\pi)$ *the unique BBL projector such that* $(\mathbf{1} - \rho_d(\pi))\boldsymbol{D} = \boldsymbol{D}(\mathbf{1} - \rho_d(\pi)) = \boldsymbol{D}$. *Then,* Ker $\boldsymbol{D} = \mathcal{W}_{\boldsymbol{D}} \oplus$ Range $\rho_d(\pi)$, *with*

$$\mathrm{W}_{\boldsymbol{D}} = \mathrm{Span}\{\Phi_{z_\ell, v}|m\rangle \mid m = 1, \ldots, d_0;\ \ell = 1, \ldots, n;\ v = 1, \ldots, s_{m\ell}\}. \tag{5.26}$$

Proof Since $\boldsymbol{D}\Psi = \sum_{m=1}^{d_0} |m\rangle \boldsymbol{g}_m \{\psi_{jm}\}_{j\in\mathrm{Z}}$, a sequence is annihilated by $\boldsymbol{D}$ if and only if it is of the form $\Psi = \sum_{m=d_0+1}^{d} |m\rangle\{\psi_{jm}\}_{j\in\mathrm{Z}}$, or $\Psi = \sum_{m=1}^{d_0} |m\rangle\{\psi_{jm}\}_{j\in\mathrm{Z}}$ and $\boldsymbol{g}_m\{\psi_{jm}\}_{j\in\mathrm{Z}} = 0$ for all $m \leq d_0$. According to Eq. (5.25),

$$\boldsymbol{g}_m = \prod_{\ell=0}^{n} (\boldsymbol{V} - z_\ell)^{s_{m\ell}}$$

and so its kernel can be determined by inspection with the help of Lemma 5.9. □

The space Range $\rho_d(\pi)$ is an uncountably infinite-dimensional space. Since we have not specified a topology on any space of sequences, there is no easy way to describe a basis for it. Fortunately, for the purpose of investigating the kernel of $A_N = \boldsymbol{P}_{L,R}\boldsymbol{A}|_{\mathcal{V}_{L,R}}$, we only need to characterize explicitly $\boldsymbol{F}^{-1}\boldsymbol{\Pi}_0$, where $\boldsymbol{\Pi}_0$ denotes the subspace of sequences in Ker $\rho_d(\pi)$ of finite support. The sequences

$$\Delta_i|m\rangle = \{\delta_{ij}|m\rangle\}_{j\in\mathrm{Z}}, \quad i \in \mathrm{Z}, \quad m = d_0 + 1, \ldots, d,$$

where δ_{ij} denotes the Kronecker delta, form a basis of $\boldsymbol{\Pi}_0$. For convenience, we will call the space $\boldsymbol{F}^{-1}\mathrm{W}_{\boldsymbol{D}} \oplus \boldsymbol{F}^{-1}\boldsymbol{\Pi}_0 \equiv \mathrm{Ker}_c\, \boldsymbol{A} \subset \mathrm{Ker}\, \boldsymbol{A}$ the *countable kernel* of $\boldsymbol{A}$. Putting things together, we obtain a complete and explicit description of $\mathrm{Ker}_c\, \boldsymbol{A}$. We will write $\boldsymbol{F}^{-1} = \rho_d(F^{-1}(w))$, with

$$F^{-1}(w) \equiv \sum_{s=0}^{\deg F^{-1}} w^s \hat{f}_s, \quad \deg F^{-1} \in N, \quad \hat{f}_s \in \mathbb{C}_d.$$

Corollary 5.12 *The countable kernel of* $\boldsymbol{A}$ *is spanned by the sequences*

$$\boldsymbol{F}^{-1}\Delta_i|m\rangle = \sum_{s=0}^{\deg F^{-1}} \Delta_{i-s}\hat{f}_s|m\rangle, \quad m = d_0+1,\ldots,d;\ i \in Z,$$

$$\boldsymbol{F}^{-1}\Phi_{z_\ell,v}|m\rangle = \sum_{x=1}^{s_{m\ell}} \Phi_{z_\ell,x}[(F^{-1}, z_\ell, s_{m\ell})]_{xv}|m\rangle, \quad m = 1,\ldots,d_0;$$

$$\ell = 1,\ldots,n;\ v = 1,\ldots,s_{m\ell}\,.$$

Proof Since $\text{Ker}_c\, \boldsymbol{A} = \boldsymbol{F}^{-1}\text{W}_{\boldsymbol{D}} \oplus \boldsymbol{F}^{-1}\boldsymbol{\Pi}_0$, the claim follows from computing the action of $\boldsymbol{F}^{-1} = \rho_d(F^{-1}(w)) = \sum_{s=0}^{\deg F^{-1}} \boldsymbol{V}^s\hat{f}_s$ on the explicit bases of W_D and $\boldsymbol{\Pi}_0$. □

If $d_0 = d$, then $\boldsymbol{A}$ is regular and $\text{Ker}\,\boldsymbol{A}$ is finite-dimensional. In this case, $\text{Ker}\,\boldsymbol{A}$ may be determined *without* explicitly computing the Smith factorization of $\boldsymbol{A}$.

Theorem 5.13 *Let* $\boldsymbol{A} = \rho_d(A(w, w^{-1}))$ *be regular with Smith normal form* $\boldsymbol{D} = \rho_d(D(w))$, *and write*

$$\det A(w, w^{-1}) = c w^{dp} \prod_{m=1}^{d} g_m(w) \equiv c\, w^{dp} \prod_{\ell=0}^{n} (w - z_\ell)^{s_\ell},$$

where $s_\ell = \sum_{m=1}^{d} s_{m\ell}$ *is the multiplicity of* z_ℓ *as a root of* $\det A \neq 0$, *and* $s_0 = 0$ *if* $z_0 = 0$ *is not a root of* $\det A$. *Then,*

1. $\text{Ker}\,\boldsymbol{A} = \bigoplus_{\ell=1}^{n} \text{Ker}\,[(A, z_\ell, s_\ell)]$.
2. $\dim \text{Ker}\, A(z_\ell, z_\ell^{-1}) \leq \dim \text{Ker}\,[(A, z_\ell, s_\ell)] = s_\ell$, *for any* $\ell = 1,\ldots,n$.

Furthermore, if the inequality in (ii) saturates, then $\text{Ker}\,[(A, z_\ell, s_\ell)] = \text{Span}\{\Phi_{z_\ell,1}|u_s\rangle\}_{s=1}^{s_\ell}$, *where* $\{|u_s\rangle\}_{s=1}^{s_\ell}$ *is a basis of* $\text{Ker}\, A(z_\ell, z_\ell^{-1})$.

Proof

(i) We know an explicit basis of $\text{Ker}\,\boldsymbol{A}$ from Corollary 5.12. By grouping together the basis vectors associated with each non-zero root z_ℓ, we obtain the decomposition

$$\text{Ker}\,\boldsymbol{A} = \bigoplus_{\ell=1}^{n} \mathcal{W}_{z_\ell}, \quad \mathcal{W}_{z_\ell} = \boldsymbol{F}^{-1}\text{Span}\,\{\Phi_{z_\ell,v}|m\rangle \,\big|\, v = 1,\ldots,s_{m\ell}\}_{m=1}^{d},$$

so that $\dim \mathcal{W}_{z_\ell} = \sum_{m=1}^{d} s_{m\ell} = s_\ell$. Moreover, by Lemma 5.10 (iii), $\mathcal{W}_{z_\ell} \subset \boldsymbol{F}^{-1}\mathcal{T}_{z_\ell,s_\ell} = \mathcal{T}_{z_\ell,s_\ell}$. By construction, $\boldsymbol{A}$ has no kernel vectors in $\mathcal{T}_{z_\ell,s_\ell}$ other than the ones in $\mathcal{W}_{z_\ell}$. Hence, $\mathcal{W}_{z_\ell} = \operatorname{Ker} \boldsymbol{A} \cap \mathcal{T}_{z_\ell,s_\ell} = \operatorname{Ker}[(A, z_\ell, s_\ell)]$. We conclude that $\dim \operatorname{Ker}[(A, z_\ell, s_\ell)] = s_\ell$, the multiplicity of $z_\ell \neq 0$ as a root of $\det A$.

(ii) Because $g_1(w)|g_2(w)|\dots|g_d(w)$, a non-zero root z_ℓ of $\det A(w, w^{-1})$ will appear for the first time in one of the $g_m(w)$, say $g_{m_\ell}(w)$, and reappear with equal or greater multiplicity in every $g_m(w)$ with $d \geq m > m_\ell$. In particular, $g_m(z_\ell) \neq 0$ if $m < m_\ell$, and vanishes otherwise. Hence, $\dim \operatorname{Ker} D(z_\ell) = \dim \operatorname{Ker} A(z_\ell, z_\ell^{-1}) = d - m_\ell + 1$. The number m_ℓ is as small as possible whenever z_ℓ is a root with multiplicity one of each one of the $g_{m_\ell}(w), \dots, g_d(w)$, in which case $m_\ell = d - s_\ell + 1$. This shows that $\dim \operatorname{Ker} A(z_\ell, z_\ell^{-1}) \leq s_\ell = \dim \operatorname{Ker}[(A, z_\ell, s_\ell)]$.

Suppose next that $\dim \operatorname{Ker} A(z_\ell, z_\ell^{-1}) = s_\ell$, and let $\{|u_s\rangle\}_{s=1}^{s_\ell}$ denote a basis of this space. Then, $\boldsymbol{A}\Phi_{z_\ell,1}|u_s\rangle = \Phi_{z_\ell,1}A(z_\ell, z_\ell^{-1})|u_s\rangle = 0$, $s = 1, \dots, s_\ell$. Since $\operatorname{Ker}[(A, z_\ell, s_\ell)]$ is s_ℓ-dimensional, it follows that these linearly independent sequences span this space. □

Finite-Support Solutions

According to Theorem 5.7, if the principal coefficients of the BBT matrix $A_N = \boldsymbol{P}_{L,R}\boldsymbol{A}|_{\mathcal{V}_{L,R}}$ fail to be invertible (which can happen even for $d = 1$), the bulk solution space $\mathcal{M}_{L,R} = \operatorname{Ker} P_B A_N$ contains, but need *not* be contained in $\boldsymbol{P}_{L,R}\operatorname{Ker} \boldsymbol{A}$. The solutions of the bulk equation that are *not* in $\boldsymbol{P}_{L,R}\operatorname{Ker} \boldsymbol{A}$ were referred to as emergent before. We aim to establish a structural characterization of $\mathcal{M}_{L,R}$. Our strategy will be to characterize the spaces $\mathcal{M}_{L,\infty}$ and $\mathcal{M}_{-\infty,R}$ first, and then proceed to establish their relationship to $\mathcal{M}_{L,R}$.

Theorem 5.14 *If the BBL matrix $\boldsymbol{A}$ is regular, then there exist spaces $\mathcal{F}_L^+ \subseteq \mathcal{V}_{L,\overline{L}}$ and $\mathcal{F}_R^- \subseteq \mathcal{V}_{\overline{R},R}$ such that*

$$\mathcal{M}_{L,\infty} = \boldsymbol{P}_{L,\infty}\operatorname{Ker} \boldsymbol{A} \oplus \mathcal{F}_L^+, \quad \text{and} \quad \mathcal{M}_{-\infty,R} = \boldsymbol{P}_{-\infty,R}\operatorname{Ker} \boldsymbol{A} \oplus \mathcal{F}_R^-,$$

with $\overline{L} = L + \sigma - 1$, $\overline{R} = R - \sigma + 1$, and $\sigma = d(q' - p') - \dim \operatorname{Ker} \boldsymbol{A}$.

Proof We will prove the claim for $\mathcal{M}_{L,\infty}$ only; the reasoning is the same for $\mathcal{M}_{-\infty,R}$.

We will begin by establishing that $\mathcal{M}_{L,\infty}$ is finite-dimensional. Since $\dim \boldsymbol{P}_{L,R}\mathcal{M}_{L,\infty}$ is necessarily finite-dimensional if $R - L$ is finite, the question becomes whether there are non-zero sequences in $\mathcal{M}_{L,\infty}$ annihilated by $\boldsymbol{P}_{L,R}$. The answer is in the negative provided $R - L \geq \tau$. We will reason by contradiction. Suppose Ψ satisfies $\boldsymbol{P}_{L-p',\infty}\boldsymbol{A}\Psi = 0 = \boldsymbol{P}_{L,R}\Psi$, and assume also, without loss of generality, that Ψ vanishes for all $j \leq R$, that is, $P_{-\infty,R}\Psi = 0$. Then,

$$A\Psi = P_{-\infty,L-p'-1}A\Psi = \sum_{r=p}^{q} a_r V^r P_{-\infty,L+(r-p')-1}\Psi = 0$$

because $L+(r-p')-1 \le L+(q-p')-1 \le L+(q'-p') \le R$. By construction, the translates $\{\Psi_n = V^n\Psi\}_{n\in\mathbb{Z}}$ are linearly independent and satisfy $A\Psi_n = 0$, implying that $\dim \operatorname{Ker} A = \infty$. But this is in contradiction with the regularity of A, see Theorem 5.13. Hence, if $\Psi \in \mathcal{M}_{L,\infty}$ is a non-zero sequence and $R - L \ge \tau$, then $P_{L,R}\Psi \ne 0$, and so $\dim \mathcal{M}_{L,\infty} = \dim P_{L,R}\mathcal{M}_{L,\infty} < \infty$, as was to be shown. The necessary condition $R - L \ge \tau$ yields in fact a stronger result. Combining nesting with Theorem 5.8, we conclude that $\dim \mathcal{M}_{L,\infty} = \dim P_{L,R}\mathcal{M}_{L,\infty} \le \dim \mathcal{M}_{L,R} = d\tau$.

While $\mathcal{M}_{L,\infty}$ is finite dimensional just like $\mathcal{M}_{L,R}$, what is special about this space is that it is an invariant subspace of a translation-like transformation, the "unilateral shift" $P_{L,\infty}V$. To see that this is the case, just notice that

$$P_{L-p',\infty}AP_{L,\infty}V\mathcal{M}_{L,\infty} = VP_{L+1-p',\infty}AP_{L+1,\infty}\mathcal{M}_{L,\infty} = 0$$

because of the nesting property $P_{L+1,\infty}\mathcal{M}_{L,\infty} \subseteq \mathcal{M}_{L+1,\infty}$. Hence, $P_{L,\infty}V\mathcal{M}_{L,\infty} \subseteq \mathcal{M}_{L,\infty}$, as was to be shown. Since $\mathcal{M}_{L,\infty}$ is finite-dimensional, it can be decomposed into the direct sum of generalized eigenspaces of $P_{L,\infty}V|_{\mathcal{M}_{L,\infty}}$. Accordingly, let us write $\mathcal{M}_{L,\infty} = \mathcal{N} \oplus \mathcal{F}_L^+$, where $P_{L,\infty}V|_{\mathcal{F}_L^+}$ is nilpotent and $P_{L,\infty}V|_{\mathcal{N}}$ is invertible.

The space $\mathcal{F}_L^+$, that is, the generalized kernel $P_{L,R}V|_{\mathcal{M}_{L,\infty}}$, is a subspace of $\mathcal{V}_{L,\overline{L}}$. The reason is that, if $\Psi \in \mathcal{F}_L^+$, then there is a smallest positive integer v, its "rank," such that $(P_{L,\infty}V)^v\Psi = 0$. The rank v is necessarily bounded above by

$$\sigma \equiv \dim \mathcal{F}_L^+ = \dim \mathcal{M}_{L,\infty} - \dim P_{L,\infty}\operatorname{Ker} A = d\tau - \dim \operatorname{Ker} A.$$

As a consequence, if $\Psi \in \mathcal{F}_L^+$, then

$$P_{L,\infty}V^\sigma\Psi = (P_{L,\infty}V)^\sigma\Psi = 0 \quad \text{and so } \Psi \in \mathcal{V}_{L,\overline{L}}\text{ , with } \overline{L} = L + \sigma - 1.$$

The space $\mathcal{N}$, that is, the direct sum of all the generalized eigenspaces of $P_{L,R}V|_{\mathcal{M}_{L,\infty}}$ associated with non-zero eigenvalues, coincides with $P_{L,\infty}\operatorname{Ker} A$. To see that this is the case, let $\Psi_n = (P_{L,\infty}V)^{-n}\Psi \in \mathcal{N}$ for any positive integer n and $\Psi \in \mathcal{N}$. Since $V^n\Psi_n \in V^n\mathcal{M}_{L,\infty} = \mathcal{M}_{L-n,\infty}$, and $P_{L,\infty}V^n\Psi_n = (P_{L,\infty}V)^n\Psi_n = \Psi$, we conclude that $\Psi \in P_{L,\infty}\mathcal{M}_{L-n,\infty}$ for any n. It follows that $\Psi \in P_{L,\infty}\mathcal{M}_{-\infty,\infty} = P_{L,\infty}\operatorname{Ker} A$. The reverse inclusion $P_{L,\infty}\operatorname{Ker} A \subseteq \mathcal{N}$ holds because

$$P_{L,\infty}VP_{L,\infty}\operatorname{Ker} A = P_{L,\infty}P_{L-1,\infty}V\operatorname{Ker} A = P_{L,\infty}\operatorname{Ker} A.$$

□

This theorem suggests that $\mathcal{M}_{L,R}$ consists of three qualitatively distinct pieces: bulk solutions associated with $\boldsymbol{P}_{L,R}\mathrm{Ker}\,\boldsymbol{A}$, bulk solutions localized near L, and bulk solutions localized near R. The latter two types are the emergent solutions of finite support. Here is the precise confirmation of this picture.

Theorem 5.15

1. *If* $A_N = \boldsymbol{P}_{L,R}\boldsymbol{A}|_{\mathcal{V}_{L,R}}$ *is regular, then* $\mathcal{M}_{L,R} = \mathrm{Span}\big(\boldsymbol{P}_{L,R}\mathcal{M}_{-\infty,R} \cup \boldsymbol{P}_{L,R}\mathcal{M}_{L,\infty}\big)$.
2. *If* $N \geq 2\sigma + \tau$, *then* $\mathcal{M}_{L,R} = \boldsymbol{P}_{L,R}\mathrm{Ker}\,\boldsymbol{A} \oplus \mathcal{F}_L^+ \oplus \mathcal{F}_R^-$.

Proof

(i) The inclusion $\mathrm{Span}\big(\boldsymbol{P}_{L,R}\,\mathcal{M}_{L,\infty} \cup \boldsymbol{P}_{L,R}\,\mathcal{M}_{-\infty,R}\big) \subseteq \mathcal{M}_{L,R}$ holds true because of the nesting property, Eq. (5.19). The task is to establish the reverse inclusion. Let us begin by showing that

$$\boldsymbol{P}_{L,R}\,\mathrm{Ker}\,\boldsymbol{A} = \boldsymbol{P}_{L,R}\,\mathcal{M}_{-\infty,R} \cap \boldsymbol{P}_{L,R}\,\mathcal{M}_{L,\infty}.$$

Again, because of nesting, $\boldsymbol{P}_{L,R}\,\mathrm{Ker}\,\boldsymbol{A} \equiv \boldsymbol{P}_{L,R}\mathcal{M}_{-\infty,\infty} \subseteq \boldsymbol{P}_{L,R}\,\mathcal{M}_{-\infty,R} \cap \boldsymbol{P}_{L,R}\,\mathcal{M}_{L,\infty}$. In order to prove the reverse inclusion, take

$$\{|\chi_j\rangle\}_{j=L}^{R} \in \boldsymbol{P}_{L,R}\mathcal{M}_{L,\infty} \cap \boldsymbol{P}_{L,R}\mathcal{M}_{-\infty,R} \subseteq \mathcal{M}_{L,R}$$

arbitrary. By definition, there exist sequences $\Psi_1 = \{|\psi_{1j}\rangle\}_{j=L}^{\infty} \in \mathcal{M}_{L,\infty}$ and $\Psi_2 = \{|\psi_{2j}\rangle\}_{j=-\infty}^{R} \in \mathcal{M}_{-\infty,R}$ such that $\boldsymbol{P}_{L,R}\Psi_1 = \boldsymbol{P}_{L,R}\Psi_2 = \{|\chi_j\rangle\}_{j=L}^{R}$. Let Ψ denote the unique sequence with $\boldsymbol{P}_{-\infty,R}\Psi = \Psi_1$ and $\boldsymbol{P}_{L,\infty}\Psi = \Psi_2$. Then,

$$\boldsymbol{A}\Psi = (\boldsymbol{P}_{-\infty,R-q'} + \boldsymbol{P}_{L-p',\infty} - \boldsymbol{P}_{L-p',R-q'})\boldsymbol{A}\Psi = 0,$$

confirming that $\{|\chi_j\rangle\}_{j=L}^{R} = \boldsymbol{P}_{L,R}\Psi \in \boldsymbol{P}_{L,R}\mathrm{Ker}\,\boldsymbol{A}$. The immediate but important consequence is that

$$\dim \mathrm{Span}\big(\boldsymbol{P}_{L,R}\mathcal{M}_{-\infty,R} \cup \boldsymbol{P}_{L,R}\mathcal{M}_{L,\infty}\big) = \dim \mathcal{M}_{-\infty,R} + \dim \mathcal{M}_{L,\infty} - \dim \mathrm{Ker}\,\boldsymbol{A}.$$

In particular, this dimension is independent of L, R (provided $R-L \geq q'-p'$).

The next step is to show that there exists an $\infty > R_0 \geq R$ such that

$$\boldsymbol{P}_{L,R}\mathcal{M}_{L,\widetilde{R}} = \boldsymbol{P}_{L,R}\mathcal{M}_{L,\infty}, \quad \forall \widetilde{R} \geq R_0.$$

At an intuitive level, this result is very appealing: it states that a measurement on sites L to R cannot tell $\mathcal{M}_{L,\widetilde{R}}$ apart from $\mathcal{M}_{L,\infty}$ if $\widetilde{R}$ is large enough. Let $\delta(n) = \dim \boldsymbol{P}_{L,R}\mathcal{M}_{L,R+n}$, $n \geq 0$ (in particular, $\delta(0) = d(q'-p')$). If $n_2 \geq n_1$, then, by nesting, $\boldsymbol{P}_{L,R}\mathcal{M}_{L,R+n_2} = \boldsymbol{P}_{L,R}\boldsymbol{P}_{L,R+n_1}\mathcal{M}_{L,R+n_2} \subseteq$

$\boldsymbol{P}_{L,R}\mathcal{M}_{L,R+n_1}$, showing that $\delta(n_2) \leq \delta(n_1)$. Since δ is a non-decreasing function bounded below, there exists n_0 such that $\delta(n) = \delta(n_0) \equiv \delta_0$ for all $n \geq n_0$. Let $\widetilde{R} \geq R + n_0$, so that $\dim \boldsymbol{P}_{L,R}\mathcal{M}_{L,\widetilde{R}} = \delta_0$ independently of $\widetilde{R}$. By nesting, $\boldsymbol{P}_{L,R}\mathcal{M}_{L,\infty} \subseteq \boldsymbol{P}_{L,R}\mathcal{M}_{L,\widetilde{R}}$. Hence, we would like to establish that $\delta_0 = \dim \boldsymbol{P}_{L,R}\mathcal{M}_{L,\infty} = \dim \mathcal{M}_{L,\infty}$.

Thanks to the special properties of R_0, now we can prove a special instance of our general claim, the equality

$$\mathcal{M}_{L,R_0} = \text{Span}\big(\boldsymbol{P}_{L,R_0}\mathcal{M}_{-\infty,R_0} \cup \boldsymbol{P}_{L,R_0}\mathcal{M}_{L,\infty}\big). \tag{5.27}$$

By definition of R_0, if $\Psi \in \mathcal{M}_{L,R_0}$, then $\boldsymbol{P}_{L,R}\Psi \in \boldsymbol{P}_{L,R}\mathcal{M}_{L,\infty}$. This means that there is $\Upsilon \in \mathcal{M}_{L,\infty}$ such that $\boldsymbol{P}_{L,R}\Upsilon = \boldsymbol{P}_{L,R}\Psi$. Let us decompose

$$\Psi = \Psi_1 + \Psi_2, \quad \text{with} \quad \Psi_1 \equiv \boldsymbol{P}_{L,R_0}\Upsilon, \ \ \Psi_2 = \Psi - \boldsymbol{P}_{L,R_0}\Upsilon.$$

Since $\Psi_1 \in \boldsymbol{P}_{L,R_0}\mathcal{M}_{L,\infty}$ by construction, the only thing left to do is show that $\Psi_2 \in \boldsymbol{P}_{L,R_0}\mathcal{M}_{-\infty,R_0}$. By nesting, $\boldsymbol{P}_{L,R_0}\Upsilon \in \mathcal{M}_{L,R_0}$ and so $\Psi_2 \in \mathcal{M}_{L,R_0}$. In particular, $\boldsymbol{P}_{-\infty,L-1}\Psi_2 = 0$. Hence,

$$\boldsymbol{P}_{-\infty,R_0-q'}\boldsymbol{A}\Psi_2 = \boldsymbol{P}_{-\infty,L-p'-1}\boldsymbol{A}\Psi_2 = \sum_{r=q}^{p} a_r \boldsymbol{V}^r \boldsymbol{P}_{L,L+r-p'-1}\Psi_2 = 0,$$

because $\boldsymbol{P}_{L,R}\Psi_2 = \boldsymbol{P}_{L,R}\Psi - \boldsymbol{P}_{L,R}\Upsilon = 0$ (by the way Υ was chosen), and $L + r - p' - 1 < R$. It follows that $\Psi_2 \in \mathcal{M}_{-\infty,R_0}$, and, more importantly, since $\boldsymbol{P}_{L,R_0}\Psi_2 = \Psi_2$, it also follows that $\Psi_2 \in \boldsymbol{P}_{L,R_0}\mathcal{M}_{-\infty,R_0}$. This concludes the proof of Eq. (5.27). The general claim follows from this special instance, because the dimension of the span in question is independent of L, R, and $\dim \mathcal{M}_{L,R} = d(q' - p')$ by Theorem 5.8.

(ii) Because of the lower bound on N, $\mathcal{F}_L^+ = \boldsymbol{P}_{L,R}\mathcal{F}_L^+$ and $\mathcal{F}_R^- = \boldsymbol{P}_{L,R}\mathcal{F}_R^-$. Using Theorem 5.15 and the direct sum decompositions in Theorem 5.14, the only additional result required to prove the corollary is $\{\boldsymbol{P}_{L,R}\text{Ker}\,\boldsymbol{A} \oplus \mathcal{F}_L^+\} \cap \mathcal{F}_R^- = \{0\}$. By contradiction, assume that there exists some non-zero vector Ψ in this intersection. Then $\Psi \in \mathcal{F}_R^-$ implies $\boldsymbol{P}_{L,\bar{R}-1}\Psi = 0$, which can be split into two conditions $\boldsymbol{P}_{L,\bar{L}}\Psi = 0$, $\boldsymbol{P}_{\bar{L}+1,\bar{R}-1}\Psi = 0$. Since $\Psi \in \boldsymbol{P}_{L,R}\text{Ker}\,\boldsymbol{A} \oplus \mathcal{F}_L^+$, we may also express Ψ as $\Psi = \Psi_1 + \Psi_2$, where $\Psi_1 \in \boldsymbol{P}_{L,R}\text{Ker}\,\boldsymbol{A}$ and $\Psi_2 \in \mathcal{F}_L^+$. Now since $\boldsymbol{P}_{\bar{L}+1,\bar{R}-1}\Psi_2 = 0$, therefore the second of the two equations implies $\boldsymbol{P}_{\bar{L}+1,\bar{R}-1}\Psi_1 = 0$. The lower bound on N implies that $\bar{R} - 1 - (\bar{L} + 1) + 1 > \tau$, so that $\Psi_1 = 0$. Then the first equation leads to $\Psi_2 = 0$, so that $\Psi = \Psi_1 + \Psi_2 = 0$, which is a contradiction. □

According to Theorem 5.15 *(ii)*, one can obtain a basis of the bulk solution space $\mathcal{M}_{L,R}$ by combining the bases of the three subspaces $\boldsymbol{P}_{L,R}\text{Ker}\,\boldsymbol{A}$, $\mathcal{F}_L^+$, and $\mathcal{F}_R^-$. Since $\mathcal{F}_L^+ \subseteq \mathcal{V}_{L,\bar{L}}$, in order to find a basis of $\mathcal{F}_L^+$, we only need to find one for the kernel of

$$\boldsymbol{K}_+ \equiv \boldsymbol{P}_{L-p',R-q'}\boldsymbol{A}|_{\mathcal{V}_{L,\overline{L}}}.$$

Similarly, a basis of $\mathcal{F}_R^+$ can be computed from one of the kernel of

$$\boldsymbol{K}_- \equiv \boldsymbol{P}_{L-p',R-q'}\boldsymbol{A}|_{\mathcal{V}_{\overline{R},R}}.$$

Consider the BBT transformation

$$A_{(\overline{L}+1,\overline{R}-1)} \equiv \boldsymbol{P}_{\overline{L}+1,\overline{R}-1}\boldsymbol{A}|_{\mathcal{V}_{\overline{L}+1,\overline{R}-1}},$$

and let $\overline{P}_B$ denote the projector on its bulk. Then $\overline{P}_B\boldsymbol{A} = \overline{P}_B\boldsymbol{A}\boldsymbol{P}_{\overline{L}+1,\overline{R}-1}$, which means that the bulk of $A_{(\overline{L}+1,\overline{R}-1)}$ is outside the range of $\boldsymbol{A}|_{\mathcal{V}_{L,\overline{L}}}$. Therefore,

$$\boldsymbol{K}_+ = P_B\boldsymbol{A}|_{\mathcal{V}_{L,\overline{L}}} = \overline{P}_-\boldsymbol{A}|_{\mathcal{V}_{L,\overline{L}}}, \tag{5.28}$$

where $\overline{P}_- \equiv \boldsymbol{P}_{L-p',\overline{L}-p'}$ is the projector on $\mathcal{V}_{L-p',\overline{L}-p'}$. Similarly,

$$\boldsymbol{K}_- = P_B\boldsymbol{A}|_{\mathcal{V}_{\overline{R},R}} = \overline{P}_+\boldsymbol{A}|_{\mathcal{V}_{\overline{R},R}}, \quad \overline{P}_+ \equiv \boldsymbol{P}_{\overline{R}-q',R-q'}. \tag{5.29}$$

Let us point out in closing that, if $\boldsymbol{A}$ is singular, the kernels of $\boldsymbol{K}^+$ and $\boldsymbol{K}^-$ are still contained in the bulk solution space. Together with Theorem 5.7, this implies that

$$\mathcal{M}_N \supseteq \operatorname{Span}\big(\boldsymbol{P}_N \operatorname{Ker} \boldsymbol{A} \cup (\mathcal{F}_1^+ \oplus \mathcal{F}_N^-)\big) \quad (\boldsymbol{A} \text{ singular}).$$

5.2.3 *An Exact Result for the Boundary Equation*

As we already remarked in Sect. 5.2.1, the boundary equation is generally associated with an unstructured matrix, the boundary matrix (Definition 5.5). While its solution thus relies in general on numerical methods, there is one exact result that follows from our work so far. We isolate it here.

Theorem 5.16 *If* $A_N = \boldsymbol{P}_N\boldsymbol{A}|_{\mathcal{V}_N}$ *is regular, then* $\dim \operatorname{Ker} P_B A_N = \dim \operatorname{Range} P_\partial$, *and the boundary matrix of the corner-modified BBT matrix* $A_{tot} = A_N + W$ *is square independently of the corner modification* W.

Proof The boundary matrix is the matrix of the compatibility map $B = P_\partial(A_N + W)|_{\operatorname{Ker} P_B A_N}$, see Sect. 5.2.1. Hence, the number of rows of the boundary matrix is determined by the dimension of the range of P_∂, which is $d\tau$. The number of columns is determined by the dimension of the kernel of $P_B A_N$, which, by Lemma 5.8, is also $d\tau$ if A_N is regular. □

As its proof indicates, this result leverages strongly properties of "regular" BBT matrices. The boundary matrix is the matrix form of the compatibility map of the method of kernel determination by projectors, and, in general, one can only expect the compatibility map to induce rectangular matrices. For example, as far as we know, a singular BBT matrix might yield a rectangular boundary matrix.

5.2.4 Multiplication of Corner-Modified Banded Block-Toeplitz Matrices

In general, the product of two BBT matrices is not a BBT matrix (see Ref. [4] for a lucid discussion of this point). However, the product of two corner-modified BBT matrices is again a corner modified BBT matrix, provided they are large enough. As a consequence, the problem of determining the generalized kernel of a corner-modified BBT matrix $A_{\rm tot}$ may turn out to be equivalent to that of determining the kernel of $A_{\rm tot}^{v}$ in the same class.

Theorem 5.17 *Let $A_{tot,i} = A_{N,i} + W_i$, $i = 1, 2$, denote corner-modified BBT matrices, with $A_{N,i} = \boldsymbol{P}_N \boldsymbol{A}_i|_{\mathcal{V}_N}$ of bandwidth (p_i, q_i) and $2N - 1 > q_1 - p_1 + q_2 - p_2 + 1$. Then, $A_{tot,1}A_{tot,2} = A_N + W$ is a corner-modified BBT matrix of bandwidth $(p_1 + p_2, q_1 + q_2)$, and $A_N = \boldsymbol{P}_N \boldsymbol{A}_1 \boldsymbol{A}_2|_{\mathcal{V}_N}$.*

Proof Let us begin by showing that

$$P_B A_{\rm tot,1} A_{\rm tot,2} = P_B(A_{N,1} + W_1)(A_{N,2} + W_2) = P_B A_{N,1} A_{N,2}, \tag{5.30}$$

where P_B is the bulk projector for bandwidth $(p_1 + p_2, q_1 + q_2)$. In particular, $P_B P_{B,1} = P_B$, where $P_{B,1}$ is the bulk projectors for bandwidth (p_1, q_1). Hence, $P_B W_1(A_{N,2}+W_2) = P_B P_{B,1} W_1(A_{N,2}+W_2) = 0$ because $P_{B,1} W_1 = 0$. Moreover, since

$$P_B A_{N,1} P_{\partial,2}|\psi\rangle = P_B \sum_{j=1}^{-p'_2-p'_1} |j\rangle|\phi_j\rangle + P_B \sum_{N-q'_2-q'_1+1}^{N} |j\rangle|\phi_j\rangle = 0$$

for arbitrary $|\psi\rangle$, one concludes that $P_B A_{N,1} P_{\partial,2} = 0$ and $P_B A_{N,1} P_{\partial,2} W_2 = P_B A_{N,1} W_2 = 0$ as well. The next step is to show that

$$P_{B,1} A_{N,1} A_{N,2} = P_{B,1} \boldsymbol{A}_1 \boldsymbol{A}_2|_{\mathcal{V}_N}. \tag{5.31}$$

By definition, $P_{B,1} A_{N,1} A_{N,2} = P_{B,1} \boldsymbol{P}_N \boldsymbol{A}_1 \boldsymbol{P}_N \boldsymbol{A}_2|_{\mathcal{V}_N} = P_{B,1} \boldsymbol{A}_1 \boldsymbol{P}_N \boldsymbol{A}_2|_{\mathcal{V}_N}$. However,

$$P_{B,1} \boldsymbol{A}_1 \boldsymbol{A}_2|_{\mathcal{V}_N} - P_{B,1} \boldsymbol{A}_1 \boldsymbol{P}_N \boldsymbol{A}_2|_{\mathcal{V}_N} = P_{B,1} \boldsymbol{A}_1(\boldsymbol{1} - \boldsymbol{P}_N)\boldsymbol{A}_2|_{\mathcal{V}_N} = 0.$$

The reason is that, on the one hand, the sequences in the range of $(\mathbf{1} - \boldsymbol{P}_N)\boldsymbol{A}_2|_{\mathcal{V}_N}$ necessarily vanish on sites 1 to N. On the other hand, acting with $\boldsymbol{A}_1$ on any such sequence produces a sequence that necessarily vanishes on sites $1 - p_1'$ to $N - q_1'$. Such sequences are annihilated by $P_{B,1}$. Combining Eqs. (5.30) and (5.31), we conclude that

$$P_B \boldsymbol{A}_1 \boldsymbol{A}_2|_{\mathcal{V}_N} = P_B A_{N,1} A_{N,2} = P_B A_{\text{tot},1} A_{\text{tot},2},$$

and so the bulk of $A_{\text{tot},1} A_{\text{tot},2}$ coincides with the bulk of the BBT matrix $A_N = \boldsymbol{P}_N \boldsymbol{A}_1 \boldsymbol{A}_2|_{\mathcal{V}_N}$. □

Corollary 5.18 *Let $A_{tot,1} = A_N + W$ denote a corner-modified BBT transformation of bandwidth (p, q) and $A_N = \boldsymbol{P}_N \boldsymbol{A}|_{\mathcal{V}_N}$. Then, $A_{tot,1}^v = A_N^{(v)} + W_v$ is a corner-modified BBT matrix as long as $2N - 1 > v(q - p) + 1 \geq 1$. The bandwidth of $A_N^{(v)} \equiv \boldsymbol{P}_N \boldsymbol{A}^v|_{\mathcal{V}_N}$ is (vp, vq), and W_v a corner modification for this bandwidth.*

5.3 An Ansatz for the Eigenvectors of Block-Toeplitz Operators on Hilbert Space

5.3.1 Ansatz for Finite-Dimensional Hilbert Space

Based on the analysis in Sect. 5.2, we will now formulate an exact eigenvalue-dependent ansatz for the eigenvectors of a given corner-modified BBT transformation. Any eigenvector of a corner-modified BBT transformation J, corresponding to eigenvalue ϵ, is a kernel vector of the transformation $A_{\text{tot}} \equiv J - \epsilon$, which is also a corner-modified BBT transformation. Therefore it suffices to derive an ansatz for kernel vectors of A_{tot}. In fact, an ansatz for generalized eigenvectors of J of rank $v > 1$ can be obtained similarly, since $A_{\text{tot}}^v = (J - \epsilon)^v$ is also a corner-modified BBT matrix, as shown in Corollary 5.18.

It is proved in Lemma 5.7 that if the principal coefficients are invertible, then $\mathcal{M}_N$ is simply the projection of the kernel of $\boldsymbol{A}$. For non-invertible principal coefficients instead, $\mathcal{M}_N$ is a direct sum of the three subspaces $\boldsymbol{P}_N \text{Ker}\, \boldsymbol{A}$, $\mathcal{F}_1^+$, and $\mathcal{F}_N^-$ (Corollary 5.15). We proceed to constructing a basis of each of these subspaces. According to Lemma 5.13,

$$\text{Ker}\, \boldsymbol{A} = \bigoplus_{\ell=1}^{n} \text{Ker}\, \boldsymbol{A} \cap \mathcal{T}_{z_\ell, s_\ell},$$

where $\{z_\ell\}_{\ell=1}^n$ are *non-zero* roots of the characteristic equation of $G(z)$ and $\{s_\ell\}_{\ell=1}^n$ their multiplicities, respectively. Each of the subspaces $\mathcal{T}_{z_\ell, s_\ell}$ is invariant under the action of G, and $\mathrm{G}|\mathcal{T}_{z_\ell,s_\ell}$ has representation $[(G, z_\ell, s_\ell)]$ in the canonical basis given in Lemma 5.10. Therefore, a block-vector $|u\rangle = \left[|u_1\rangle \dots |u_{s_\ell}\rangle\right]^\mathsf{T}$, $\{|u_v\rangle \in \mathbb{C}^d\}_{v=1}^{s_\ell}$, belonging to the kernel of $[(G, z_\ell, s_\ell)]$, represents the sequence

$$\Psi_{\ell,s} \equiv \sum_{v=1}^{s_\ell} \Phi_{z_\ell,v}|u_v\rangle \in \operatorname{Ker} \boldsymbol{A} \cap \mathcal{T}_{z_\ell,s_\ell}$$

in the kernel of G. Its projection on $\mathbb{C}^N \otimes \mathbb{C}^d$, namely

$$|\psi_{\ell,s}\rangle \equiv \boldsymbol{P}_N\Psi_{\ell,s} = \sum_{v=1}^{s_\ell} |z_\ell, v\rangle|u_v\rangle \in \mathcal{M}_N^{\pm,\infty}, \quad |z_\ell, v\rangle = \sum_{j=1}^{N} j^{(v-1)} z^j |j\rangle,$$

is a solution of the bulk equation. A basis $\mathcal{B}_\pm^{(\ell)} \equiv \{|\psi_{\ell,s}\rangle\}$ of $\boldsymbol{P}_N \operatorname{Ker} \boldsymbol{A}$, corresponding to root z_ℓ, may thus be inferred from a basis $\{|u_{\ell,s}\rangle\}_{s=1}^{s_\ell}$ of $\operatorname{Ker}[(G, z_\ell, s_\ell)]$, where

$$|u_{\ell,s}\rangle = \left[|u_{\ell,s,1}\rangle \dots |u_{\ell,s,s_\ell}\rangle\right]^\mathsf{T}, \quad \{|u_{\ell,s,v}\rangle \in \mathbb{C}^d\}_{v=1}^{s_\ell} \ \forall \ell, s. \tag{5.32}$$

Then a basis of $\boldsymbol{P}_N \operatorname{Ker} \boldsymbol{A}$ is given by $\mathcal{B}_\pm = \bigcup_{\ell=1}^{n} \mathcal{B}_\pm^{(\ell)}$, with this basis being stored in the form of vectors $\left\{\{|u_{\ell,s}\rangle\}_{s=1}^{s_\ell}\right\}_{\ell=1}^{n}$.

If the principal coefficients of $\boldsymbol{A}$ are not invertible, one needs to additionally obtain bases of $\mathcal{F}_1^+$ and $\mathcal{F}_N^-$. It is proved in Sect. 5.2.2 that the kernel of $\mathcal{F}_1^+$ ($\mathcal{F}_N^-$) coincides with $\operatorname{Ker} \boldsymbol{K}_+$ ($\operatorname{Ker} \boldsymbol{K}_-$), with $\boldsymbol{K}_+$ and $\boldsymbol{K}_-$ acting as in Eqs. (5.28)–(5.29). In the standard bases $\{|j\rangle\}_{j=1}^{\sigma}$ and $\{|j-p'\rangle\}_{j=1}^{\sigma}$ of the subspaces $\mathcal{V}_{1,d\tau}$ and Range $\bar{P}_-$, respectively, $\boldsymbol{K}_+$ is a square block matrix of size $d^2\tau \times d^2\tau$, with block-entries

$$[\boldsymbol{K}_+]_{jj'} = \begin{cases} a_{j-j'+p'} & \text{if } j' \le j \le j'+\tau \\ 0 & \text{otherwise} \end{cases}, \quad 1 \le j, j' \le \sigma,$$

Every vector $|u\rangle = \left[|u_1\rangle \dots |u_\sigma\rangle\right]^\mathsf{T}$, with $\{|u_j\rangle \in \mathbb{C}^d\}_{j=1}^{\sigma}$, in the kernel of $\boldsymbol{K}_+$ provides corresponding solution of the bulk equation, namely,

$$|\psi\rangle = \sum_{j=1}^{v} |j\rangle|u_j\rangle \in \mathcal{F}_1^+.$$

Accordingly, a basis $\mathcal{B}_+ \equiv \{|\psi_{0s}\rangle\}_{s=1}^{s_+}$ of $\mathcal{F}_1^+$ may be stored as the basis $\{|u_s^+\rangle\}_{s=1}^{s_+}$ of $\operatorname{Ker} \boldsymbol{K}_+$, where $s_+ \equiv \dim(\operatorname{Ker} \boldsymbol{K}_+)$ and

$$|u_s^+\rangle = \left[|u_{s,1}^+\rangle \dots |u_{s,\sigma}^+\rangle\right]^\mathsf{T}, \quad \{|u_{s,j}^+\rangle \in \mathbb{C}^d\}_{j=1}^{\sigma} \ \forall s. \tag{5.33}$$

Similarly, a basis of $\mathcal{F}_N^-$ can be obtained from a basis of $\boldsymbol{K}_-$, with entries

$$[\boldsymbol{K}_-]_{jj'} = \begin{cases} a_{j-j'+q'} & \text{if } j \le j' \le j+\tau \\ 0 & \text{otherwise} \end{cases}, \quad 1 \le j, j' \le \sigma.$$

In this case, each $|u\rangle = \left[|u_1\rangle \dots |u_\sigma\rangle\right]^\mathsf{T}$, $\{|u_j\rangle \in \mathbb{C}^d\}_{j=1}^{\sigma}$, in the kernel of $\boldsymbol{K}_-$ represents the solution

$$|\psi\rangle = \sum_{j=1}^{\sigma} |N - \sigma + j\rangle|u_j\rangle \in \mathcal{F}_N^-$$

of the bulk equation. Then a basis $\mathcal{B}_+ \equiv \{|\psi_{(n+1)s}\rangle\}_{s=1}^{s_-}$ of $\mathcal{F}_N^-$ is stored as the basis $\{|u_s^-\rangle\}_{s=1}^{s_-}$ of Ker $\boldsymbol{K}_-$, where $s_- \equiv \dim(\text{Ker}\,\boldsymbol{K}_-)$ and

$$|u_s^-\rangle = \left[|u_{s,1}^-\rangle \dots |u_{s,\sigma}^-\rangle\right]^\mathsf{T}, \quad \{|u_{s,j}^-\rangle \in \mathbb{C}^d\}_{j=1}^{\sigma}\ \forall s. \tag{5.34}$$

If the principal coefficients of $\boldsymbol{A}$ are invertible, then both $\mathcal{B}_+$ and $\mathcal{B}_-$ are empty.

Now it follows that any kernel vector of $A_{\text{tot}} = J - \epsilon$ may be expressed as a linear combination of the form

$$|\epsilon\rangle = \sum_{\ell=0}^{n+1}\sum_{s=1}^{s_\ell} \alpha_{\ell s}|\psi_{\ell s}\rangle, \tag{5.35}$$

where the complex coefficients $\alpha_{\ell s} \in \mathbb{C}$ are parameters to be determined and

$$|\psi_{\ell s}\rangle = \sum_{v=1}^{s_\ell} |z_\ell, v\rangle|u_{\ell s v}\rangle \in \boldsymbol{P}_N\text{Ker}\,\boldsymbol{A}\,, \quad \ell = 1, \dots, n,$$

$$|\psi_{0s}\rangle = \sum_{j=1}^{d\tau} |j\rangle|u_{sj}^+\rangle \in \mathcal{F}_1^+\,,$$

$$|\psi_{(n+1)s}\rangle = \sum_{j=1}^{d\tau} |N - d\tau + j\rangle|u_{sj}^-\rangle \in \mathcal{F}_N^-\,,$$

for basis vectors described in Eqs. (5.32)–(5.34).

Remark 5.19 Corollary 5.15 applies only to those cases where the matrix Laurent polynomial under consideration is regular. Therefore, the ansatz in Eq. (5.35) is *provably complete* only for those corner-modified BBT matrices $A_N + W$, where the associated matrix Laurent polynomial $A(w, w^{-1}) - \epsilon$ is regular for every ϵ, which is usually the case. If $A(w, w^{-1}) - \epsilon$ is singular for some ϵ, our analysis suggests that $\boldsymbol{P}_N\text{Ker}\,\boldsymbol{A}$ and $\mathcal{F}_1^+ \oplus \mathcal{F}_N^-$ are subspaces of $\mathcal{M}_N$; however, they need *not* span the entire $\mathcal{M}_N$. Such cases are important but rare, and typically correspond to some

Table 5.1 Structural characterization of the bulk solution space, depending on the invertibility of the principal coefficients and regularity of the corresponding matrix polynomial. Table adapted with permission from Ref. [5]. Copyrighted by IOP Publishing

	$a_{p'}, a_{q'}$	
$G(w)$	Invertible	Non-invertible
Regular	$\mathcal{M}_N = \boldsymbol{P}_N \operatorname{Ker} \boldsymbol{A}$	$\mathcal{M}_N = \boldsymbol{P}_N \operatorname{Ker} \boldsymbol{A} \oplus \mathcal{F}_1^- \oplus \mathcal{F}_N^+$
Singular	–	$\mathcal{M}_N \supseteq \operatorname{Span}(\boldsymbol{P}_N \operatorname{Ker} \boldsymbol{A} \cup (\mathcal{F}_1^- \oplus \mathcal{F}_N^+))$

exactly solvable limits. In these cases, ϵ is a highly degenerate eigenvalue, with $\mathcal{O}(N)$ eigenvectors of the form given in Corollary 5.12 that have *finite support in the bulk*. In free-fermionic Hamiltonians as considered in Chaps. 2–4, dispersionless energy bands form for such eigenvalues. A summary of our results on the structural characterization of $\mathcal{M}_N$ is given in Table 5.1.

If the principal coefficients of the associated matrix Laurent polynomial $G(w)$ are invertible, both the second and third terms in the ansatz of Eq. (5.35) vanish. Irrespective of the invertibility of the principal coefficients, a simplification in the first term occurs if $s_\ell = \dim \operatorname{Ker} G(z_\ell)$ for some ℓ, where s_ℓ is the algebraic multiplicity z_ℓ as a root of $\det G(w)$ (recall that $s_\ell = \dim \operatorname{Ker} [(G, z_\ell, s_\ell)]$). In these cases, Lemma 5.13 implies that each of the $|\psi_{\ell,s}\rangle$ in Eq. (5.35) is of the form

$$|\psi_{\ell,s}\rangle = |z_\ell, 1\rangle |u_{\ell s1}\rangle \quad \forall s, \tag{5.36}$$

and there are no contributions from terms $|z_\ell, v\rangle |u_{\ell s1}\rangle$ with $v > 1$. We call vectors of the form in Eq. (5.36) "exponential solutions," because the amplitude $\big|\langle j|\langle m|\psi_{\ell s}\rangle\big| = \big|\langle m|u_{\ell s1}\rangle z_\ell^j\big| \propto |z_\ell^j|$ varies exponentially with j for any value of $m = 1, \ldots, d$. If z lies on unit circle, then these solutions correspond to plane waves, with amplitude that is independent of j. Otherwise, the amplitude has an exponential behavior that depends on the magnitude of z. The condition $s_\ell = \dim \operatorname{Ker} G(z_\ell)$ is satisfied under generic situations by all roots z_ℓ. In those special situations where $s_\ell > \dim \operatorname{Ker} G(z_\ell)$, one must allow for the possibility of $|\psi_{\ell,s}\rangle$ in Eq. (5.35) to describe "power-law solutions," whose amplitude varies with j as

$$\big|\langle j|\langle m|\psi_{\ell s}\rangle\big| = \Big|\sum_{v=1}^{s_\ell} \langle m|u_{\ell s v}\rangle j^{(v-1)} z_\ell^j\Big| \propto |j^{s_\ell - 1} z_\ell^j| \quad \forall m.$$

If the principal coefficients of the matrix Laurent polynomial $G(w)$ are not invertible, the contributions to the ansatz in Eq. (5.35) that belong to $\mathcal{F}_1^+$ and $\mathcal{F}_N^-$ are *finite-support solutions*. This refers to the fact that, for all m, their amplitude

$$\big|\langle j|\langle m|\psi_{0s}\rangle\big| = \begin{cases} \big|\langle m|u_{sj}^+\rangle\big| & \text{if } 1 \le j \le d\tau \\ 0 & \text{if } j > d\tau \end{cases}, \tag{5.37}$$

for $j > d\tau$ in the case of $\mathcal{F}_1^+$, and $j < N - d\tau$ for $\mathcal{F}_N^-$, respectively. The support of these solutions clearly does not change with N.

5.3.2 *Ansatz for Infinite-Dimensional Hilbert Space*

We will now construct an ansatz for eigenvectors of the linear transformations of $\mathcal{V}_{1,\infty} = \boldsymbol{P}_{1,\infty}\mathcal{V}_d^S$ of the form $A = \boldsymbol{P}_{1,\infty}\boldsymbol{A}|_{\mathcal{V}_{1,\infty}}$. The task is to compute the square-summable sequences in the kernel of A or some closely related corner-modified version of A. We will make this problem precise after some preliminaries.

Elements of $\mathcal{V}_{1,\infty}$ can be seen as half-infinite sequences. We will use the letter $\Upsilon \in \mathcal{V}_{1,\infty}$ to denote one such sequence, and write $\Upsilon \equiv \{|\upsilon_j\rangle\}_{j\in N}$. If $\boldsymbol{A}$ has bandwidth (p, q), with $p \le q$, then A_∞ is induced by the "infinite-downwards" square array

$$A_\infty = \begin{bmatrix} a_0 & \dots & a_q & & & 0 & \ddots \\ \vdots & \ddots & \ddots & \ddots & & & \ddots \\ a_p & \ddots & a_0 & \ddots & a_q & & \\ & \ddots & \ddots & \ddots & \ddots & \ddots & \\ & & a_p & \ddots & \ddots & & \\ 0 & & & \ddots & & & \\ \ddots & \ddots & & & & & \end{bmatrix}.$$

We will call A_∞ an *infinite* BBT matrix or IBBT for short. The transformation induced by A_∞ is an IBBT transformation.

Definition 5.20 Let $p' = \min(p, 0)$ and $q' = \max(0, q)$ for integers $p \le q$. The projector

$$P_B\Upsilon \equiv \{|\upsilon_j'\rangle\}_{j\in N}, \quad |\upsilon_j'\rangle = \begin{cases} 0 & \text{if } j = 1, \dots, -p' \\ |\upsilon_j\rangle & \text{if } -p' < j \end{cases}$$

is the "right bulk projector" for bandwidth (p, q). The projector

$$Q_B\Upsilon \equiv \{|\upsilon_j'\rangle\}_{j\in N}, \quad |\upsilon_j'\rangle = \begin{cases} 0 & \text{if } j = 1, \dots, q' \\ |\upsilon_j\rangle & \text{if } q' < j \end{cases}$$

is the "left bulk projector." The corresponding left and right boundary projectors are $P_\partial = I - P_B$ and $Q_\partial = I - Q_\partial$, respectively.

With this definition, it follows that if $p' = 0$ ($q' = 0$), then $P_B = I$ ($Q_B = I$).

Definition 5.21 A linear transformation $A_{\text{tot},1}$ of $\mathcal{V}_{1,\infty}$ is a "corner-modified IBBT transformation" if there exists an IBBT transformation $A_\infty = P_{1,\infty}\boldsymbol{A}|_{\mathcal{V}_{1,\infty}}$, necessarily unique, such that $P_B A_{\text{tot},1} = P_B A_\infty$. $A_{\text{tot},1}$ is "symmetrical" if, in addition, $A_{\text{tot},1} Q_B = A_\infty Q_B$.

Lemma 5.22 *If the principal coefficient $a_{p'}$ of A is invertible, then* $\text{Ker}\, P_B A = \boldsymbol{P}_{1,\infty}\text{Ker}\, \boldsymbol{A}$. *Otherwise,* $\boldsymbol{P}_{1,\infty}\text{Ker}\, \boldsymbol{A} \subseteq \text{Ker}\, P_B A$.

Proof See the proof of Theorem 5.7. In contrast to the situation for A_N, the principal coefficient $a_{q'}$ plays no role here. As noted, if $p' = 0$, then $P_B = I$. □

From here onwards, we denote the solution space of the relevant bulk equation by $\mathcal{M} = \text{Ker}\, P_B A$. Since A is a linear transformation of an infinite-dimensional vector space, $\mathcal{M}$ may also, in principle, be infinite-dimensional if the principal coefficient a'_p is not invertible. The following Lemma describes $\mathcal{M}$ explicitly, showing in particular that it is finite dimensional. However, there is no guarantee that dim Range P_∂ should match $\mathcal{M}$. The proof of Theorem 5.8 breaks down for IBBT transformations, and so one may expect that the boundary matrix will be rectangular in general.

We will recall the first result of Theorem 5.14 before introducing the space of square-summable sequences: if the polynomial part of $\boldsymbol{A}$ is regular, then

$$\mathcal{M}_{1,\infty} = \mathcal{F}_1^+ \oplus \boldsymbol{P}_{1,\infty}\text{Ker}\, \boldsymbol{A}\ ,$$

where $\mathcal{F}_1^+ \in \mathcal{V}_{1,\sigma}$ for any $N > \tau + 2\sigma$. If the principal coefficient $a_{p'}$ of A is invertible, then $\mathcal{F}_1^+ = \{0\}$. The subspace

$$\mathcal{H} = \left\{ \{|\upsilon_j\rangle\}_{j\in N} \mid \sum_{j=1}^{\infty} \langle \upsilon_j|\upsilon_j\rangle < \infty \right\} \subset \mathcal{V}_{1,\infty}$$

is the Hilbert space of square-summable sequences. We will denote square-summable sequences as $|\Upsilon\rangle \in \mathcal{H}$, so that $\langle \Upsilon_1|\Upsilon_2\rangle = \sum_{j\in N}\langle \upsilon_{1,j}|\upsilon_{2,j}\rangle < \infty$. Our final task is to compute a basis of Ker $A_{\text{tot}} \cap \mathcal{H}$ for an arbitrary corner-modified IBBT transformation. Physically, these states correspond to normalizable, bound states.

Lemma 5.23 *Let $z \in \mathbb{C}$, $z \neq 0$, and $s \in \mathbb{N}$. If $|z| < 1$, then $\boldsymbol{P}_{1,\infty}\mathcal{T}_{z,s} \subset \mathcal{H}$.*

Proof Let $\boldsymbol{P}_{1,\infty}\Phi_{z,v}|m\rangle = \{j^{(v-1)}z^j|m\rangle\}_{j\in N} = \{|\phi_j\rangle\}_{j\in N}$. Then, since $\langle m|m\rangle = 1$, the l^2-norm of this sequence would be given by the series $\sum_{j\in N}\langle\phi_j|\phi_j\rangle = \sum_{j\in N}(j^{(v-1)})^2|z|^{2j}$ if convergent. The limit

$$\lim_{j\to\infty}\left|\frac{((j+1)^{(v-1)})^2|z|^{2(j+1)}}{(j^{(v-1)})^2|z|^{2j}}\right| = |z|^2$$

and so, by the ratio test, the series converges if $|z| < 1$ and diverges $|z| > 1$. It is immediate to check that it also diverges if $|z| = 1$, in which case the series is attempting to sum a non-decreasing sequence of strictly positive numbers. □

Theorem 5.24 *If $\boldsymbol{A}$ is regular, the space of the square-summable solutions of the bulk equation is given by*

$$\mathcal{M}_{1,\infty} \cap \mathcal{H} = \mathcal{F}_1^+ \oplus \boldsymbol{P}_{1,\infty} \bigoplus_{|z_\ell|<1} \text{Ker}\,\boldsymbol{A} \cap \mathcal{T}_{z_\ell, s_\ell}.$$

Proof The sequences in $\mathcal{F}_1^+$ have finite support, and so they are square-summable. Then $\mathcal{F}_1^+ \subset \mathcal{H}$ implies that $\mathcal{M}_{1,\infty} \cap \mathcal{H} = \mathcal{F}_1^+ \oplus \left(\boldsymbol{P}_{1,\infty}\text{Ker}\,\boldsymbol{A} \cap \mathcal{H}\right)$. For every $\Psi \in \text{Ker}\,\boldsymbol{A}$,

$$\boldsymbol{P}_{1,\infty}\Psi = \sum_{\ell=1}^{n}\sum_{v=1}^{s_\ell}\sum_{m=1}^{d} \alpha_{\ell,v,m}\boldsymbol{P}_{1,\infty}\Phi_{z_\ell,v}|m\rangle \equiv \{|\psi_j\rangle\}_{j\in N},$$

so that $|\psi_j\rangle = \sum_{m=1}^{d}|m\rangle\sum_{\ell=1}^{n} y_{\ell,m}(j)z_\ell^j$, with $y_{\ell,m}(j) = \sum_{v=1}^{s_\ell}\alpha_{\ell,v,m}j^{(v-1)}$ polynomials in j of degree at most s_ℓ. The sequence $\boldsymbol{P}_{1,\infty}\Psi$ cannot be square-summable unless $\lim_{j\to\infty}\langle\psi_j|\psi_j\rangle = 0$, which in turn implies $\lim_{j\to\infty}|\psi_j\rangle = 0$. Hence, for any $m = 1, \ldots, d$,

$$\lim_{j\to\infty}\langle m|\psi_j\rangle = 0 \quad \Longrightarrow \quad \lim_{j\to\infty}\sum_{\ell=1}^{n} y_{\ell,m}(j)z_\ell^j = 0.$$

The necessary condition for square-summability can be met if and only if $\alpha_{\ell,s,m} = 0$ whenever $|z_\ell| \geq 1$, for all s, m. □

Based on the above characterization of the solution space $\mathcal{M}_{1,\infty} \cap \mathcal{H}$, the ansatz for the kernel vector of A_{tot} in the space of square-summable sequences may be written by suitably truncating the general ansatz presented in Eq. (5.35),

$$|\epsilon\rangle = \sum_{\substack{|z_\ell|<1,\\ \ell=0}}\sum_{s=1}^{s_\ell}\alpha_{\ell s}|\psi_{\ell s}\rangle, \quad \alpha_{\ell s} \in \mathbb{C}.$$

References

1. P.R. Halmos, *Finite-Dimensional Vector Spaces* (Courier Dover Publications, Mineola, 2017)
2. B. Mourrain, V.Y. Pan, Multivariate polynomials, duality, and structured matrices. J. Complex. **16**, 110–180 (2000). https://doi.org/10.1006/jcom.1999.0530
3. I. Gohberg, P. Lancaster, L. Rodman, *Matrix Polynomials* (Academic Press, New York, 1982)

4. M. Fardad, The operator algebra of almost Toeplitz matrices and the optimal control of large-scale systems, in *2009 American Control Conference* (IEEE, Piscataway, 2009), pp. 854–859
5. E. Cobanera, A. Alase, G. Ortiz, L. Viola, Exact solution of corner-modified banded block-Toeplitz eigensystems. J. Phys. A: Math. Theor. **50**, 195204 (2017). https://doi.org/10.1088/1751-8121/aa6046

Chapter 6
Summary and Outlook

The main goal of the work presented in this thesis is to contribute to the understanding of the physics of topological phases of matter and, in particular, of the phenomenon of bulk-boundary correspondence observed in them. We made good progress towards this goal by formulating a generalization of Bloch's theorem to fermionic lattice systems with non-trivial boundary, by applying it to various prototypical models of topological insulators and superconductors, and also by employing a matrix factorization based approach to prove the bulk-boundary correspondence for all symmetry classes in 1D systems.

6.1 The Generalized Bloch Theorem and Its Applications

The generalization of Bloch's theorem presented in Chap. 2 is applicable to clean systems of independent fermions on a lattice, subject to BCs that are arbitrary—other than respecting the finite-range nature of the overall Hamiltonian. Our theorem leverages the corner-modified block-Toeplitz structure of the single-particle Hamiltonian to provide exact, analytical expressions for all the energy eigenvalues and eigenstates of the system, and consistently recovers the ones derived from the standard Bloch's theorem for periodic BCs. This generalization is already sufficient to study all toy models as well as realistic tight-binding models of topological materials, as evidenced by our analysis of many of the known models in Chap. 3. The fact that topology manifests itself at low energy scales makes the assumption of finite range justified. Within our framework, BCs for D-dimensional systems must be imposed on two parallel hyperplanes, but are otherwise arbitrary. Therefore, the generalized Bloch theorem yields highly-effective tools for analyzing systems subject to anything from pristine terminations to surface relaxation, reconstruction, and disorder. A theoretically simple extension to interfaces between multiple bulks in practice adds a great deal of utility to the theorem, as it allows us to study arbitrary

A. Alase, *Boundary Physics and Bulk-Boundary Correspondence in Topological Phases of Matter*, Springer Theses, https://doi.org/10.1007/978-3-030-31960-1_6

junctions, including interface modes resulting from putting in contact two exotic topologically non-trivial bulks.

The real strength of the generalized Bloch theorem lies in the fact that it leverages the symmetries of the system, including the translation invariance mildly broken by the boundary. As a key component to this theorem, one obtains an exact *structural ansatz*, close in spirit to the Bethe ansatz, for all (regular) energy eigenstates in *dispersive* bands. This ansatz is easy to construct since it depends only on the energy eigenvalue and the bulk properties of the Hamiltonian. The individual components of this ansatz reflect translation invariance in a way we have made precise and are, as such, determined by the analytic continuation of the Bloch Hamiltonian, as we showed. Besides, as can be witnessed in many of the applications, further symmetries including those considered in the tenfold classification can be easily incorporated to provide a further structure to this ansatz. This is one of the main features of the theorem that make it much more powerful as an algebraic tool than some other existing approaches such as the transfer matrix and Green's function methods, apart from having a well-defined class of BCs that can be handled within the framework. While our generalized Bloch theorem makes no prediction about the singular energy values which correspond to *dispersionless* (flat) bands of eigenstates, we have provided a prescription for identifying such energy values without diagonalizing the full Hamiltonian, and showed how they necessarily enter the physical energy spectrum irrespective of the BCs. In such singular cases, we have further provided a procedure to effectively obtain a (possibly over-complete) basis of *perfectly localized states* using an analytic continuation of the Bloch Hamiltonian, and explicitly illustrated such a procedure in the Kitaev's Majorana chain Hamiltonian at its sweet spot.

The generalized Bloch theorem properly accounts for two types of energy eigenstates that do not exist once translation invariance is imposed via Born–von Karman (periodic) BCs: perfectly localized energy states and localized energy states whose exponential decay exhibits a power-law prefactor. Our generalized Bloch theorem predicts the existence, under specific (non-generic) conditions, of edge states that decay exponentially in space with a *power-law prefactor*. Such exotic states were previously believed to arise only in systems with long-range couplings. In our framework, their origin may be traced back to the description of the system's eigenstates in terms of *non-unitary* representations of translational symmetry "outside Hilbert space"—again capturing the fact that such a symmetry is only mildly broken by the BCs, in a precise sense. Notably, we have shown how the emergence of zero-energy Majorana modes with a linear prefactor is possible in the paradigmatic Kitaev chain by proper Hamiltonian tuning *on* the so-called "circle of oscillations." Their "critical" spatial behavior separates the theoretically observed Majorana wavefunction oscillations inside such a circle from the simple exponential decay outside.

Yet another strength of the generalized Bloch theorem is its ability to access in an exact manner the systems of semi-infinite extent in one direction. All the work in this thesis points to the fact that such systems are the best candidates to rigorously study bulk-boundary correspondence. They have well-defined bulk states

(states with non-normalizable wavefunctions) because of their infinite extent, yet they also show topological features on the boundary. In some sense, it is only for such systems that the topological invariant is a well-defined quantum mechanical observable, since it can formulated as the expectation value of a compact operator relating to the localized energy states on the boundary. This was our motivation behind the derivation of the "indicator for the bulk-boundary correspondence" using semi-infinite systems. This indicator diverges if and only if the system hosts a bound mode. This indicator leverages the other key component to our generalized Bloch theorem, the *boundary matrix*, and is unique in the sense that, unlike most other indicators in the literature, it combines information from *both* the bulk and the boundary. The utility of this indicator is seen from our analysis of the 4π-periodic Josephson effect in a model of a s-wave topological superconductor. In the process, we show how, interestingly, the 4π-periodicity that distinguishes a topologically non-trivial response is *not* accompanied by a fermionic parity switch in this system. We have provided a physical explanation of this behavior by exhibiting a decoupling transformation, which maps the relevant Hamiltonian to two uncoupled "virtual" wires—each undergoing a parity switch.

The mathematical results presented in Chap. 5 are essential for the formulation of the generalized Bloch theorem. Much of the analysis is devoted to finding a *complete* basis of the bulk solution space. It is for this reason that we can be sure in our treatments of topological Hamiltonians that we have not missed any solution, except in those cases where an incomplete ansatz is used deliberately to simplify calculations. As we already mentioned, the generalized Bloch theorem may be thought of as bestowing *exact solvability* in the same sense as the algebraic Bethe ansatz does: the linear-algebraic task of diagonalizing the single-particle Hamiltonian is mapped to one of solving a small (independent of the number of sites) system of *polynomial* equations. While in general, if the polynomial degree is higher than four, the roots must be found numerically, whenever this polynomial system can be solved analytically, one has managed to solve the original linear-algebraic problem analytically as well. In fact, fully analytical solutions are less rare than one might think, and either emerge in special parameter regimes, or by suitably adjusting the BCs. The use of the Smith normal form might lead the way towards bringing more tools from the theory of polynomials to applications in condensed-matter physics. From a computational standpoint, we expect that the diagonalization algorithms emerging from our approach will be useful for large-scale electronic calculations in both one and higher dimensions, possibly in conjunction with perturbative approaches for incorporating interactions.

Looking afresh at the transfer matrix approach from the generalized Bloch theorem's perspective yields a remarkable result: the generalized eigenvectors of the transfer matrix, whose role, as we emphasized, has been appreciated only recently, describe energy eigenstates with power-law corrections to an otherwise exponential behavior. An explicit example is seen, again, in the semi-infinite Kitaev's chain with open BCs, precisely in the same circle-of-oscillations parameter regime that hosts power-law zero-energy Majorana modes. While, in this way, our method provides a natural inversion-free alternative to the standard transfer matrix approach, the connections we have identified point to further possibilities for fruitfully combining the two approaches. In particular, since the bulk-boundary separation we propose remains useful in the presence of bulk disorder, one may envision a hybrid approach for solving disordered systems subject to arbitrary BCs, by employing transfer-matrix techniques to handle the resulting bulk equation. Interestingly, from the standpoint of computing energy levels, our bulk-boundary separation is in many ways complementary to the transfer matrix method. While the latter can handle bulk disorder (at a computational cost), it does not lend itself to investigating the space of arbitrary BCs in a transparent way. On the contrary, our generalized Bloch theorem can handle arbitrary BCs efficiently, as long as the bulk respects translational invariance—with arbitrary (finite-range) disorder on the boundary being permitted. Our generalized Bloch theorem further suggests a way to extend the transfer matrix approach to a disordered bulk and arbitrary BCs.

The applications of the generalized Bloch theorem presented in Chap. 3 are important for several reasons. First, they have all been treated uniformly to get exact expressions for the energy eigenstates, analytically to the extent possible. Such analysis serves as a sanity check in some cases (e.g., the SSH model), while the solutions are new in other cases. Table 3.1 summarizes all the systems that we have solved so far by our techniques, where exact analytic solutions were largely unknown prior to our findings, to the best of our knowledge. Since the solutions are analytic, it is possible to read off the exact dependence of the localized energy states on the parameters of the model as well as BCs in some cases. They therefore serve as excellent starting points for investigation of the bulk-boundary correspondence. 2D models have been historically important in the developments in topological phases. In our analysis of the 2D p-wave superconductor, we provide the first provable and completely analytic treatment of chiral edge states. We also showed that it is possible to *analytically* determine Andreev bound states for an idealized SNS junction, and that for large values of superconducting gap, all normal modes of the metallic part transform into Andreev bound states. With an eye toward applications in synthetic quantum matter, we have also used the generalized Bloch theorem to engineer a quasi 1D Hamiltonian that supports a perfectly localized, robust zero-energy mode, notwithstanding the lack of chiral and charge-conjugation protecting symmetries. More importantly, the existence of power-law modes would not have been unveiled without our mathematical formalism. We also find power-law modes on the surface of the $p + ip$ chiral superconductor as part of our closed-form full calculation of the surface states, see Sect. 3.2.3. We have also included applications to other 2D systems, such as the full closed-form diagonalization of graphene ribbons for zigzag-bearded and armchair surface terminations. While the edge modes for zigzag-bearded graphene have been computed before in closed-form, the closed-form band states appear to be new in the literature. It seems a distinctive feature of the generalized Bloch theorem that *both* edge and bulk bands can be treated analytically on equal footing. Finally, we investigated in detail the Majorana flat bands of the gapless s-wave topological superconductor. There, we find an extensive contribution of the surface MFB to the 4π-periodic component of the Josephson current, which would serve as a smoking gun for experimental detection should a candidate material realization be identified.

Towards a deeper understanding of the bulk-boundary correspondence in topological insulators and superconductors, an important next step is to study the robustness of edge states against boundary perturbations. It is natural to start by asking how certain symmetries of the system influence the nature of the proposed indicator, or the boundary matrix from which the indicator itself is derived. This can possibly lead to identifying a symmetry principle which dictates the bulk-boundary correspondence, as well as an interpretation at the basic dynamical-system level in terms of stability theory. Likewise, the framework we have developed may also serve as a concrete starting point for rigorously deriving an effective boundary theory for lattice systems and for exploring generalizations of the concept of Wannier function [1] in the presence of arbitrary BCs.

The advantages of our algorithm extend straightforwardly to D-dimensional quadratic Hamiltonians, $D > 1$, if the standard procedure of imposing periodic BCs in $D-1$ directions is employed: in this way, the model reduces to a poly-sized set of 1D lattices subject to arbitrary BCs, to which our algorithm applies. We expect our approach to further elucidate physical phenomena that take place at the surfaces [2–4]. On the one hand, this prompts the question of whether the concept of a Wannier function may also be generalized for arbitrary BCs. On the other hand, the procedure for incorporating BCs is more involved, calling for separate investigation. For more general BCs, e.g., two open directions, the bulk-boundary separation goes through essentially unchanged: for example, the range of the boundary projector consists of a hyper-cubic surface layer of thickness determined by the bulk structure of the system. The challenge in higher dimensions is solving the bulk equation explicitly and in full generality. It is a worthy challenge, because it would yield insight into the plethora of corner states that can appear in such systems [5–7]. While special cases may still be able to be handled on a case-by-case basis, in general we see little hope of using the same mathematical techniques (crucially, the Smith decomposition) that work so well in our setup. The analytic continuation of the Bloch Hamiltonian then becomes a matrix-valued analytic function of D complex variables. The passage from one complex variable to several makes a critical difference.

Beyond equilibrium scenarios, our approach should prove advantageous to evaluate in closed-form the unitary propagator $\exp(-i\widehat{H}t)$ describing free evolution under arbitrary BCs [8], and to diagonalize the Floquet propagator describing periodically driven fermionic systems [9, 10]. Since our algorithm does not exploit the Hermiticity of the Hamiltonian, a further direction of investigation is the application to open Fermi systems obeying quadratic Lindblad master equations [11], with the potential to shed light onto bulk-boundary correspondence in engineered topological phases far from equilibrium [12]. Lastly, while we have focused on fermions in this thesis, the general foundation of our method laid out in Chap. 5 is equally valid for bosons and immediately applicable to non-Hermitian effective Hamiltonians with non-trivial boundaries, as arising in semi-classical models of open quantum systems in various contexts [13–17].

6.2 Matrix-Factorization Approach to Bulk-Boundary Correspondence

In Chap. 4, we presented an outline of a proof of the known bulk-boundary correspondence in 1D systems using matrix Wiener–Hopf factorization technique. Our proofs apply for tight-binding Hamiltonians on semi-infinite lattices, with disorder of compact support. This assumption means that our results apply for finite systems which have disorder near one of the edges, but with possibly larger penetration depth than the localized energy states. First of all, for symmetry classes other than AIII, the proofs of the conjectures are not known or explicitly provided in the existing literature to the best of our knowledge. In particular, our analysis covers the case of p-wave Majorana chain of Kitaev, proposed almost two decades ago and used as a prototypical example of a topological superconductor, yet without a rigorous proof of bulk-boundary correspondence conjecture in the literature.

More importantly, our approach and methodology provide hope of revealing new aspects of the bulk-boundary correspondence. In the Introduction, we mentioned the need of an approach that is based on a technique of solving boundary value problems. The Wiener–Hopf factorization is precisely one such tool. As an intermediate step in our proof, we define a new way of spectral flattening of a finite-range tight-binding Hamiltonian using Wiener–Hopf factorization. Although this comes at the cost of a change in wavefunctions of the energy states in the bulk energy bands, we show that the relevant topological invariant is preserved under this transformation. The advantage is that the flattened Hamiltonian is also finite range, and therefore easier to handle analytically. We also extended the notion of spectral flattening to semi-infinite systems. Operationally, such a Hamiltonian can have only emergent (finite support) bulk solutions at all regular energies, therefore any localized energy state irrespective of BCs must be finitely supported. The boundary invariant of a semi-infinite system does not change under this definition of spectral flattening.

We have also presented a rigorous extension of the bulk-boundary correspondence for interfaces, provided that the bulks forming the interface, and the junction joining them, all belong to the same symmetry class. The topological invariant of the interface then depends completely on the invariants of the individual bulks. It is not yet clear, however, whether there is any interesting physics in interfaces formed by bulks belonging to distinct symmetry classes. If it is the case, new approaches need to be explored to formulate and prove the bulk-boundary correspondence in those cases.

To solidify various claims about the "protection" of localized energy states in SPT phases, we define the concept of stability of zero modes, and also discuss rigorously how it relates to symmetries and topological invariants. In particular, we find that not all zero modes are stable, even if they are hosted by a system with a non-trivial topological invariant. We provide and prove an exact criterion for the stability of zero modes of topological origin. We also define the sensitivity of zero modes to perturbations, and provide bounds on the sensitivity in terms of generalized condition number of the Hamiltonian. These bounds are demonstrated with the example of SSH model. In the future, this can form a bridge between the literature on SPT phases and system theory.

The rigorously proved bulk-boundary correspondence also clarifies the role of unitary symmetries in the classification of topological phases. In the tenfold classification and related works, it is suggested that whenever a system satisfies a unitary symmetry, one must block-diagonalize the Hamiltonian and proceed with the topological analysis of the reduced blocks. Our stability analysis of a model of s-wave topological superconductor presents a different picture; whether the topological invariant should be computed from the full Hamiltonian or the reduced block depends on whether the perturbations (e.g., arising from disorder) of interest obey the same unitary symmetry and therefore they have the same block structure. Our proof of the bulk-boundary correspondence does not assume the absence of unitary symmetries. In light of this result, a fresh assessment of the tenfold classification itself might be necessary.

The analysis of 1D systems naturally raises the question on whether a similar analysis can be performed for topological systems in higher spatial dimension. The answer is not yet clear, although one would need to look at analogous factorizations of families of matrix polynomials for this purpose. Some such factorizations have been studied in the context of loop groups [18]. Since the original Wiener–Hopf factorization is applicable to any matrix polynomial without the constraint of Hermiticity, there is also a possibility that similar treatment can be extended to bosonic and effective non-Hermitian Hamiltonians [17], as well as open quantum systems described by Lindblad master equations [11].

Finally, the stability analysis of zero modes prompts the question as to whether a similar analysis for intrinsic topological phases, such as the fractional quantum Hall state, would yield qualitatively different results. The sensitivity to perturbations can therefore be a great tool to compare intrinsic and SPT phases, in order to either differentiate among them operationally, or perhaps treat them on an equal footing.

References

1. D. Vanderbilt, R. Resta, Quantum electrostatics of insulators: polarization, Wannier functions, and electric fields. Contemp. Concepts Condens. Matter Sci. **2**, 139–163 (2006). https://doi.org/10.1016/S1572-0934(06)02005-1
2. L. Gor'kov, Surface and superconductivity, in *Recent Progress in Many-body Theories*, ed. by J.A. Carlson, G. Ortiz (2006), pp. 3–7
3. L. Isaev, G. Ortiz, I. Vekhter, Tunable unconventional Kondo effect on topological insulator surfaces. Phys. Rev. B **92**, 205423 (2015). https://doi.org/10.1103/PhysRevB.92.205423
4. K. Binder, D. Landau, Critical phenomena at surfaces. Phys. A Stat. Mech. Appl. **163**, 17–30 (1990). https://doi.org/10.1016/0378-4371(90)90311-F
5. W.A. Benalcazar, B.A. Bernevig, T.L. Hughes, Quantized electric multipole insulators. Science **357**, 61–66 (2017). https://doi.org/10.1126/science.aah6442
6. K. Hashimoto, X. Wu, T. Kimura, Edge states at an intersection of edges of a topological material. Phys. Rev. B **95**, 165443 (2017). https://doi.org/10.1103/PhysRevB.95.165443

7. F.K. Kunst, G. van Miert, E.J. Bergholtz, Lattice models with exactly solvable topological hinge and corner states. Phys. Rev. B **97**, 241405 (2018). https://doi.org/10.1103/PhysRevB.97.241405
8. S. Hegde, V. Shivamoggi, S. Vishveshwara, D. Sen, Quench dynamics and parity blocking in Majorana wires. New J. Phys. **17**, 053036 (2015). https://doi.org/10.1088/1367-2630/17/5/053036
9. T. Kitagawa, E. Berg, M. Rudner, E. Demler, Topological characterization of periodically driven quantum systems. Phys. Rev. B **82**, 235114 (2010). https://doi.org/10.1103/PhysRevB.82.235114
10. A. Poudel, G. Ortiz, L. Viola, Dynamical generation of Floquet Majorana flat bands in s-wave superconductors. Europhys. Lett. **110**, 17004 (2015). https://doi.org/10.1209/0295-5075/110/17004
11. T. Prosen, Third quantization: a general method to solve master equations for quadratic open Fermi systems. New J. Phys. **10**, 043026 (2008). https://doi.org/10.1088/1367-2630/10/4/043026
12. S. Diehl, E. Rico, M.A. Baranov, P. Zoller, Topology by dissipation in atomic quantum wires. Nat. Phys. **7**, 971 (2011). https://doi.org/10.1038/nphys2106
13. I. Mandal, Exceptional points for chiral Majorana fermions in arbitrary dimensions. Europhys. Lett. **110**, 67005 (2015). https://doi.org/10.1209/0295-5075/110/67005
14. A. Tayebi, T.N. Hoatson, J. Wang, V. Zelevinsky, Environment-protected solid-state-based distributed charge qubit. Phys. Rev. B **94**, 235150 (2016). https://doi.org/10.1103/PhysRevB.94.235150
15. D. Leykam, S. Flach, Y.D. Chong, Flat bands in lattices with non-Hermitian coupling. Phys. Rev. B **96**, 064305 (2017). https://doi.org/10.1103/PhysRevB.96.064305
16. M. Gluza, C. Krumnow, M. Friesdorf, C. Gogolin, J. Eisert, Equilibration via gaussification in fermionic lattice systems. Phys. Rev. Lett. **117**, 190602 (2016). https://doi.org/10.1103/PhysRevLett.117.190602
17. I. Rotter, A non-Hermitian Hamilton operator and the physics of open quantum systems. J. Phys. A Math. Theor. **42**, 153001 (2009). https://doi.org/10.1088/1751-8113/42/15/153001
18. G. Giorgadze, G. Khimshiashvili, Factorization of loops in loop groups. Bull. Georgian Natl. Acad. Sci. **5**, 35 (2011)

Curriculum Vitae: Abhijeet Alase

Personal Data

Postdoctoral Associate
Institute for Quantum Science and Technology
University of Calgary
Email: alase.abhijeet@gmail.com

Research Interests

Topological insulators and superconductors, topological quantum computation, quantum information science.

Education

09/2013–02/2019	**Dartmouth College**, USA Ph.D. in Physics, Advisor: Prof. Lorenza Viola
08/2009–05/2013	**Indian Institute of Technology, Bombay**, India B. Tech. in Engineering Physics

Publications

- A. Alase, E. Cobanera, G. Ortiz, and L. Viola, "Matrix-factorization approach to bulk-boundary correspondence and stability of zero modes", in preparation.

A. Alase, *Boundary Physics and Bulk-Boundary Correspondence in Topological Phases of Matter*, Springer Theses, https://doi.org/10.1007/978-3-030-31960-1

- E. Cobanera, A. Alase, G. Ortiz, and L. Viola, "Generalization of Bloch's theorem for arbitrary boundary conditions: Interfaces and topological surface band structure", Phys. Rev. B **98**, 245423 (2018).
- A. Alase, E. Cobanera, G. Ortiz, and L. Viola, "Generalization of Bloch's theorem for arbitrary boundary conditions: Theory", Phys. Rev. B. **96**, 195133 (2017). Selected as **Editor's suggestion**. Featured as a **Synopsis** in *Physics*.
- E. Cobanera, A. Alase, G. Ortiz, and L. Viola, "Exact solution of corner-modified banded block-Toeplitz eigensystems", J. Phys. A: Math. Theor. **50**, 195204 (2017). **Highlights collection.**
- A. Alase, E. Cobanera, G. Ortiz, and L. Viola, "Exact solution of quadratic fermionic Hamiltonians for arbitrary boundary conditions", Phys. Rev. Lett. **117**, 076804 (2016).

Honors and Awards

- **The Physics and Astronomy Graduate Research Award**, Dartmouth College (2018) .
- **MITACS Globalink** intern in Institute for Quantum Computing, Waterloo (2012).
- Invited participant in **National Initiative for Undergraduate Science** camp, HBCSE (2010).
- Ranked one among the top 50 candidates in **National Physics Olympiad** (2009).
- Awarded **National Talent Search** scholarship by Government of India (2007).

Professional Societies and Service

- Reviewer for **Mathematical Reviews** published by American Mathematical Society.
- Member of the **American Physical Society**.

CPSIA information can be obtained
at www.ICGtesting.com
Printed in the USA
LVHW012132011219
639064LV00001B/41/P

9 783030 319595